Northumbria, 500–1100

Creation and Destruction of a Kingdom

This book deals with the development of Northumbria from the beginning of its history to the end, thereby offering a unique opportunity to study the rise and fall of a kingdom rather than just focusing on its period of greatness.

It examines the mechanisms of ethnic, political, social and religious change which, beginning after the end of the Roman Empire, welded the large and disparate area between the Humber and the Firth of Forth into one of the most powerful kingdoms of early medieval England, and those which led to its disintegration and the emergence in its place of the political structures of northern England and southern Scotland.

These mechanisms are set in a wider European context so that the history of Northumbria is seen as paradigmatic for an understanding of state formation and religious and cultural change in the early medieval world. In doing so, the book explores the consequences for our understanding of the past of the characteristics and value of available source-materials. Attention is focused particularly on the interpretation of archaeological and art-historical material as it relates to written source-material, and the extent to which narrative sources were shaped by sectional interests and created imagined visions of the past.

DAVID ROLLASON is Professor of History, University of Durham. His many publications include *The Mildrith Legend: A Study in Early Medieval Hagiography in England* (1982), *St Cuthbert, his Cult and his Community to AD 1200* (ed. 1989, with Gerald Bonner and Clare Stancliffe), *Saints and Relics in Anglo-Saxon England* (1989), *Anglo-Norman Durham 1093–1193* (ed. 1994, with Margaret Harvey and Michael Prestwich), *Sources for York History before 1100* (ed. 1998, with Derek Gore and Gillian Fellows-Jensen), *Symeon of Durham, On the Origins and Progress of this the Church of Durham* (ed. 2000) and *Bede and Germany* (Jarrow Lecture, 2002).

Northumbria, 500–1100

Creation and Destruction of a Kingdom

DAVID ROLLASON

CAMBRIDGE UNIVERSITY PRESS
Cambridge, New York, Melbourne, Madrid, Cape Town, Singapore, São Paulo

Cambridge University Press
The Edinburgh Building, Cambridge CB2 8RU, UK

Published in the United States of America by Cambridge University Press, New York

www.cambridge.org
Information on this title: www.cambridge.org/9780521813358

First published 2003
This digitally printed version 2007

A catalogue record for this publication is available from the British Library

ISBN 978-0-521-81335-8 hardback
ISBN 978-0-521-04102-7 paperback

For Jon, Will and Ed

Contents

Illustrations

Figures

Maps

Preface

This book is principally the product of teaching a special subject on the kingdom of Northumbria to undergraduate students in the University of Durham History Department, although it also draws on the experience of teaching extramural students in Birmingham with which my career as a teacher began. The book's contents and approach have been guided by my perceptions of the interests and requirements of both groups. I have tried to emphasize throughout what seem to me to be the crucial elements of studying history. First, close attention to the sources and an awareness of what they do or do not contribute to our understanding of the past. The book tries wherever possible to expose the reader to those sources and to engage in close consideration of them. The footnotes aim not to survey the entire bibliography available, which is immense, but rather to direct the reader to where they may consult further sources for themselves, and the modern works which are (in my view) most helpful in understanding them. Many of those readers whose previous experience has been in modern history need introducing to the importance of the full range of source-materials, whether chronicles, documents, coins, art-objects, buildings or whatever. The depth of understanding made possible by embracing this range is one of the most exciting features of early medieval history as it has developed in recent years, although the approach is in fact equally applicable to other periods. I have already gone some way along the road to presenting the written sources for a fair amount of Northumbrian history in the original language and in translation with detailed commentary in my book (with Derek Gower and Gillian Fellows-Jensen), *Sources for York History before 1100* (Archaeology of York 1;

York: York Archaeological Trust, 1998). This might be regarded as almost a companion volume to the present book, and I hope readers will forgive me for frequently citing it as the most convenient guide to many of the sources and further study (I have invariably given alternative references for those without access to it).

Secondly, the importance of geography and the inter-relationship in the past as in the present between human activity and the physical space in which it is undertaken. This is not a geography book and it does not aspire to the sophistication of the approaches developed by historical geography, nor to the wonderful sense of contact with the past landscape which is so evident in a book like Charles Phythian-Adams's *The Land of the Cumbrians: A Study in British Provincial Origins, AD 400–1120* (London: Scolar, 1996). The book is nevertheless founded on the notion that geography really matters in history, and that variations in space as well as time even within a relatively restricted area such as the kingdom of Northumbria are of immense importance to understanding how past societies worked. For that reason, the book is well endowed with maps and with a whole chapter devoted to the frontiers and heartlands of the kingdom of Northumbria. I have been persistent, perhaps irritatingly so, in giving locations by the counties as they were before the reorganization of 1974, which I hope will provide a suitable frame of reference. None of this, however, is a substitute for poring over maps, and indeed for exploring at first hand the areas of northern England and southern Scotland with which this book is concerned.

Thirdly, the importance of history in the wider scheme of things. Although an understanding of the past has to be founded on rigorous attention to detail and above all on criticism of sources to the highest possible levels of technical sophistication, that in itself is not enough. H. S. Offler's opinion, expressed in his University of Durham inaugural lecture in 1958, that 'historian's history' is 'the only sort that matters' will not do in the twenty-first century (Offler (1958)). History cannot be just about dotting the 'i's and crossing the 't's of the historical record, important as that may be. For history to hold its place not only in the public eye but also alongside its peers among academic disciplines, it needs to show that it is illuminating matters of fundamental importance to understanding human activity and organization. For that reason, this book aims to propose an agenda for the history of Northumbria which places it in a wider context. I have chosen to relate that history, crudely perhaps, to the overall question of the mechanisms underlying the formation and dissolution of states. Hence the title of my book which seeks to convey quite precisely what I have opted to deal with and what to leave to others. My sense of the need to be always aware of the wider context of any historical subject and period owes much to my students, especially my extramural students. But it also owes much to Bill Davies at the Cambridge University Press who, in addition to invaluable support and encouragement, was rightly sceptical of publishing a book which was 'just about Northumbria' and spurred me on to think more generally about the importance of the subject.

My thanks for support, erudition and friendship go to those who have helped me so generously with information, discussion, sharing of results, and above all to Rosemary Cramp, Simon Keynes, David Palliser, Charles Phythian-Adams and my wife Lynda for reading parts or the whole of the book in draft. As with all my work in Durham, I am indebted to the enthusiastic support of the University Library (especially Beth Rainey and Ian Doyle) and the Cathedral Library (especially Roger Norris), to the support of my colleagues in the History Department (especially in covering me during research leave) and to all my colleagues in the AHRB Centre for North-East England History for the stimulus and new ideas which its conferences, seminars and projects have given me.

The book is proudly and affectionately dedicated to my three sons, with whom my wife and I have shared so many happy rambles in Northumbria.

David Rollason

Acknowledgments

For permission to reproduce illustrations, the author is very grateful to the following: the National Monuments Record, the York Archaeological Trust, English Heritage, Dr Nancy Edwards, Her Majesty's Stationery Office, Hull City Museums, Ms Susan Hirst, Four Courts Press, Dr David Longley, Professor Philip Rahtz, Tees Archaeology, Cambridge University Press, the University of Durham, National Museums of Scotland, the Ministry of Defence, the Biblioteca Medicea Laurenziana, the British Library, Durham Cathedral, the Board of Trinity College Dublin, the Royal Irish Academy, the British Museum and the Fitzwilliam Museum. For sections of text from translated sources, the author gratefully acknowledges the permission of Cambridge University Press, Oxford University Press, Eyre and Spottiswoode, Llanerch Publishers, D. S. Brewer, York Archaeological Trust, Messrs Dent, the Borthwick Institute of the University of York, and Penguin Books.

Abbreviations

Æthelweard	Alastair Campbell (ed. and trans.), *The Chronicle of Æthelweard* (Nelson's Medieval Texts; London: Thomas Nelson and Sons, 1962).
Alcuin, *BKSY*	Peter Godman (ed. and trans.), *Alcuin, The Bishops, Kings, and Saints of York* (Oxford Medieval Texts; Oxford: Clarendon Press, 1982).
Alcuin, *Letters*	Ernst Dümmler (ed.), *Epistolae Karolini Aevi, Tomus II* (Monumenta Germaniae Historica, Epistolarum Tomus IV, Karolini Aevi II; Berlin: Weidmann, 1895); Stephen Allott (trans.), *Alcuin of York c. AD 732 to 804: His Life and Letters* (York: Ebor Press, 1974). Cited by the numbers in Dümmler's edition (concordance in Allott, p. 173).
Aneirin, *Gododdin*	Modern Welsh and English translation: A. O. H. Jarman (trans.), *Aneirin: Y Gododdin. Britain's Oldest Heroic Poem* (Welsh Classics 3; Llandysul, 1988). Brittonic reconstruction and English translation: John T. Koch, *The Gododdin of Aneirin: Text and Context in Dark-Age North Britain* (Cardiff: University of Wales Press, 1997), pp. 1–129. English translation: Kenneth Jackson, *The Gododdin: The Oldest Scottish Poem* (Edinburgh: Edinburgh University Press, 1969).
Annals Ulster	Seán Mac Airt and Gearóid Mac Niocaill (eds.), *The Annals of Ulster (to A.D. 1131): Part I: Text and Translation* (Dublin: Dublin Institute for Advanced Studies, 1983).

Anon, *Life Cuth.*	Anonymous, *Life of St Cuthbert*; Bertram Colgrave (ed. and trans.), *Two Lives of St Cuthbert: A Life by an Anonymous Monk of Lindisfarne and Bede's Prose Life* (Cambridge: Cambridge University Press, 1940, repr. 1985), pp. 60–139.
ASC	*Anglo-Saxon Chronicle*, cited *sub anno* (*s.a.*) and, where necessary, by the conventional manuscript sigla, A–G; ed. J. Earle and C. Plummer, *Two of the Saxon Chronicles Parallel*, 2 vols. (Oxford: Oxford University Press, 1892–9); trans. (with corrected dates) D. Whitelock, D. C. Douglas and S. I. Tucker, *The Anglo-Saxon Chronicle: A Revised Translation* (Cambridge: Cambridge University Press, 1961), and in *EHD I*, pp. 145–61. The D version cited from G. P. Cubbin (ed.), *MS D* (The Anglo-Saxon Chronicle: A Collaborative Edition 6; Cambridge: D. S. Brewer, 1996).
Asser	W. H. Stevenson (ed.), *Asser's Life of King Alfred together with the Annals of Saint Neot's Erroneously Ascribed to Asser* (Oxford: Oxford University Press, 1959), and Simon Keynes and Michael Lapidge (trans.), *Alfred the Great: Asser's Life of King Alfred and Other Contemporary Sources* (Penguin Classics; Harmondsworth: Penguin, 1983), pp. 67–120.
Bede, *Continuations*	*Continuations* to Bede's *Ecclesiastical History of the English People*. Edition: Plummer, *Bede*, I, pp. 361–3. Edition and translation: Bertram Colgrave and R. A. B. Mynors, *Bede's Ecclesiastical History of the English People* (Oxford Medieval Texts; Oxford: Clarendon Press, 1969), pp. 572–7. Translation: Judith McClure and Roger Collins (trans.), *Bede: The Ecclesiastical History of the English People, The Greater Chronicle, Bede's Letter to Egbert* (Oxford: Oxford University Press, 1994), pp. 296–8. Cited *sub anno* (*s.a.*).
Bede, *Eccles. Hist.*	Bede, *Ecclesiastical History of the English People.* Edition: Plummer, *Bede*, I, pp. 5–360. Edition and translation: Bertram Colgrave and R. A. B. Mynors, *Bede's Ecclesiastical History of the English People* (Oxford Medieval Texts; Oxford: Clarendon Press, 1969); the translation from the latter is published with additional notes in Judith McClure and Roger Collins, *Bede: The Ecclesiastical History of the English People, The Greater Chronicle, Bede's Letter to Egbert* (Oxford: Oxford University Press, 1994).

Bede, *Hist. Abbots* — Bede, *History of the Abbots of Monkwearmouth and Jarrow.* Edition: Plummer, *Bede*, I, pp. 364–87. Translation: J. F. Webb and D. F. Farmer, *The Age of Bede: Bede, Life of Cuthbert; Eddius Stephanus, Life of Wilfrid; Bede, Lives of the Abbots of Wearmouth and Jarrow, with the Voyage of St Brendan*, 2nd edn (Penguin Classics; Harmondsworth: Penguin, 1983), pp. 183–208.

Bede, *Letter* — Bede, *Letter to Bishop Egbert.* Edition: Plummer, *Bede*, I, pp. 405–23. Translation: Judith McClure and Roger Collins, *Bede: The Ecclesiastical History of the English People, The Greater Chronicle, Bede's Letter to Egbert* (Oxford: Oxford University Press, 1994), pp. 343–57.

Bede, *Life Cuth.* — Edition and translation: Bertram Colgrave, *Two Lives of St Cuthbert: A Life by an Anonymous Monk of Lindisfarne and Bede's Prose Life* (Cambridge: Cambridge University Press, 1940, repr. 1985), pp. 141–307.

Bede, *Met. Life Cuth.* — Werner Jaager (ed.), *Bedas metrische Vita sancta Cuthberti* (Palaestra 198; Leipzig: Mayer and Müller, 1935).

Bede, *Reckoning* — Bede, *The Reckoning of Time.* Edition: C. W. Jones, *Bedae opera de temporibus* (Cambridge, Mass.: Mediaeval Academy of America, 1943). Translation: Faith Wallis, *Bede: The Reckoning of Time* (Translated Texts for Historians 29; Liverpool: Liverpool University Press, 1999).

Beowulf — Edition: Friederich Klaeber, *Beowulf and the Fight at Finnsburg*, 3rd edn (Boston: D. C. Heath, 1950). Translation: Michael Alexander, *Beowulf: A Verse Translation* (Penguin Classics; Harmondsworth: Penguin, 1973).

Boldon Book — David Austin (ed. and trans.), *Boldon Book: Northumberland and Durham* (Chichester: Phillimore, 1982).

Boniface, *Letters* — Michael Tangl (ed.), *Die Briefe des heiligen Bonifatius und Lullus* (Epistolae Selectae in usum scholarum ex Monumentis Germaniae Historicis separatim editae 1; Berlin: Weidmann, 1916); Ephraim Emerton (trans.), *The Letters of Saint Boniface*, 2nd edn (Records of Western Civilisation; New York: Columbia University Press, 2000).

Domesday Yorks. — Margaret L. Faull and Marie Stinson (eds. and trans.), *Yorkshire*, 2 vols. (Domesday Book 30; Chichester: Phillimore, 1986).

EHD I — Dorothy Whitelock (trans.), *English Historical Documents, I, c. 500–1042*, 2nd edn (London: Eyre and Spottiswoode, 1979).

Encyclopaedia ASE — Michael Lapidge, John Blair, Simon Keynes, et al. (eds.), *The Blackwell Encyclopaedia of Anglo-Saxon England* (Oxford: Blackwell, 1998).

Gildas, *Ruin* — Michael Winterbottom (ed. and trans.), *Gildas, The Ruin of Britain and Other Works* (Chichester: Phillimore, 1978).

Hist. Cuth. — Ted Johnson South (ed. and trans.), Historia de Sancto Cuthberto: *A History of Saint Cuthbert and a Record of his Patrimony* (Anglo-Saxon Texts 3; Cambridge: D. S. Brewer, 2002).

Hist. Kings — *History of the Kings.* Edition: Thomas Arnold, *Symeonis monachi Opera omnia*, 2 vols. (Rolls Series; London: Longmans, 1882–5), II, pp. 1–283. Translation: J. Stevenson, *The Church Historians of England, III, part ii, containing the Historical Works of Symeon of Durham* (London: Seeleys, 1855), pp. 425–617 (facsimile reprint, Felinfach: Llanerch, 1987). Cited *sub anno* (*s.a.*), but also with page references from Arnold's edition.

Hugh Chanter — Charles Johnson, M. Brett, C. N. L. Brooke, et al. (eds. and trans.), *Hugh the Chanter: The History of the Church of York 1066–1127*, 2nd edn (Oxford Medieval Texts; Oxford: Clarendon Press, 1990).

John Worcs. — R. R. Darlington, P. McGurk and Jennifer Bray (eds. and trans.), *The Chronicle of John of Worcester, Volume II, The Annals from 450 to 1066* (Oxford Medieval Texts; Oxford: Clarendon Press, 1995).

Life Ceolf. — *Life of Ceolfrith* (also known as Anonymous, *History of the Abbots of Monkwearmouth and Jarrow*). Edition: Plummer, *Bede*, I, pp. 388–404. Translation: *EHD I*, no. 155.

Life Greg. — Bertram Colgrave (ed. and trans.), *The Earliest Life of Gregory the Great by an Anonymous Monk of Whitby* (Kansas: University of Kansas Press, 1968; repr. Cambridge: Cambridge University Press, 1985).

Life Willibald — O. Holder Egger, 'Vita Willibaldi episopi', in G. Waitz (ed.), *Scriptores* (Monumenta Germaniae Historica; Hanover: Hahn, 1887), XV.1, pp. 86–106; C. H. Talbot (trans.), *The Anglo-Saxon Missionaries in Germany being the*

	Lives of SS Willibrord, Boniface, Sturm, Leoba, and Lebuin, together with the Hodoeporicon of St Willibald and a Selection of the Correspondence of St Boniface (London: Sheed and Ward, 1954).
Nennius, *Hist. Britons*	Edition and translation: John Morris, *Nennius: British History and the Welsh Annals* (Chichester: Phillimore, 1980). Edition from one manuscript: D. N. Dumville, *The Historia Brittonum III: The 'Vatican' Recension* (Cambridge: D. S. Brewer, 1985).
Orderic	Marjorie Chibnall (ed. and trans.), *The Ecclesiastical History of Orderic Vitalis*, 6 vols. (Oxford Medieval Texts; Oxford: Clarendon Press, 1969–80).
Plummer, *Bede*	Charles Plummer (ed.), *Venerabilis Baedae Opera historica*, 2 vols. (Oxford: Clarendon Press, 1896).
RFA	*Royal Frankish Annals.* Edition: F. Kurze, *Annales regni Francorum qui dicuntur Annales Laurissenses maiores et Einhardi* (Monumenta Germaniae Historica Series Rerum Germanicarum VI; Hanover: Hahn, 1895). Translation: Bernhard Walter Scholz and Barbara Rogers, *Carolingian Chronicles: Royal Frankish Annals and Nithard's Histories* (Ann Arbor: University of Michigan Press, 1972).
Rog. Hove.	William Stubbs (ed.), *Chronica Rogeri de Houedene*, 4 vols. (Rolls Series 51; London: Longman, 1868–71).
Siege of Durham	*The Siege of Durham and the Probity of Earl Uhtred.* Edition: Thomas Arnold, *Symeonis monachi Opera omnia*, 2 vols. (Rolls Series; London: Longmans, 1882–5), I, pp. 215–20. Translation: Christopher J. Morris, *Marriage and Murder in Eleventh-Century Northumbria: A Study of 'De Obsessione Dunelmi'* (Borthwick Paper 82; York: Borthwick Institute of Historical Research, 1992), pp. 1–5.
Sources	David Rollason with Derek Gower and Gillian Fellows-Jensen, *Sources for York History before 1100* (Archaeology of York 1; York: York Archaeological Trust, 1998).
Stephen, *Life Wilf.*	Bertram Colgrave (ed. and trans.), *The Life of Bishop Wilfrid by Eddius Stephanus* (Cambridge: Cambridge University Press, 1927; repr. 1985).
Symeon, *Origins*	David Rollason (ed. and trans.), *Symeon of Durham, Libellus de exordio atque procursu istius hoc est Dunelmensis ecclesie*

	(On the Origins and Progress of this the Church of Durham) (Oxford Medieval Texts; Oxford: Clarendon Press, 2000), cited by book and chapter.
Tacitus, *Germania*	Edition and translation: M. Hutton, R. M. Ogilvie, E. H. Warmington et al., *Tacitus in Five Volumes*, 5 vols. (Loeb Classical Library; London: Heinemann, 1925–70). Translation: H. Mattingly, *Tacitus: On Britain and Germany* (Penguin Classics; Harmondsworth: Penguin, 1948).
Taliesin, *Poems*	Edition: Ifor Williams, *The Poems of Taliesin* (Medieval and Modern Welsh Series 3; Dublin: Dublin Institute for Advanced Studies) (English version by J. E. Caerwyn Williams). Translation: M. Pennar, *The Poems of Taliesin* (Felinfach: Llanerch, 1988).
Welsh Annals	*Annales Cambriae*, in John Morris (ed. and trans.), *Nennius: British History and the Welsh Annals* (Chichester: Phillimore, 1980), text, pp. 85–91, trans., pp. 45–9.
Wendover, *Flowers*	Roger of Wendover, *Flowers of History*. Edition: Henry Coxe, *Rogeri de Wendover, Chronica sive flores historiarium*, 4 vols. (English Historical Society; London: Longman, 1841–2). Translated extracts: *EHD I*, no. 4.
William Malms., *History*	Edition and translation: R. A. B. Mynors, R. M. Thomson and Michael Winterbottom, *William of Malmesbury, Gesta Regum Anglorum, The History of the English Kings* (Oxford Medieval Texts; Oxford: Clarendon Press, 1998).

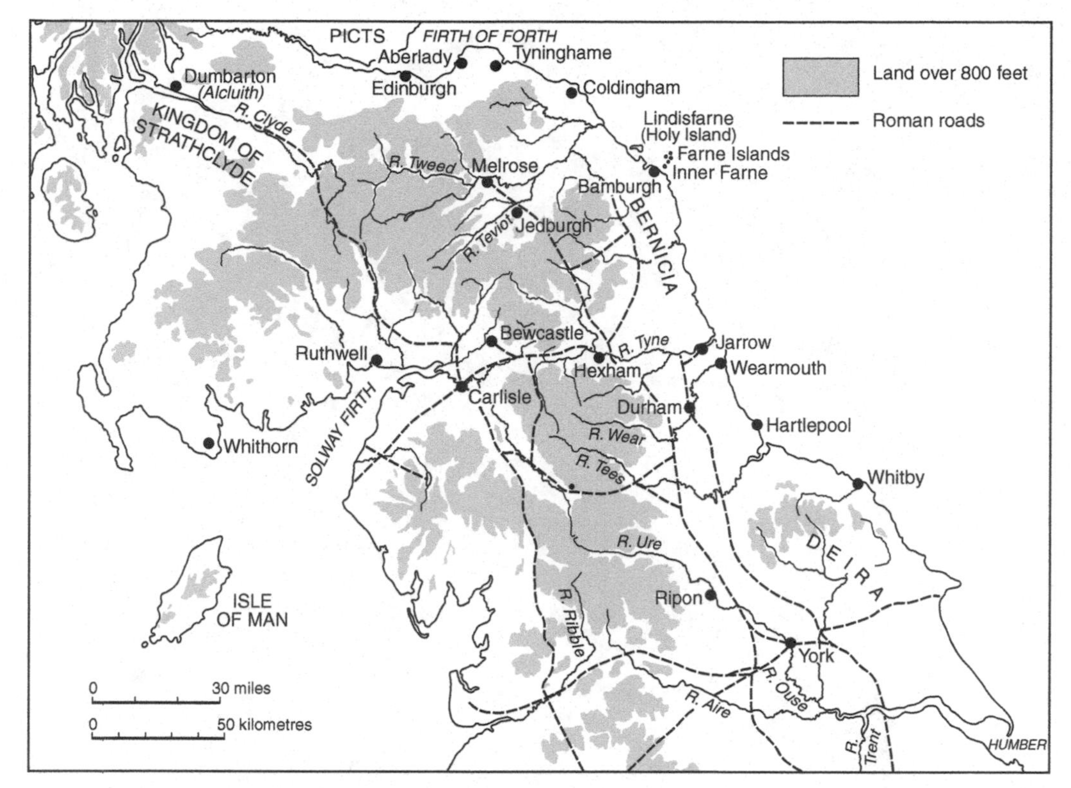

Northumbria: map for general reference

PART I

The kingdom of Northumbria

CHAPTER 1

Kingdoms, peoples and nations: Northumbria in context

Understanding the nature of political organization is a central preoccupation of historical research. Why do states form? Why are people willing to accept their authority, and how do they come to identify themselves as members of those states? In the formation of people's attitudes to the said states, what are the respective roles of ethnic origin, religion and culture? More prosaically, what are the mechanisms which hold states together and what are the processes which cause them to fragment or to coalesce? What are the relative roles of bureaucratic and judicial machinery on the one hand, and ideologies and their associated rituals on the other? Such questions are not the preserve of any one period or geographical area of historical research and they underlie much of the work that has been carried out in recent – and less recent – times, notably, for example, by historians of the modern period concerned with nationalism and the origins of nations. The fluidity of political structures in the middle ages, however, has encouraged historians to pose similar questions of that period, particularly in the context of the development of national identity, or of the relationship between belief in Christianity and affiliation to political structures, especially in circumstances of military and ideological expansion.[1]

Within the middle ages, the centuries from the end of the Roman Empire in western Europe to the rise of recognizable European states around 1100

[1] For nationalism and its possible pre-modern antecedents, see, for example, Anthony D. Smith (1986), and, for a lucid critique of recent scholarship, Anthony D. Smith (1998), pp. 57–95. For the middle ages, see, for example, the excellent series of articles by R. R. Davies (1994, 1995, 1996, 1997) and, specifically for the role of Christianity, Bartlett (1993), pp. 5–23.

offer a particularly promising field of study. A new political map of western Europe emerged within and beyond the former territory of that Empire, a map dominated by a multiplicity of political units which were predominantly kingdoms ruled by kings. The emergence of these units and their evolution into the embryonic forms of the states which were to dominate western European history until the modern period raise the questions outlined above in acute form. For the dissolution of the western part of the Roman Empire, which ceased to exist formally with the deposition of its last emperor in 476, was a long and complex process, more perhaps a transformation of European society and political organization than the cataclysmic process it was once regarded as being. It was accompanied by the presence in western Europe of substantial numbers of barbarian people (or 'Germans' as they are sometimes termed), who came from the lands to the east and north of the Roman Empire, and gave their names to the new kingdoms of western Europe, usually described as 'barbarian': the kingdom of the Franks, the kingdoms of the Lombards, the kingdom of the Visigoths and so on.[2] It is of central importance to understanding the history of these kingdoms, how far they had really been created by, and owed their characteristics to, these barbarians. To pursue this, it is necessary to investigate whether the barbarians were so numerous or powerful that they formed the dominant group in these kingdoms; or whether the original inhabitants remained the predominant element in the population but began to think of themselves as Goths or Lombards or Franks, thus transforming their Roman identity into a new barbarian identity. Historians have recently leaned towards the second of these possibilities, and this has caused them to consider the processes by which such transformations of identity might have occurred – what part was played in them by religion and intellectual and material culture, for example, and the extent to which people were in a position to make a conscious choice of their identity as either Roman or barbarian and to assume the appropriate trappings of this identity. These historical considerations are not limited to the period of the origins of the barbarian kingdoms, for western Europe was subject to other developments which pose similar problems of interpretation: the Viking raids and invasions of the ninth and tenth centuries, for example, raise issues of the manner by which the ethnically and religiously distinct Vikings, who were pagans from Scandinavia, were absorbed into Christian western European society and their contribution to it.[3]

[2] See, for example, Wolfram (1997).

[3] Geary (2002) is an accessible discussion of this approach, which also relates the issues to recent and contemporary political history, and provides additional bibliographical references. Its assertions

The early middle ages also offer a rich field for studying the origins of political structures and the ideologies which underpinned them, for the kingdoms which succeeded to the Roman Empire in the west were in important respects new political structures – even the office of king as it emerged in these centuries was a new one, and had certainly not been in general use amongst the Romans.[4] Moreover, the conversion of the Roman Empire from 'classical' paganism to Christianity in the fourth century, and the conversion of the barbarian groups from 'Germanic' paganism to Christianity after their settlement within the Empire or the former Empire from the fifth century onwards represented major cultural and religious shifts in western Europe, against the background of which the political changes we have been sketching occurred.[5] It is of considerable importance to understand the role of the new beliefs and of the institutions and political ideology of the Christian church in the development of political organization and of people's allegiance to the new political structures. Important also to understand are the mechanisms by which those political structures functioned in a period when, even if the written word may have been more important than was once thought, bureaucratic forms of government were much less developed than in later centuries, and political power may have rested much more on ideological foundations, on personal relationships and on the structures and processes of organizations other than the state, particularly the Christian church.[6]

It is of course feasible and rewarding to pursue these issues for Europe as a whole, or for one of the emerging nations of western Europe, such as France, Germany or England. There are, however, strong reasons for narrowing the scope to a smaller kingdom which proved as things turned out to have no future in the political map of western Europe. This makes it possible to consider the reasons for the failure and destruction of such a kingdom as well as for its formation and earlier success. Because the geographical scope is narrower, it also makes it possible to examine the functioning of political processes in relation to particular conditions and circumstances; and to relate the kingdom to a defined geographical region.

The suitability of the kingdom of Northumbria for this can best be appreciated from a brief sketch of its history across the period under consideration

about the history of Britain, however, need to be treated cautiously. A useful summary of the essential elements of the approach is in Pohl (1997).

4 Wallace-Hadrill (1971).

5 For general treatments, see Peter Brown (1997) and Fletcher (1997), which is notably lucid.

6 On literacy, see McKitterick (1990), and Clanchy (1979).

here, that is broadly from the end of the Roman Empire in Britain to the definitive absorption of the kingdom's former territory into the kingdom of England on the one hand and the kingdom of Scots on the other.[7] The emergence of Northumbria occurred at some time after the end of the Roman Empire in Britain (early fifth century), and in connection with the arrival in Britain of the barbarian groups who came to be known as the English. Obscure as this period is, we can at least discern that Northumbria appeared first as two kingdoms, that of Deira to the south of the River Tees and that of Bernicia to the north of it (map, p. xxvii). The earliest king of Bernicia, as Bede laconically informs us, was a certain Ida who began his reign in 547 and from whom 'the Northumbrian royal family trace their origin'. This is consistent with the materials, partly narrative, partly genealogical, which a British scholar known as Nennius compiled in the early ninth century, and with surviving ninth-century genealogies. These name Ida and also his successors as kings of Bernicia, giving a list of kings who are really no more than names until the time of Æthelfrith (d. 616), who appears to have expelled the heir to Deira, the future King Edwin, and to have ruled over a united Deira and Bernicia, forming for the first time the kingdom of Northumbria (fig. 1).[8]

The earliest king of Deira to find a mention in Bede's pages was an equally obscure figure, the father of Edwin, Ælle, whom Bede mentions only because he figured in the famous story of how Pope Gregory the Great, finding English slave-boys for sale in Rome and being told amongst other things that the king of their country was Ælle, declared that they should henceforth be taught to sing Alleluia – and thus, so the story goes, was inspired to undertake the conversion of the English people to Christianity.[9] In the period from 616 to 633, Edwin, king of Deira, ruled over Bernicia as well as his own kingdom, thus uniting Northumbria under Deiran rule; but on his death in battle in 633 the fortunes of his line suffered an irreversible decline, and it was the Bernician kings who again dominated Deira in their turn and forged the united kingdom of Northumbria which Bede knew. This process, Bede believed, began with Oswald (634–42), by whose efforts 'the kingdoms of Deira and Bernicia, which had up to this time been at strife with one another, were peacefully united and became one people'.[10] In fact, as Bede

[7] For general surveys, see Kirby (1991), pp. 77–112 and 142–62; and Yorke (1990), pp. 72–99.

[8] Bede, *Eccles. Hist.*, bk 5 ch. 24 (note that Bede does not state that 547 was the starting-point of the kingdom of Bernicia itself); Nennius, *Hist. Britons*, para. 57; for genealogies, Dumville (1976), pp. 30–7.

[9] Bede, *Eccles. Hist.*, bk 2 ch. 1.

[10] Bede, *Eccles. Hist.*, bk 3 ch. 6.

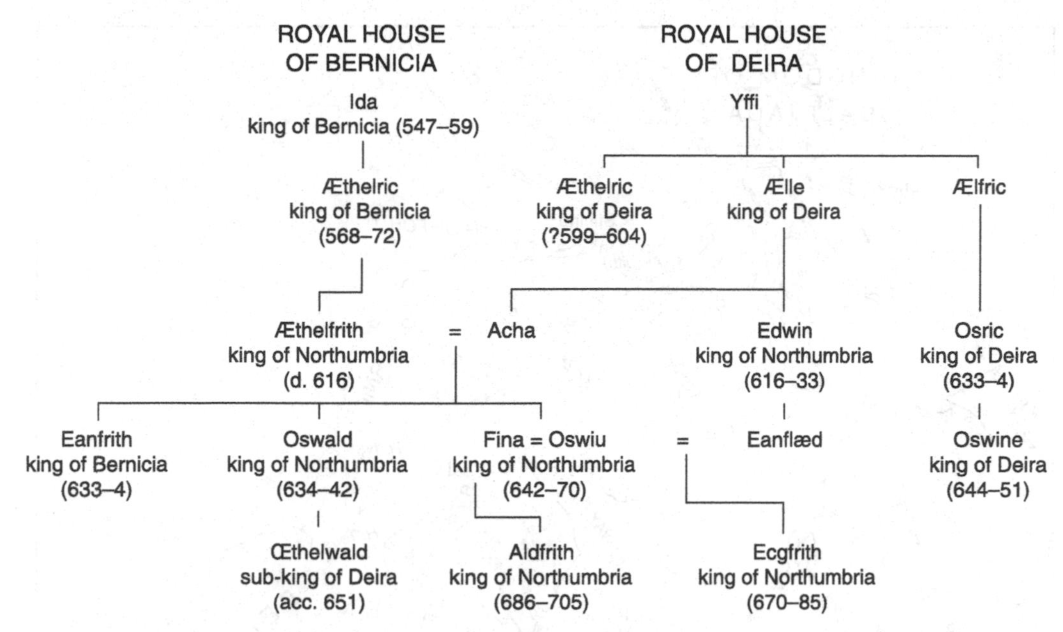

FIGURE 1 Simplified genealogical chart of the early kings of Deira and Bernicia
The dates of reigns are in many cases subject to discussion (see *Sources*, pp. 46–9, and Kirby (1991), pp. 61–112), and are especially problematic for Ida, Æthelric, Yffi, Ælle and Ælfric who belong to the sixth century. Note the marriage alliances between the two houses, involving Ælle's daughter Acha and Edwin's daughter Eanflæd. Note also the relationship between Ecgfrith and Aldfrith, the latter being termed illegitimate (*nothus*) by Bede (*Life Cuth.*, ch. 24); his mother was Irish and his name among the Irish was Fland, and Fína which was his mother's name (Plummer (1896), II, p. 263).

himself shows, a series of kings continued to rule in Deira after Oswald: Oswine, who was of the house of Edwin and was killed at the behest of Oswald's Bernician successor Oswiu in 651; Alhfrith, son of Oswiu, who disappears from history, presumed dead, after 664; and Œthelwald, a son of Oswald, who was killed in battle in 655. From then on, it appears that Deira and Bernicia were one kingdom, although we must be alert to the possibility that the influence of what had been was still felt in the united kingdom of Northumbria. In Bede's time, at any rate, such tensions could apparently be regarded as of the past, and at the end of his *Ecclesiastical History*, he was able to give a complacent picture of the state of his kingdom and of England in general: these were, he tells us, 'favourable times of peace and prosperity'.[11] Moreover, Northumbria was definitively Christian: the first conversion under

[11] Bede, *Eccles. Hist.*, bk 5 ch. 23.

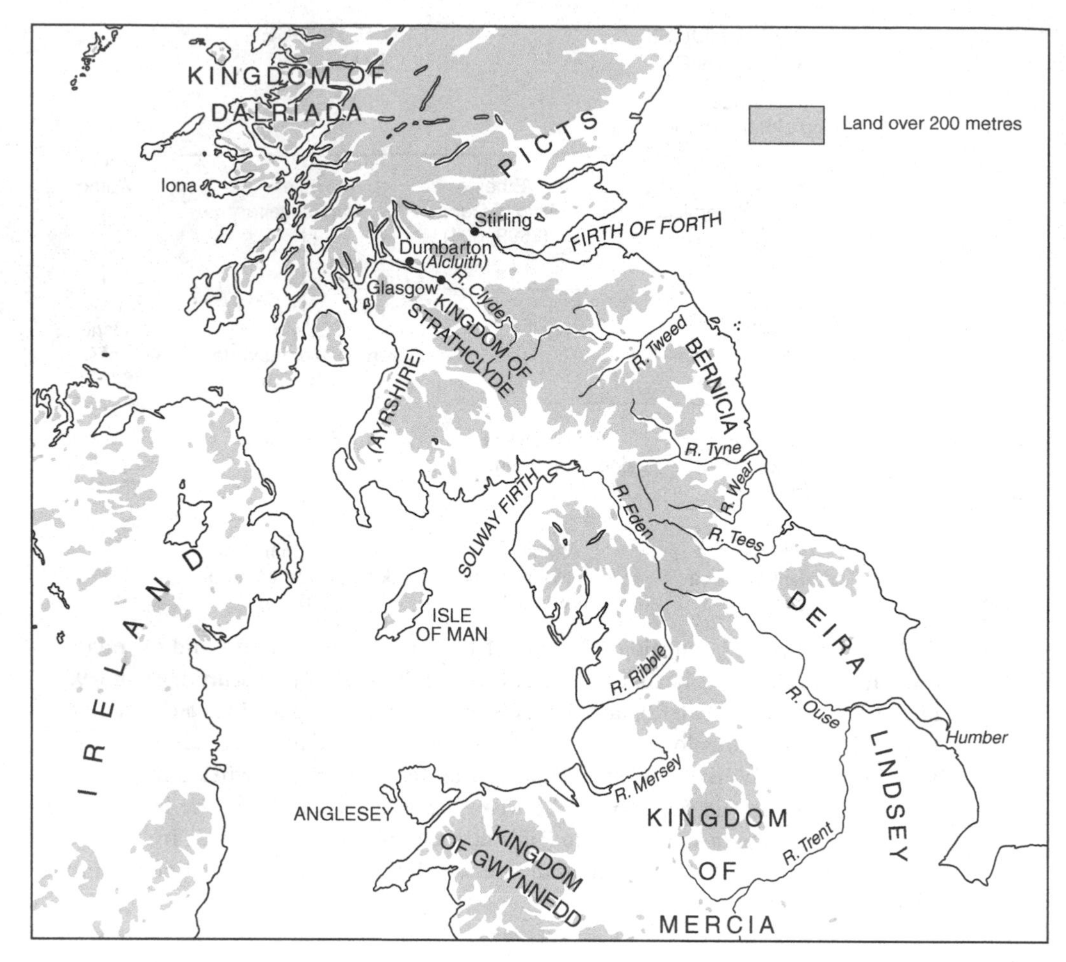

MAP 1 Map of northern central Britain, showing principal rivers, land over 200 metres, and the rough location of the principal kingdoms mentioned in the text
Note the location of modern Ayrshire to the north of the Solway Firth, and of Iona founded by Columba in 563.

Edwin had been brought to an end by his death; but Christianization had been resumed by Oswald and there had been no further apostasy.

At the height of its territorial power at least from the mid-seventh century to the ninth, the kingdom of Northumbria extended from the Humber in the south, northwards on the east side of the country to the Firth of Forth, westwards to the Irish Sea and the southern coastlands of the Solway Firth, and northwards of that, at least transiently, to take in modern Ayrshire (map 1). Moreover, according to Bede, its kings claimed at certain times dominance over southern England and even over other peoples of Britain to the north and west. Here then was a kingdom which not only extended its power very widely, but did so at the periphery of the power and organization of the former Roman Empire, the northern frontier of which, represented by Hadrian's Wall, bisected the territory of Northumbria, although Roman influence and organization had also extended to the north of that frontier. The potential for studying the processes by which this kingdom emerged in the aftermath of the end of Roman power is thus enhanced by the fact that we have the possibility of examining those processes simultaneously in areas which had lain both inside and outside the Roman Empire; and of exploring the relationship of incoming barbarians to both the Romanized and the non-Romanized native inhabitants of Britain, with its implications for the emergence of the peoples of the British Isles: English, British, Picts and Scots. Moreover, the conversion of Northumbria to Christianity in this period, and the growth of a notable Christian culture of which Bede himself was the foremost representative, offer the opportunity to consider the role of Christianity and Christian culture in the making of the kingdom.

Ultimately, however, Northumbria was unsuccessful as a kingdom. Already in the eighth century, the favourable times extolled by Bede had come to an end. On the death of King Aldfrith in 705, there was uncertainty over the succession to the kingship, and the man who eventually succeeded, Osred I, was violently done to death. There followed a fairly stable period, notably the reign of Eadberht (737/8–58), but the late eighth century was characterized by assassinations, exiles and coup d'états, with some kings reigning for very short periods, in the case of Osbald (796) for only a month before he was driven out. This pattern of political instability appears to have continued into the early ninth century, but beyond that we have no firm information as the chronology of the reigns of ninth-century kings is uncertain. In 866/7 York fell to a Viking army, and the kingdom of Northumbria was fractured into three areas: first, south of the River Tees, which was for nearly a century ruled by Viking kings; secondly, the area between the River Tyne and the Firth of Forth which fell to the lot of English rulers based at the old royal centre of Bamburgh, and the developing ecclesiastical lordship of the religious community which, formerly on the island monastery of Lindisfarne, now centred around the body of its principal saint, Cuthbert, wherever that might be taken; and thirdly Cumbria, the north-west portion of Northumbria, which emerged as a separate political unit, although whether an independent one or as a part of the kingdom of Strathclyde, which was centred on Glasgow and Dumbarton, is open to question. Here then we have the opportunity not only to investigate the nature of Viking influence on Northumbria and to consider again in this context the questions of ethnicity and cultural identity, but we can explore too the processes by which the kingdom was fractured and the implications of that fracturing for culture, political and ecclesiastical organization, and the identity of its inhabitants. Moreover, the processes of change continued with the expulsion in 954 of the last Viking king of York, Eric (sometimes identified with Eric Bloodaxe), and the consolidation – at least after a fashion – of the power in Northumbria of the kings ruling southern England, the English dynasty of Wessex and, for a time, the Danish dynasty of Cnut. The kings of Wessex, however, could effectively rule only a part of the old kingdom of Northumbria. In the north-west Cumbria had been absorbed into the kingdom of Scotland after the death of its last king Owain the Bald in or around 1018; and in the north-east the area north of the River Tweed (known as Lothian) had passed into the hands of the kings of Scotland. The pressure in the late eleventh and early twelfth centuries of the power of the kings of England from the south and that of the kings of Scotland from the north led to the creation of a frontier which was the Scottish Border in embryo. The creation of this frontier offers us our final opportunity in the context of this book to explore the processes of the making and destruction of political units in the early middle ages.

The following questions concerning the history of Northumbria will dictate the form of the subsequent chapters. First, what constituted the kingdom of Northumbria: what its frontiers were at various periods, what they signified, and how they related – if at all – to the frontiers along the lines of which the kingdom was dismembered. Secondly, the ethnic origins and sense of their own identity of the inhabitants of Northumbria. Did they constitute a new ethnic group or an amalgam of English ('Germanic') incomers with the native British population? Did they interact later with, or were they displaced by, the new groups which influenced Northumbria, the Vikings and the Normans? How far, if at all, did their identity as Northumbrians survive the changes and political fracturing of the ninth, tenth and eleventh centuries? Thirdly, the role of culture and religion in the kingdom of Northumbria: the relationship between any pre-existing Roman or British culture and the incoming English or 'Germanic' culture; and above all the role of the development of an imported Christian culture in the assimilation of incoming groups, whether the English of the post-Roman period or the Vikings of later centuries. Fourthly and finally, the framework of power which made it possible for the kingdom to be held together: the machinery of that power in terms of royal officers and governmental mechanisms, as well as the ideology and attitudes which underpinned it.

This is an ambitious programme, and we must be realistic about what can be achieved in the light of the evidence available to us. The written evidence is inevitably exiguous and fragmentary, given not only the remoteness of the period, but also the lack of institutions to preserve written records, and the relative lack of development of the bureaucratic organs of the state. Moreover, it is in some respects deficient when compared with the evidence for southern and midland England in the period before the Norman Conquest. For those areas, and especially for Kent and Mercia, the important sources are predominantly charters, that is to say documents recording the grant of lands and privileges, mostly to churches, and in this period exclusively by kings.[12] These charters, although they present problems, notably in establishing their authenticity and reliability, are of immense value for studying the succession of kings in the various kingdoms, the administrative machinery available to them, the personnel of their courts who were generally listed as witnesses to the charters, the extent of the privileges which kings were able to grant (exemption from service to the king and from taxation, for example) and also for the land itself, what it produced and how it was cultivated. The evidence of the charters bearing on this last point in particular is naturally patchy, but given the remoteness of the period the richness of these documents is amazing. They are also frequently our only witnesses to the existence of particular churches in southern and midland England. A whole group of monasteries in the west midlands, for example, would be wholly unknown to us had charters making grants to them not survived, albeit in the form of copies.[13] By contrast there are no surviving charters whatsoever from pre-Viking Northumbria apart from an almost certainly spurious charter of King Ecgfrith, and what appear to be summaries of a handful of charters relating to the church of St Cuthbert, contained in the text of an eleventh-century narrative from Durham, the *History of St Cuthbert*.[14] There are in fact very few charters from Northumbria at any period before the Norman Conquest: some documents pertaining to the lands of the church of York towards the end of our period, a single charter of King Athelstan (924–39), a single document (of the eleventh century)

[12] Sawyer (1968), nos. 1–226 and *passim*. For a classic discussion, see Stenton (1955); see also Scharer (1982).

[13] Sims-Williams (1990), pp. 115–76.

[14] *Hist. Cuth.* (see Hart (1975), pp. 117–42) and Sawyer (1968), no. 66 (charter of King Ecgfrith).

from Cumbria and three rather suspicious charters associated with the Yorkshire lands of Peterborough Abbey.[15] There is reason to suppose that charters were composed at least in seventh-century Northumbria, but it must be presumed that the disruption of ecclesiastical life in the Viking period meant that church archives were not preserved, except in part for the church of St Cuthbert which was somewhat less disrupted than other Northumbrian ecclesiastical establishments.[16]

Another source which is relatively rich for the south and midlands comprises the law-codes of the kings. For Kent, we have three from the seventh century; for Wessex, we have the important code of King Ine (688–726x?) and the laws of King Alfred (871–99). There are difficulties with interpreting the character and purposes of these codes; but they are unquestionably an extremely valuable source for social structure, royal government and justice, the working of the land (clauses in Ine's code provide the earliest evidence bearing on the origin of the open field system of agriculture) and so on. For Northumbria, we have no such codes and indeed no trace of legislation at all before the early eleventh century, when we have two documents of rather minor importance associated with Archbishop Wulfstan of York (1002–23), one called the *Northumbrian Priests' Law*, the other the *Law of the North People*. If the great Northumbrian kings of the pre-Viking period had written law-codes (and they need not have done), nothing whatever has survived.[17]

These gaps in the source-materials for Northumbria have a very considerable effect on our ability to fill in the history of that kingdom. For example, we can never be precise about land-holdings (charters in the south and midlands often have boundary clauses which permit very detailed reconstruction of estates);[18] we can never have any sharp focus on the distribution of royal estates (there is nothing like the will of King Alfred in Wessex which shows in detail the extent of the royal lands at a particular point in time);[19]

[15] The York documents are: Robertson (1956), nos. LIV and LXXXIV, and Sawyer (1968), nos. 712 (Sherburn-in-Elmet), 716 (Newbald), 968 (Patrington); the Athelstan charter is no. 407 (Amounderness), the Cumbria document no. 1243 (writ of Gospatric), and the Peterborough documents nos. 68 (Howden), 451 (Beverley), 681 (Howden and Old Drax).

[16] See below, pp. 175–6.

[17] The laws are ed. and trans. Attenborough (1922), and Robertson (1925); trans. also in *EHD I*, nos. 29–53; for editions and detailed commentary, see Liebermann (1903–16) and Hector Munro Chadwick (1905), pp. 76–114. For modern scholarship, see Wormald (1998) and, on open fields, Aston (1983). For *Law of the North People*, see below, p. 244; for the *Northumbrian Priests' Law*, see below, pp. 170, 237, 244.

[18] *Encyclopaedia ASE*, entry 'charter bounds' and references therein; see also Hooke (1998), pp. 84–102.

[19] Trans. Keynes and Lapidge (1983), pp. 173–8 (with map), and *EHD I*, no. 96; ed. Harmer (1914), no. 11.

we have no systematic picture of the social structure of Northumbria (such as is provided, albeit in idealized form, by law-codes such as that of Ine); and we are missing the considerable possibilities which exist in the south and midlands for documenting churches such as those of the west midlands mentioned above.

We need not despair, however, of writing this book. If we have virtually no charters and no law-codes for Northumbria, we do have a much richer collection of historical and hagiographical texts at least for the pre-Viking period than exists for the south and the midlands. Aside from some annals in the *Anglo-Saxon Chronicle* of very doubtful reliability and concerning mostly Kent and Mercia, and some Kentish hagiography of possibly eighth-century date, the south and midlands have nothing to offer before the time of King Alfred when the *Anglo-Saxon Chronicle* becomes an important source, and writings such as Asser's *Life of King Alfred* appear. For Mercia, we have absolutely no sources of this type at all, apart from the *Life* of the hermit of Crowland, Guthlac, written between perhaps 730 and 740 by Felix, who may have been a monk of Repton; this work has something to say of the Mercian kings, particularly Æthelbald, but not a great deal.[20]

By contrast, Northumbria was extremely productive in Latin prose and verse, and it is from the Northumbrian writer Bede, principally his *Ecclesiastical History of the English People*, completed in 731, that the narrative history of the southern and midland kingdoms down to that date has to be largely reconstructed; but the most detailed information it provides is about Northumbria itself.[21] In addition to this work, we possess a series of hagiographical texts of the first half of the eighth century: two prose *Lives of St Cuthbert* (bishop of Lindisfarne, d. 687) by a monk of Lindisfarne and by Bede;[22] a metrical *Life* of the same saint by Bede;[23] a *Life of Bishop Wilfrid* (bishop of York, founder of the monasteries of Ripon and Hexham; d. 714) by Stephen of Ripon;[24] and a *Life of Pope Gregory the Great*, written at Whitby and incorporating information about Northumbria, principally about the Northumbrian king Edwin (616–33).[25] Further, we have two accounts of the joint monasteries of Monkwearmouth and Jarrow, one by an anonymous monk of those monasteries and usually known as the *Life of Ceolfrith*, and one by Bede, entitled the *History of the Abbots of Monkwearmouth and Jarrow*.[26]

[20] On *ASC*, see Sims-Williams (1980), pp. 1–41; on Kentish hagiography, David Rollason (1982), pp. 15–18, 33–40; *Life of Guthlac* is ed. and trans. Colgrave (1956); for authorship, see pp. 15–19, for Æthelbald, pp. 164–7.

[21] Bede, *Eccles. Hist.* [22] Anon., *Life Cuth.*; Bede, *Life Cuth.*

[23] Bede, *Met. Life Cuth.*; on this work, see Lapidge (1989).

[24] Stephen, *Life Wilf.* [25] *Life Greg.* [26] *Life Ceolf.*; Bede, *Hist. Abbots.*

Finally, Bede himself composed a chronicle (the *Greater Chronicle*), as part of his work *The Reckoning of Time*, which was primarily about performing chronological calculations.[27]

We should notice at once that this corpus of texts casts light chiefly on the history of seventh-century and to a lesser extent early eighth-century Northumbria. For the sixth century and before, we have available only various British writings which are, from a strictly historical viewpoint, of very unsatisfactory character. Gildas, *On the Ruin and Conquest of Britain*, which was written before 600, possibly in Wales, does not specifically concern Northumbria at all and is indeed written in a very generalized way.[28] Nennius, *History of the Britons*, which was compiled in the early ninth century, possibly in Wales, possibly in north Britain, does contain information about Northumbria in the sixth and more particularly in the seventh century. It is, however, very hard to evaluate this information, which is sometimes at variance with that given by Bede. It has been argued that Nennius absorbed without significantly modifying a variety of early sources, and that parts of his work are therefore of very high authority. It seems more likely, however, that, despite his modest avowal in his preface that he has simply heaped together what he found, Nennius was producing his *History* by modifying, in an attempt to collate and synchronize it, information from various sources, some perhaps reliable, some perhaps not.[29] Even more problematic is a series of British poems apparently composed in north Britain but preserved in Wales. Those which appear directly to concern the area of Northumbria are *The Gododdin* attributed to Aneirin, which is a series of elegies for British heroes fallen at an engagement at a place called *Catraeth*, usually identified as Catterick in Yorkshire; and the poems of Taliesin, which eulogize the martial activities of a king of Rheged called Urien, notably a battle he fought against English forces in an area identified as Wensleydale. As we shall see, these poems offer considerable potential for studying the developing hostility of the British to the English; but, literary in form and of uncertain date, they are probably unusable as guides to the early history of Northumbria.[30] Also preserved in later medieval Wales were genealogies of dynasties of British rulers apparently active in north Britain in the earlier part of our period: these are equally difficult to interpret, partly because of the late date of

[27] Bede, *Reckoning*; trans. (sixth age only) in McClure and Collins (1994), pp. 307–40.

[28] Gildas, *Ruin*.

[29] Nennius, *Hist. Britons*, on which see Dumville (1975–6, 1986, 1989a). Jackson (1963b) argued that Nennius was drawing on a British historical compilation of the late seventh century, but this has now been disproved by Dumville.

[30] Aneirin, *Gododdin*, Taliesin, *Poems*; for discussion, see Dumville (1977); see further below, pp. 101–3.

the manuscripts in which they have been preserved, but also because they too are essentially literary in form, their content and structure influenced by later rulers seeking legitimation for their own political claims.[31] The result of all this is that available information for the sixth century – and for the northern and western neighbours of Northumbria at any period – is very fragmented and unsatisfactory, whereas that for the seventh is much richer.

The Northumbrian sources detailed above tell us a little about the early eighth century, although Bede is notably more reticent about events after the late seventh. After he laid down his pen in 731, however, the character of the Northumbrian sources changes markedly. In book 5, chapter 24 of his *Ecclesiastical History*, Bede included a series of annals, which he had compiled (those concerned with Northumbria extended from 547 to 731); and similar annals had appeared in Bede's chronicle in his *The Reckoning of Time*.[32] This appears to have inaugurated a process of keeping annals of Northumbrian history, for it is clear that such annals were preserved from 732 until perhaps 806. These annals, known here as the Northern Annals, have not survived in their original form, but their content can be partially surmised from their preservation to a greater or lesser extent, first in the so-called Moore manuscript of Bede's *Ecclesiastical History* (which has annals for 731–4), and then in the following later works: continuations to Bede's *Ecclesiastical History* found in twelfth-century and later manuscripts and comprising annals from 732 to 766;[33] the northern recension of the *Anglo-Saxon Chronicle*;[34] Symeon of Durham's early twelfth-century *On the Origins and Progress of this the Church of Durham*,[35] the somewhat later twelfth-century *History of the Kings*, which was associated in the twelfth century with the same writer and may well have been compiled by him;[36] the *Chronicle of Melrose*;[37] and the work of Roger of Hoveden in the late twelfth century. This last writer incorporated wholesale into his own history a work called the *History after the Death of Bede* which seems to have been compiled in Durham largely on the basis of an early version of the *History of the Kings*, perhaps in the 1120s.[38] Comparison between the sources listed above makes it possible to reconstruct

[31] See, for example, Miller (1975). [32] See above, n. 27.
[33] Bede, *Continuations*. [34] *ASC* (D and E). [35] Symeon, *Origins*, bk 2.
[36] *Hist. Kings*, *s.a.*732–802 (pp. 30–68, with interpolations). On authorship, see Peter Hunter Blair (1963) and, for new evidence pointing to Symeon as author, David Rollason (1998), p. 10; Norton (1998b), pp. 101–2; and Story (1998), pp. 212–13.
[37] Facsimile Anderson, Orr and Anderson (1936).
[38] Rog. Hove. On the *History after the Death of Bede*, see Piper (1998), pp. 302–3. In the absence of detailed study, the question of the relationship between this text and *Hist. Kings* remains problematic; for a different view, see Gransden (1974), pp. 225–6.

with some confidence part at least of the content of the annals from 732 to 806. That they were based on authentic information is shown in part by the details which they contain (for example, the names of otherwise unknown Northumbrian princes and aristocrats), which are unlikely to be the work of a later compiler, in part by the accuracy of their recording of astronomical information which can only have been known by a contemporary. A good example is the account given under the year 756 of how a 'bright star', now known to be the planet Jupiter, passed behind the eclipsed moon; this corresponds to a 'transit' of Jupiter which can be reconstructed by astronomical computation as having occurred not admittedly in 756 but in 755 (the chronology of the annals must be slightly out). The exact place of composition of these annals has been debated, but the level of detail relating to York and the indications in their Latin style that their composition may have been influenced by Alcuin point to their having been written at York. Some scholars refer to them as the 'York Annals'.[39] It is possible that the tradition of keeping annals was continued in Northumbria after 806, but if so all that has been preserved relating to the ninth century are some annals of dubious accuracy in Roger of Wendover's *Flowers of History*, together with isolated annals in other sources such as Symeon of Durham's *On the Origins and Progress of this the Church of Durham*.[40]

The chronological pattern of these sources is of immense importance for understanding what can or cannot be said about Northumbrian history. For the sixth century, we are hardly informed at all. For the seventh, we have the information supplied by the early eighth-century group of histories and hagiographies. From 731 to the early ninth century, our principal sources are annalistic – laconic and to a large extent lacking in the sophistication of the early eighth-century texts. For the ninth century, we are very poorly informed indeed; and even for such momentous events as the conquest of York by the Vikings in 866/7 we have to depend either on southern sources, principally Asser's *Life of Alfred*, or on twelfth-century sources such as Symeon's *On the Origins and Progress of this the Church of Durham*.[41]

For particular centres, however, we have a little more material at our disposal. York was the original home of Alcuin, the scholar who rose to prominence at the court of the Frankish ruler Charlemagne. Aside from his

[39] See *Encyclopaedia ASE*, entry 'Northern Annals' and references therein, especially Peter Hunter Blair (1963). See also Story (2003), pp. 126–33. For discussion of the 756 annal, I am grateful to Richard Stephenson, Department of Physics, University of Durham.

[40] Wendover, *Flowers*; Symeon, *Origins*, for example, bk 2 ch. 3.

[41] *Sources*, p. 71, where the date of the Vikings' landing in East Anglia is erroneously given as 866 (cf. David Rollason (forthcoming)).

possible influence on the Northern Annals, he certainly composed at the end of the eighth century a long Latin poem, embodying a versification of parts of Bede's *Ecclesiastical History*, together with some original material, and some very valuable information, especially on York in that period. Moreover, his letters have survived and several of them concern York and events more widely in Northumbria.[42] An unidentified monastery connected in some way with Lindisfarne, possibly Bywell on the River Tyne, possibly Crayke near York, produced in the early ninth century a Latin poem about its own history, the *De abbatibus*, by a certain monk of the church called Æthelwulf.[43]

From the Viking conquest of York onwards, our sources are exiguous and in many respects unsatisfactory. The *Anglo-Saxon Chronicle* becomes more useful, especially the sections of the D and E versions of this work which relate more to the north than to the south and Wessex. But the record is nevertheless a thin one, and it is only marginally enriched by the annals contained in the two sections of the *History of the Kings*, a laconic and in many ways obscure chronicle running from 888 to 957, and another, based on the early twelfth-century chronicle of John of Worcester, running from 848 to 1118.[44] As we proceed through the eleventh century, Anglo-Norman writers such as John of Worcester begin to have some information of interest;[45] and the writings of Durham itself assume some importance – from the remarkable text called the *History of St Cuthbert* produced in Durham, probably in the eleventh century,[46] to the historical writing produced at Durham in the Anglo-Norman period, especially Symeon of Durham's *On the Origins and Progress of this the Church of Durham*. None the less, the record is a sketchy one, and the lack of a body of charters is a serious handicap.

If the written sources are uneven, it is all the more important that we should in our investigations exploit to the full the non-written evidence which, although it naturally presents considerable problems of interpretation and has unevennesses of its own, is nevertheless remarkably rich for Northumbria. Archaeological evidence includes sites of European significance which have a direct bearing on the agenda we have laid out. The excavations carried out in recent decades at York offer very considerable

[42] *Encyclopaedia ASE*, entry 'Alcuin of York'; Alcuin, *BKSY* (for comment, see Bullough (1981), pp. 339–59); and Alcuin, *Letters*.

[43] Æthelwulf; for the location, see John Blair (1992), at pp. 227–8; for arguments in favour of Crayke, see Lapidge (1996), p. 96.

[44] *Hist. Kings*; see Peter Hunter Blair (1963), pp. 104–11.

[45] John Worcs.

[46] *Hist. Cuth.* An original core dating to the mid-tenth century was identified by Craster (1954), but Johnson South (2002) convincingly throws doubt on this; see also *Sources*, p. 22.

potential: in the case of those under York Minster, for understanding the transition from Roman to barbarian rule in the principal centre of Roman power in northern Britain;[47] in the case of excavations in the more easterly part of the city, especially the sites of Coppergate and Fishergate, extremely well-preserved archaeological deposits have permitted considerable insight into the nature and character of Viking activity in York.[48] Equally remarkable is a series of ecclesiastical sites excavated in the area of the kingdom of Northumbria, notably those at Bede's own monasteries of Monkwearmouth and Jarrow but also those at the former British church at Whithorn in Galloway, at Hartlepool in County Durham, at Whitby in Yorkshire and to a lesser extent at St Abb's Head near Coldingham in southern Scotland.[49] Alongside these must be set a series of secular sites, including settlements at Sprouston in Roxburghshire, Thirlings in Northumberland and Wharram Percy, Cottam and West Heslerton in Yorkshire,[50] but above all the palace-site of Yeavering in the fringes of the Cheviots, representing the most complex and potentially important site of its kind to be excavated in Britain.[51] These are significant sites, but the record they offer is inevitably uneven. The same is true of the record of surviving cemeteries which can be identified for the period from the fifth to perhaps the eighth centuries, but are heavily concentrated chronologically in the sixth century and geographically in the area south of the River Tees, particularly in eastern Yorkshire. For the area north of the Tees, the record is sporadic.[52]

This, however, by no means exhausts the non-written evidence. The existence of an extraordinary series of illuminated manuscripts assigned to Northumbria, including the Lindisfarne Gospels, is well known and its importance for understanding the cultural development of the Northumbrian church in the late seventh and eighth centuries is considerable.[53] It can be set alongside metalwork and other 'mobiliary' art, notably the remarkable

[47] Phillips, Heywood and Carver (1995).

[48] Richard A. Hall (1984b, 1994); for more specialist reports, see Addyman (1976–).

[49] Survey of older excavations in Cramp (1976). For a useful update, see Webster and Backhouse (1991), pp. 108–56. Specialist studies: on Monkwearmouth and Jarrow, Cramp (1994); on Whithorn, Peter Hill (1997); on Hartlepool, Daniels (1988), on Whitby, Peers and Ralegh Radford (1943) and, for more recent interpretations, Rahtz (1995); on Coldingham, Alcock, Alcock and Foster (1986).

[50] On Sprouston, see Ian Mervin Smith (1991); on Thirlings, O'Brien and Miket (1991); on Wharram Percy, Beresford and Hurst (1991); on Cottam, Richards (2000a), and on West Heslerton, Powlesland (2000).

[51] Hope-Taylor (1977).

[52] Lucy (1999). For a useful summary of the archaeological evidence for pre-Viking Northumbria with maps, see Cramp (1988).

[53] Kendrick, Brown, Bruce-Mitford et al. (1956–60) and, for a more accessible account with illustrations, Backhouse (1981); see also below, pp. 146–57.

group of treasures recovered from the grave of St Cuthbert when it was opened in 1829.[54] An important series of surviving church buildings, including Bede's churches at Monkwearmouth and Jarrow, the almost complete church of Escomb and the near-unique crypts attributable to Bishop Wilfrid at Ripon and Hexham respectively,[55] is complemented by one of the major strengths of Northumbrian history: a body of religious carving on stone much larger than anything comparable in the south (although paralleled by areas of Mercia, notably Derbyshire).[56] This sculpture, which permits conclusions to be drawn about the distribution of ecclesiastical activity as well as about cultural influences and the transmission of artistic and iconographical ideas, includes not only two of the most sophisticated monuments of early medieval Europe, the crosses of Bewcastle and Ruthwell,[57] but also a considerable number of lesser monuments. These latter are especially common in the Viking period and form a crucial source for the progress of Christianity in areas dominated by Viking rulers.[58] The numismatic evidence, that of coins in other words, is not comparably important for Northumbria, but does offer significant insights into the history of that area in the ninth century, and above all in the period of the Viking kings from the late ninth to the mid-tenth centuries.[59] Finally, although we lack all but the most fragmentary remains of written sources in the Old English or Old Norse vernaculars, the evidence of place-names has formed an important basis for discussion of linguistic and more general cultural influences in Northumbria in both the period of its origins and that of Viking dominance.[60]

[54] Battiscombe (1956); see also Bonner, Rollason and Stancliffe (1989), pp. 231–366.
[55] Taylor and Taylor (1965–78), I, pp. 338–49, 432–46, 234–8, and Bailey (1991).
[56] Cramp (1984), Bailey and Cramp (1988) and Lang (1991); on Derbyshire, Routh (1937), Cramp (1977) and Lynda Rollason, Armstrong, Baty et al. (1996).
[57] See, for example, O Carragáin (1978) and Cassidy (1992). [58] Bailey (1980).
[59] For references, see North (1994) and Grierson and Blackburn (1986), and below, pp. 176–7, 224–7.
[60] Gelling (1997).

CHAPTER 2

The kingdom of Northumbria: frontiers and heartlands

In the time of Bede, the kingdom of Northumbria covered an area very different from the medieval and modern (pre-1974) county of Northumberland (the area between the Rivers Tyne and Tweed) or even the area between the Rivers Tees and Tweed which is the Northumbria of present-day tourist promotion. To define its territorial extent, its frontiers and the regions which composed it is the principal aim of this chapter. Deeper than this, however, the chapter also aims to explore the fundamental question of how a kingdom like that of Northumbria functioned spatially. In the context of studies of the middle ages, this question can be resolved into two subsidiary issues: that of the nature of frontiers on the one hand; and that of heartlands and peripheries on the other.

First, the question of frontiers. It is always a temptation to represent medieval states by means of lines on a map, as if their frontiers were as clearly defined and as rigorously controlled as we expect those of modern states to be. But were medieval frontiers like that? First, there is the problem of whether they were lines (linear frontiers) at all, or frontier zones, in which the transition from one kingdom to the next was a graded continuum rather than the sudden change which is implied by a line. The archaeological remains of some frontiers, notably Offa's Dyke separating Wales from England (ill. 1), the Danevirke separating Denmark from Carolingian Frankia to the south, and Hadrian's Wall, seem at first glance to leave no doubt that they were linear frontiers. In fact, even these may have functioned in a more complex way. They may have been permeable, allowing free passage of people

ILLUSTRATION 1 Offa's Dyke, looking north across the Mainstone Valley (Shropshire)
The path in the foreground runs along the now much reduced ditch, with the still massive bank rising to the right. The Dyke can be discerned clearly rising up the far side of the valley. For description, see Fox (1955), p. 129.

and goods, and thus designed to control but not to prevent movement across them and activity on either side of them. According to a classic study, the existence of English settlements west of Offa's Dyke, combined with the presence of deliberately created gaps in the Dyke, show that it was a permeable frontier, more concerned with prestige on the one hand and the control of a frontier zone on the other than with preventing free passage. In the case of Hadrian's Wall, emphasis has been laid on the interpretation of the large gateways in each of its milecastles as features to promote rather than hinder passage, and on the impact of Roman influence on a wide zone to the north of the Wall as an indicator of its permeability. Thus the distinction between a linear frontier and a frontier zone is blurred. Reasonably clear-cut examples of frontier zones have been identified for the later middle ages in the area of English settlement in eastern Ireland, the Welsh Marches and the eastern frontier of the German empire, pressing always eastwards against the Slav peoples who were its neighbours. The shifting and unstable character of such frontier zones, often in relation to the military expansion which had often created them, has provided a rich subject through which to understand the nature of medieval frontiers.[1]

The concept of a linear frontier poses the problem of whether a kingdom recognized different frontier lines for different aspects of its activities: a frontier line applying to the incidence of military service on inhabitants, for example, another for financial matters (who had to pay dues), another for ecclesiastical organization (who owed tithe payments to which church) and yet another for legal jurisdiction, separating areas where different law and legal procedures were applied. As for frontier zones, research into the Scottish border-lands in the later middle ages, for example, has shown the development of law specific to that border-land which comprised areas on either side of and had no reference to the (then linear) frontier between England and Scotland.[2] This problem of what a frontier actually defined is especially acute in the early middle ages, where historians must routinely confront not only states in the process of formation, but also (as in the later middle ages also) mosaics of quasi-independent jurisdictions, such as ecclesiastical and other liberties, and areas of authority detached from their parent kingdom. In particular, the relationship between diocesan and political boundaries has been a particular area of concern to historians. Since the conversion of the English to Christianity was generally effected through the agency of kings who were the first converts, it is often supposed that there was a direct relationship between political frontiers and diocesan boundaries. Thus the bishop of a particular kingdom, appointed as a result of the conversion of the ruler of that kingdom, would naturally have had a see which reflected political frontiers. For example, the supposed frontiers of the kingdom of the Hwicce are assumed to be those of the diocese of Worcester, those of the kingdom of the Magonsaete, those of the diocese of Hereford and so on.[3] How far, if at all, such assumptions are well founded in the case of Northumbria we must consider.

Secondly, the question of heartlands. Just as it is tempting to draw sharp lines on an historical map to demarcate medieval states, so it is tempting

[1] In general, see the stimulating treatment of the Roman frontiers by Whittaker (1994); see also Power and Standen (1999), pp. 3–6, and Pohl, Wood and Reimitz (2001). For the classic study of Offa's Dyke, see Fox (1955), pp. 279–81; more recent excavation has suggested, however, that the Dyke was a continuous barrier (*Encyclopaedia ASE*, pp. 342–3). On Hadrian's Wall, see, for example, Breeze and Dobson (2000), pp. 39–43, 46; see also Elton (1996), pp. 111–13. On Ireland, see Perros (1995) and Barry (1995). On the Welsh Marches, see R. R. Davies (1978), pp. 15–33 ('there was no well-defined boundary between England and Wales in the medieval period' (p. 15)). On the Slav frontier, see Bartlett (1993), pp. 197–242.

[2] Neville (1998). Note, however, that even frontier zones could be conceived of as having defined limits.

[3] See, for example, Sims-Williams (1990), pp. 43–4.

to shade the area so demarcated in a uniform colour suggesting that every part of the state is subject to the same level of governance and control. That this need not be the case in the middle ages is shown by research into the rule of the Carolingian kings and emperors on the continent, as their power developed and reached its apogee in the late eighth and early ninth centuries. Scholars have shown how their realm, which extended over a vast area from the English Channel to the Mediterranean and from south-east Spain eastwards to the River Elbe, can be resolved into a series of zones where the rulers enjoyed power of differing intensity. At the core of their power, quite literally, were the rulers' heartlands (*Kerngebiete*), comprising: in the west the heartland of Neustria, focused on the basin of the River Seine and its tributaries such as the Marne and extending south perhaps to the River Loire; and in the east the heartland of Austrasia, roughly a triangle bounded by the Rivers Rhine, Meuse and Moselle with an extension eastwards into the valley of the River Main. Mapping the movements (or 'itineraries', as they are called) of the Carolingian rulers shows clearly that it was in these areas that they spent most of their time, when they were not on military campaigns, and it was in these areas that they assembled their most important council meetings. Their palaces were concentrated there; and in so far as we can map the landed estates of the rulers, they too were concentrated in these heartlands, and the history of the Carolingian family tends to confirm that their ancestral lands lay there. Certainly, the most important churches which they founded or patronized were there; and their most prominent aristocratic followers had their family origins there. Beyond the heartlands lay an intermediate zone where the power of the rulers was much less intense, which was very largely without Carolingian palaces, and which the rulers seldom visited except in transit. This zone was made up of regions which were in effect sub-kingdoms (*regna*) under the control of a king (Lombardy and Aquitaine, for example) or of a governor or group of governors (as in the case of Bavaria and Saxony) whose power was comparable to that of a king. The power of the Carolingian rulers was thus in effect mediated in this zone. Beyond it lay the marches, areas organized for military defence under officials known as marquises, and which may have been more firmly under the control of the Carolingian rulers than were the *regna*, thus forming as it were a hard shell around their lands.[4]

In the context of early medieval Irish kingship, comparable gradations have been identified, but with another layer of complexity; namely, areas

[4] See Werner (1980), and Ewig (1967). On particular aspects, see Gauert (1965), Brühl (1968), Ewig (1963), Thompson (1935), Prinz (1965), Werner (1979) and Hlawitschka (1967).

which were in effect not subject to the governmental control of the ruler of the kingdom which claimed hegemony over them, but simply recognized their subservience by payment of a tribute in money or in kind.[5] Such tribute relationships are also clearly recognizable in the tenth century between the Ottonian rulers of Germany and their Slav neighbours to the east;[6] and in the story which the Norwegian traveller Ohthere told King Alfred of Wessex when he visited his court in the late ninth century. Ohthere explained to the king that the Lapps, who were his neighbours to the north, paid annually to himself and his fellow countrymen a tribute of 'the skins of beasts, the feathers of birds, whale-bone, and ship-ropes made from whale-hide and seal-skin', presumably because they had formerly been defeated by them and forced to accept responsibility for this tribute by way of making terms.[7] In this example, we have authority at its most tenuous: the imposition of what were effectively protection payments on a defeated group. Discrete frontiers and uniform levels of governance and control are the hallmarks of the modern state. Arrangements like that described by Ohthere would seem to relate to states in only an embryonic stage of development. An aim of this chapter is to ask: where did the kingdom of Northumbria stand in this spectrum of development and what light can study of it throw on the broader questions about early medieval state formation which they raise?

We must not expect too much, since it is here perhaps that we feel most keenly the lack of charters amongst our sources. A body of charter evidence, which has not survived for Northumbria, would potentially make it possible to map areas of royal influence, the places of notable assemblies and the origins of persons associated with the king recorded in witness-lists.[8] The historiographical and hagiographical sources at our disposal are poorly equipped to answer questions in any systematic way, and much of what we can learn about the kingdom of Northumbria's frontiers, heartlands and peripheries has to be gleaned from scattered clues and allusions in them. They are richest, as we have seen, for particular periods, chiefly the seventh century, and we shall inevitably find that it is easier to answer our questions for that period. References explicitly to the movements and actions of

[5] Byrne (1995); on the British Isles in general, see Charles-Edwards (1989), pp. 28–33.
[6] Reuter (1991), pp. 253–64.
[7] For the text, see Lund, Fell, Crumlin-Pedersen et al. (1984), pp. 18–22.
[8] For mapping royal power on the basis of charters, see Lemarignier (1965); for tabulation of witnesses in Anglo-Saxon charters chiefly from other parts of England, see Keynes (1998); for charters used to demonstrate the extent of royal power, see Keynes (1993).

kings and leaders, or the locations of palaces and centres of government, are nevertheless all too rare.

Fortunately, the location of churches, for which our evidence tends to be richer, may provide some supporting evidence. In the case of frontiers, where we find churches in a particular area closely connected with churches centrally placed in Northumbria this can perhaps be used as supporting evidence that the area in question was also within Northumbria. There have to be considerable reservations about this approach, since we know, for example, that the prominent Northumbrian churchman Wilfrid (d. 709) founded, and was responsible for, churches far beyond Northumbria; and others may have acted likewise.[9] A related approach is to consider the stylistic features of architecture or ecclesiastical sculpture in stone, which are often the only evidence we have for the existence of a church in our period. Where these features are characteristically Northumbrian, this can be regarded as evidence that the churches in question fell within Northumbria. But, since it is clear that styles originating in Northumbria were found elsewhere, this too must be used with due reservation and only to support other evidence.[10] As regards heartlands, it is reasonable to assume (by analogy with what we know of other kingdoms) that before the late eighth century, only kings were able to grant lands to the church so that churches were founded on royal land, or, where they were founded on the lands of aristocrats, with royal involvement and consent. Concentrations of churches may thus provide a clue to the existence of concentrations of royal lands, on which they were founded; or to concentrations of lands in the hands of aristocrats with whom the kings were on such terms as actively to participate in their endowment of churches, as well as passively consenting to it.

In attempting to define the frontiers of the kingdom of Northumbria at its greatest extent, we shall undertake a tour, beginning in the south-east and moving clockwise around the kingdom. Bede is explicit that the great inlet of the Humber, which now separates Yorkshire from Lincolnshire, was in his time the frontier of Northumbria, for he tells us in the *Ecclesiastical History* in more than one place that the meaning of the name Northumbrians was 'those who dwell to the north of the Humber'.[11] Evidence for the southern frontier of the kingdom of Northumbria westwards from the

[9] See below, pp. 133–4.

[10] Cramp (1999) attempts to use the evidence of styles in sculpture to map Northumbrian 'identity'.

[11] Bede, *Eccles. Hist.*, bk 1 ch. 15, bk 2 chs. 5, 9.

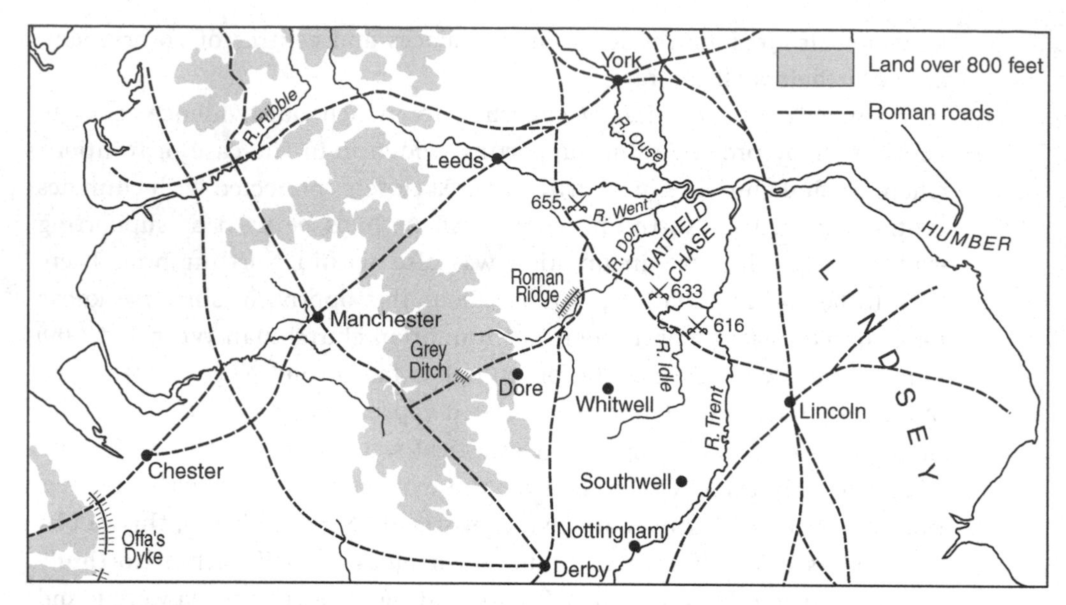

MAP 2 Map to illustrate the southern frontiers of Northumbria
Note that the Roman road running north-west from Lincoln crosses the Rivers Idle and Went, and these crossing-points are the most likely sites of the Battles of the Idle and *Winwaed*. Note too that Hatfield Chase is the name of a district rather than a settlement.

Humber, however, is fragmentary and inconclusive (map 2).[12] A poem inserted into the *Anglo-Saxon Chronicle* under the year 942 and describing the liberation of Mercia from Viking dominance by King Edmund of England gives the northern frontier of Mercia (and so by inference the southern frontier of Northumbria) as 'Dore, Whitwell Gap and Humber river'.[13] The case for arguing that the first two elements were markers of a frontier antedating the Viking period rests partly on the assumption that the point of the poem was to celebrate the re-establishment of the pre-Viking situation, partly on the fact that, according to the *Anglo-Saxon Chronicle*, already in 829 King Egbert of Wessex led an army 'to Dore against the Northumbrians; and there they offered him submission and concord'. Transactions of this type often took place on the frontier, which reinforces the identification of Dore as a point on the frontier of the kingdom of Northumbria in the early ninth century.[14] Peter Hunter Blair drew attention to two stretches of linear earthwork, the so-called Roman Ridge, which is unlikely to be Roman

[12] For what follows, see Peter Hunter Blair (1948). [13] *ASC*, *s.a.* 942.
[14] *ASC*, *s.a.* 827 (correct date 829); on homage given on frontiers, see Lemarignier (1945).

and may conceivably be contemporary with the kingdom of Northumbria (although excavation has not in fact confirmed this); and the Grey Ditch in what is now northern Derbyshire, which has been confirmed as post-Roman in date by excavation.[15] The former, however, seems unlikely to relate to any frontier including Whitwell (although it may relate to an earlier one); but the latter might be viewed as a component in a frontier extending from Dore to Manchester, which, according to an entry in the *Anglo-Saxon Chronicle* for 922, was 'in Northumbria'. We have no way of telling, however, whether this expression pertained only to the then contemporary situation, when the West Saxon kings were advancing northwards so that any earlier frontiers are likely to have been in a state of flux.[16] As for the south-west extremities of Northumbria, we know that King Æthelfrith of Northumbria (d. 616) won a victory over the British at Chester, but we have no way of knowing whether this meant that the power of his kingdom extended that far or whether he was merely engaged in a raiding expedition.[17] Possibly the River Ribble was the frontier, since it was in the middle ages the border between the see of York and that of Lichfield, and ecclesiastical arrangements might have reflected an earlier distribution of territory between the kingdom of Northumbria and that of Mercia. Given the disruption of this area in the Viking period, this is a slim basis of evidence indeed.[18] Moreover, it is clear that assumptions about the coincidence between diocesan and political boundaries have no validity for this southern frontier of Northumbria. In the later middle ages, the boundary of the diocese of York extended on the east side of the Pennines far to the south to take in Nottinghamshire, which was not as far as we know part of the kingdom of Northumbria at any time.[19] Either this was a frontier where political divisions did not match ecclesiastical ones, or the disruption caused to the ecclesiastical organization by the Viking invasions led to substantial changes in the southern boundary of the see of York – there is a possibility that it was extended to the south in the tenth and eleventh centuries to take in the important church of secular canons at Southwell which was certainly in existence by 1051 and was a possession of the archbishops of York.[20]

[15] Peter Hunter Blair (1948), pp. 119–20, 121–2. For recent excavations, see English Heritage National Monuments Record nos. 884666 (Grey Ditch) and 5040 (Roman Ridge).

[16] *ASC, s.a.* 922 (A – correct date 919); see below, pp. 219–20.

[17] Bede, *Eccles. Hist.*, bk 2 ch. 2.

[18] Peter Hunter Blair (1948), pp. 123–4. For the possibility that it was in fact the River Mersey (the name means 'boundary river'), see Kenyon (1991), p. 78.

[19] David Hill (1981), map nos. 255 and 258.

[20] William Page (1910), pp. 152–3; for a summary of more recent scholarship, see Hadley (2000), pp. 231–2. See also, below, p. 229.

Our uncertainty about this and other aspects of the southern frontier of Northumbria may be due to lack of evidence, or it may be due to the fact that we are dealing with a frontier zone rather than a linear frontier. Even so apparently clear-cut a case of a linear frontier as the Humber itself may in fact have been more of a frontier zone. For, in the late ninth century, the biographer of King Alfred, Asser, referred to York as lying on the north bank of the Humber, whereas in fact it lies a considerable distance to the north on the banks of a tributary, the River Ouse.[21] Admittedly, he may not have been well informed about the geography of Northumbria; but the possibility that the name Humber originally applied to a zone rather than just to the estuary itself is strengthened by the occurrence in a fourteenth-century document of the name *Humbreheued*, which contains the name 'Humber' but referred to part of the valley of the River Don.[22] Such a possibility is further strengthened by another observation of Bede's, to the effect that in the time of King Æthelberht of Kent (d. 616), the Humber together with the 'contiguous boundaries or territories' (*termini contigui*) was the dividing line between the southern and the northern English kingdoms, the latter of which is presumably to be identified with Northumbria.[23] It is likely from the present lie of the land that the western extremity of the Humber estuary shaded off into a low-lying marshy area where any frontier may in any case have been ill-defined. If the Grey Ditch was a frontier work of the kingdom of Northumbria, however, we might really be seeing a linear frontier in the Pennine area. Elsewhere, and so far as our evidence goes, it seems more likely that we are dealing with a frontier zone.

If the southern frontier really extended to Chester, it would appear likely that all the lands west of the Pennines were also part of Northumbria and that the western frontier extended to the coast (map 3). The only proof we have, however, is some rather disparate evidence for churches in Cumbria which were connected with the churches in what was unquestionably Northumbria to the east, and tend to confirm that the area was therefore within the kingdom of Northumbria. For example, we learn incidentally from Bede's account of a miracle worked by St Cuthbert's relics that there was a monastery 'built near the River Dacre from which it received its name'. This was almost certainly Dacre (Cumberland) near Ullswater, which is indeed on the river of the same name, and where there is ecclesiastical sculpture dated to the eighth or early ninth century; and archaeological excavations have recovered what may be the remains of an early ecclesiastical

[21] Asser, ch. 26. [22] Peter Hunter Blair (1948), p. 116. [23] Bede, *Eccles. Hist.*, bk 2 ch. 5.

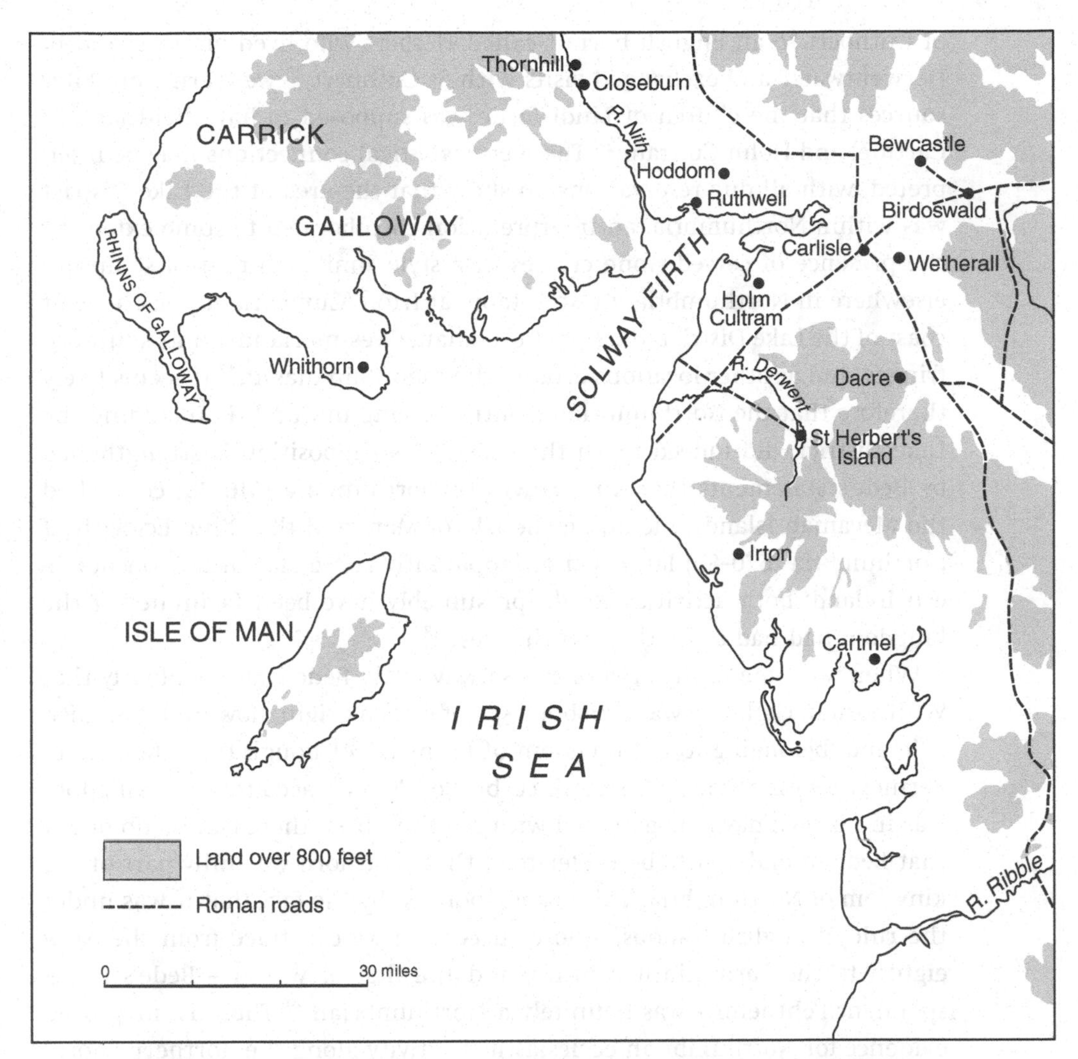

MAP 3 Map of the western and north-western areas of the kingdom of Northumbria

site. Bede was in contact with the monastery, for he knew the names of two successive abbots and he was able to relate a miracle-story from it in some detail. This involved the cure of a young man's diseased eye by some of St Cuthbert's hair which had been cut off when his coffin had been opened in 698 and the body found undecayed. This hair the monks of Lindisfarne had taken 'to give as relics to their friends', amongst whom were evidently numbered the monks of Dacre.[24] Similarly, there is a reference in the *Lives*

[24] Bede, *Eccles. Hist.*, bk 4 ch. 32 (30); for the sculpture, Bailey and Cramp (1988), pp. 90–3, and, for the excavation, Leech and Newman (1985).

of Cuthbert to an English hermit called Herbert who lived on an island in Derwentwater and exchanged visits with St Cuthbert;[25] we learn from later sources that the church of Lindisfarne was supposed to have held land at Cartmel and Holm Cultram.[26] These ecclesiastical connections may be interpreted, with all due reservations, to show that the area of the Lake District was within Northumbria, an interpretation corroborated to some extent by the presence of carved stone crosses in a style similar to that of sculpture elsewhere in Northumbria, most notably at Irton (Cumberland) on the west coast of the Lake District (ill. 2), at Heversham (Westmorland) on its southern fringes, and at other locations around the mountain massif.[27] It seems likely therefore that the Northumbrian frontier, taking in Cumbria including the Lake District, did indeed reach the coast. This supposition is strengthened by Bede's statements that King Edwin of Northumbria (616–33) controlled the Mevanian Islands, including the Isle of Man, and that King Ecgfrith of Northumbria (670–85) launched an apparently successful attack on northern Ireland: both activities would presumably have been facilitated if the kingdom had had a coastline on the west.[28]

When we come to the area of the Solway Firth, Bede states explicitly that Whithorn in Galloway was a bishop's see of British origin, now under English rule and 'belonging to the kingdom of Bernicia'. Although the reference to Bernicia seems rather archaic since by Bede's own account this kingdom was in his own day amalgamated with Northumbria, there can be no doubt that Bede intended it to be understood that Whithorn was fully part of the kingdom of Northumbria. This is corroborated by the fact that it was under the rule of English bishops, whose succession we can trace from the early eighth to the early ninth century, and the first of whom – Bede's correspondent Pehthelm – was definitely a Northumbrian.[29] There is, moreover, evidence for Northumbrian ecclesiastical activity along the northern shores and hinterland of the Solway Firth. At the Roman site of Carlisle, according to the *History of St Cuthbert*, Cuthbert, bishop of Lindisfarne (d. 687), founded

[25] Anon., *Life Cuth.*, bk 4 ch. 9; Bede, *Life Cuth.*, ch. 28.

[26] *Hist. Cuth.*, para. 6 (Cartmel), and *Hist. Kings*, *s.a.* 854 (p. 101, Holm Cultram). Doubt has been cast on the identification of *Suthgedling* in the former with the Yealands near Arnside on the coast of Lancashire, although this seems to make more sense topographically than the currently proposed identification with Gilling (Yorkshire North Riding) (Johnson South (2002), p. 138).

[27] Bailey and Cramp (1988), pp. 113–17, and p. 2 (fig. 2).

[28] Bede, *Eccles. Hist.*, bk 2 ch. 5 and bk 4 ch. 26 (24).

[29] Bede, *Eccles. Hist.*, bk 3 ch. 4 (*ad prouinciam Berniciorum pertinens*), and for Pehthelm bk 5 ch. 23. For the dates, see Fryde, Greenway, Porter et al. (1986), pp. 222–3; the suggestion there that Trumwine was briefly bishop of the see around 681 is based on lists of bishops given by John of Worcester which may themselves be based on a confusion of Whithorn with Abercorn (Plummer (1896), II, p. 224).

ILLUSTRATION 2 Irton (Cumberland), St Paul's Church, cross
Dated on stylistic grounds to the first half of the ninth century, the geometrical ornament on the face illustrated here finds parallels in the ornament of 'Insular' manuscripts, and with other carved crosses such as those at Bewcastle and Lindisfarne. It is therefore firmly within the Northumbrian cultural tradition. For description, discussion and further illustrations, see Bailey and Cramp (1988), pp. 115–17, pls. 355–64.

a religious house for women; and there too, according to his *Lives*, he was staying with King Ecgfrith's queen when he was forewarned in a vision that the king had been killed by the Picts at the Battle of *Nechtanesmere* (685).[30] Westward from Carlisle along the Solway coastlands lay Ruthwell, a completely undocumented site but with a cross very closely related to that of Bewcastle to the east and to other Northumbrian carving, notably at Jarrow in the very heart of Northumbria.[31] Just to the north-east of Ruthwell lay Hoddom, another site which is undocumented except for doubtful mentions in the late twelfth-century *Life* of the British churchman, St Kentigern (d. 612), connecting the church of Hoddom with him. Early sculpture and

[30] *Hist. Cuth.*, para. 5; Anon., *Life Cuth.*, bk 4 ch. 8; Bede, *Life Cuth.*, ch. 27; Bede, *Eccles. Hist.*, bk 4 ch. 26 (24).
[31] See below, pp. 158–9.

archaeological remains, however, are consistent with Hoddom having been a Northumbrian monastery.[32]

How far north of the Solway Firth the kingdom of Northumbria extended is not clear. It must presumably have marched with the British kingdom of Strathclyde, although the lack of sources from Strathclyde itself (we have nothing more than a genealogy and stray references in Adamnan's *Life of St Columba* and the *Annals of Ulster*) deprives us of the possibility of any precision about that kingdom's frontiers. It seems fairly clear, however, that its heartland was the Clyde Valley (map 1). Its ruler in the time of St Columba (d. 597) had his political centre at Dumbarton Rock, where early medieval enclosures of a complexity suggestive of a royal site have been excavated. It probably still had that role in Bede's time, for he refers to it as the fortified centre (*urbs*) of *Alcluith*, 'a name which in their language means "Clyde Rock" because it stands near the river of that name'. According to the *Life of St Kentigern*, Strathclyde's ecclesiastical centre was at Glasgow, and its political centre was probably Dumbarton. There is some evidence that, at least from the mid-eighth century, Northumbria impinged on the territory of Strathclyde. The continuations to Bede's *Ecclesiastical History* record a conquest of the plain of Kyle in the area of modern Ayr by King Eadberht of the Northumbrians in 750, although the entry is laconic and no detail is offered. A couple of very striking pieces of stone sculpture at Closeburn and Thornhill in Dumfries in a style comparable to Northumbrian work do something to corroborate the northward extension of Northumbrian power – or at least influence – in this area (map 3).[33]

It appears that on the east the kingdom of Northumbria extended as far northwards as the Firth of Forth, which, Bede seems clear, was in his time the boundary of Northumbria (map 5). Referring to the invasions of Roman Britain by Irish from the west and from the north, he observes: 'We call them races from over the waters, not because they dwelt outside Britain but because they were separated from the Britons by two wide and long arms of the sea, one of which enters the land from the east, the other from the west, although they do not meet.' The western arm was the Firth of Clyde, for this is the same passage in which Bede refers to *Alcluith* (Dumbarton) on the River Clyde (map 1). The eastern branch was thus the Firth of Forth, half way along which (Bede states) lay the city of *Guidi*.

[32] For the *Life*, see Jackson (1958), esp. pp. 320–1; for archaeology and sculpture, see Lowe (1991, 1993).

[33] For *Alcluith* and the sources for its history, see Alcock (1975–6) and Bede, *Eccles. Hist.*, bk 1 chs. 1, 12; for Kyle, see Bede, *Continuations*, *s.a.* 750. For the sculpture, see Romilly Allen and Anderson (1993), II, pp. 436, 449, and Cramp (1992), pp. 1–15.

ILLUSTRATION 3 Aberlady (Haddingtonshire), cross-shaft now in Edinburgh, National Museums of Scotland
Measuring 82 cm by 21 cm at the base, this shaft was found built into the manse garden in 1863. Dated to the mid-eighth century, its decoration is notably similar to that of the Lindisfarne Gospels (below, pp. 148–51; and see Cramp (1992), p. 225). Note especially the intertwined birds on the back and compare them with the birds, animals and accompanying interlace in the Lindisfarne Gospels (ill. 15). The vine-scroll on the side is equally reminiscent of sculpture at Hexham (Cramp (1984), no. 925). The cross is thus proof that the southern shores of the Firth of Forth were firmly within the orbit of Northumbrian ecclesiastical culture, as also is a sculptured cross-shaft in similar style from Abercorn near Queensferry (Linlithgow); see Cramp (1984), nos. 1434–6.

This was probably identical with a place called *Iudeu* mentioned by Nennius, and may have been Stirling (map 5). The passage quoted here referred to the Roman period; but that this frontier was still the frontier between the Picts and the Northumbrians as it had been between the Picts and the Romano-British is shown by another observation of Bede's, that the English church of Abercorn (Linlithgowshire, just to the east of Queensferry on the southern shores of the Firth of Forth) was 'close to the firth which divides the lands of the English from those of the Picts'. Evidence to corroborate the fact that Northumbria extended north to the Firth of Forth is provided by sculpture of Northumbrian origin from Abercorn itself, and also a very fine cross in Northumbrian style from Aberlady (Haddingtonshire) to the east of Edinburgh (ill. 3). In addition, it appears from the *History of the Kings* that Tyninghame a little further south-east along the coast was the site of a Northumbrian church and the hermitage of the eighth-century Northumbrian saint Balthere.[34] From there the frontier seems to have followed the coast south to the Humber, embracing the coastal or nearly coastal sites of Dunbar, Coldingham, Bamburgh, Monkwearmouth, Hartlepool and Whitby (maps 4–6).

The kingdom of Northumbria appears then to have had physical frontiers on the west (the Irish Sea), on the east (the North Sea) and on the north (the Firth of Forth), with even the southern frontier being in part defined by a physical feature (the Humber estuary). The picture we have been painting of these frontiers derives in large part from eighth-century writings, mostly earlier eighth-century writings, and we need not suppose that these were the frontiers throughout the kingdom's history. We have already noted that the frontier against Strathclyde in the north-west is said to have been extended northwards by King Eadberht in 750. It is possible that Northumbrian power only came to extend along the southern shores of the Solway Firth in the reign of King Ecgfrith (670–85),[35] or even shortly before the establishment, probably for Bishop Pehthelm (? x 731–*c.* 735), of a Northumbrian see at Whithorn. As for the period at which the kingdom was pushed northwards to the Firth of Forth, we are really in complete ignorance. A reference in the *Annals of Ulster* to the 'siege of Edinburgh' (*obsesio Etin*) in 638 is sometimes interpreted as a reference to the capture of Edinburgh by the Northumbrians, but the evidence hardly authorizes such a conclusion (map 5).[36] It has been argued that, because none of the English place-names in Lothian (between the Tweed and the Forth) were formed from elements referring to English paganism such as are found in the south, the English were not present in this area until after their conversion to Christianity. In fact, however, the distribution of such place-names is very uneven in England generally, and they are wholly absent from Northumbria south of the Tweed which certainly had an English population before the conversion; so the absence of such names from the area between the Tweed and the Forth has no validity.[37]

Most problematic of all is the period of origin of the southern frontier. The fact that Bede felt it necessary to repeat his tautologous explanation for

[34] For the texts, see: Bede, *Eccles. Hist.*, bk 1 ch. 12 (*Alcluith* and *Guidi*); bk 4 ch. 26 (24) (Abercorn); *Hist. Kings*, *s.a.* 756 (p. 41) and 941 (p. 94) (Tyninghame, and see also Alcuin, *BKSY*, lines 1319–87). On *Guidi*, see A. Graham (1959) and Jackson (1981); on sculpture from Abercorn and Aberlady, see Romilly Allen and Anderson (1993), II, pp. 418–20, 428–9.

[35] Smyth (1984), pp. 24–5. [36] See below, p. 89. [37] Hough (1997).

the name Northumbrians (i.e. meaning those dwelling north of the Humber) might suggest that the name was relatively new and therefore unfamiliar in his time and, by inference, so too was the Humber as a frontier. This would imply that at least the southern part of Northumbria had previously been linked with the lands to the south of the Humber, especially the kingdom of Lindsey in the area of modern Lincolnshire (map 2), with the Humber itself more as the link than the divide between them. Bede's account certainly suggests that King Edwin of Northumbria had considerable authority in Lindsey, for he presents Paulinus, Edwin's own bishop, as responsible for the conversion of that people, and particularly of the prefect (*prefectus*) of Lincoln which lay at the centre of it. It would appear as if Lindsey at that point was part of Edwin's kingdom and the prefect merely one of his officials. Lindsey was a bone of contention between the Northumbrian kings and the Mercian kings to the south, who eventually succeeded in incorporating it into their kingdom after they defeated the Northumbrians at the Battle of the Trent in 679. The implication is that the Northumbrians had previously ruled it. This may only have been intermittently, however, and the survival of a genealogy of Lindsey kings in a collection of genealogies drawn up in the ninth century suggests that the kingdom had an identity of its own and at most the Northumbrian kings were its overlords at times before it became part of Mercia.[38]

Although there is thus no certainty as to when the Humber emerged as a frontier, it is instructive that three crucial battles of the seventh century were fought in this area (map 2). Edwin, of the Deiran royal house, defeated the reigning king of Northumbria, Æthelfrith of the Bernician royal house, at the Battle of the River Idle (616) to become himself king of Northumbria; we cannot be sure where on the River Idle this battle took place, but it is quite likely that it was where that river was crossed by the Roman road from Lincoln to Doncaster and thus in the broad area of the Humber. Edwin himself was defeated by the British king of Gwynedd, Cædwalla, at the Battle of Hatfield in 633; if this was Hatfield Chase then it too was in the same zone. Finally, King Oswiu of Northumbria defeated the pagan king of Mercia, Penda, at a battle which Bede locates near the River *Winwaed* (655). The name cannot be identified as such with any modern place-name; but Bede states further that it was near Leeds, and it seems possible that the name is preserved in part in that of the River Went, and that the battle took place

[38] For the *prefectus*, see Bede, *Eccles. Hist.*, bk 2 ch. 16; on Lindsey, see *Encyclopaedia ASE*, entry 'Lindsey'; see also Vince (1993), esp. Sarah Foot, 'The kingdom of Lindsey', pp. 128–41 (at pp. 133–5); Eagles (1989) and Stenton (1970a).

at Wentbridge where the Roman road referred to above crosses that river north of Doncaster.[39]

We have already had occasion to comment that in the case of the Humber frontier we may be looking at a frontier zone rather than a linear boundary, and this may be true to some extent of the Firth of Forth frontier, where scholars have noted that, despite Bede's statement about its being a frontier with the Picts, there are place-names of semantically Pictish origin to the south of it, and also find-spots of massive silver jewellery-chains bearing characteristically Pictish symbols. This evidence may show that that the Firth of Forth was a frontier zone throughout which Pictish and English influence mingled. It is, however, insufficiently precise to prove it, especially when our knowledge of the chronology of the extension of Northumbrian power is so slight. The place-names and the Pictish chains may really be evidence for a period, otherwise unknown to us, when Pictish power or influence extended south of the Firth of Forth, perhaps before the frontiers of Northumbria had expanded that far.[40]

In both the north and the south, understanding the nature of the frontiers is complicated by evidence for the overlordship of Northumbrian kings over other kingdoms far beyond those frontiers. As regards the southern frontier, one of the most problematic passages in Bede's *Ecclesiastical History* relates to the seven overlords of the English. Discussing the first Christian king of the English, Æthelberht of Kent, he notes that

> he was the third English king to rule over all the southern kingdoms, which are divided from the north by the river Humber and the surrounding territory (*contiguis ei terminis*); but he was the first to enter the kingdom of heaven. The first king to hold the like sovereignty (*imperium*) was Ælle, king of the South Saxons; the second was Ceawlin, king of the West Saxons, known in their own language as Ceawlin; the third, as we have said, was Æthelberht, king of Kent; the fourth was Rædwald, king of the East Angles, who, while Æthelberht was still alive, acted as the military leader of his own people (*quartus Reduald rex Orientalium Anglorum, qui etiam uiuente Aedilbercto eidem sue genti ducatum prebebat, obtinuit*); the fifth was Edwin, king of the Northumbrians, the nation inhabiting the district north of the Humber. Edwin had still greater power and ruled over all the inhabitants of Britain, English and Britons alike, except for Kent only. He even brought under English rule the Mevanian Islands (Anglesey and Man) which lie between England and Ireland and belong to the Britons. The sixth to rule within the same bounds was Oswald, the most Christian king of

[39] Bede, *Eccles. Hist.*, bk 2 chs. 12, 20, bk 3 ch. 24.
[40] McNeill and MacQueen (1996), p. 51; Hope-Taylor (1977), p. 288.

> the Northumbrians, while the seventh was his brother Oswiu who for a time held almost the same territory. The latter overwhelmed and made tributary even the peoples (*gentes*) of the Picts and the Irish who inhabit the northern parts of Britain.[41]

Bede's words are translated into Old English in the *Anglo-Saxon Chronicle* entry for 827 describing the West Saxon king Egbert's victory over the other southern English kingdoms at the Battle of *Ellandun*. Egbert is there presented as the successor to these kings who ruled over southern England, and an Old English term *bretwalda* ('ruler of Britain') is applied to him.[42] Scholars have frequently discussed Bede's words and the implication of them, that being an overlord of the southern kingdoms (or a *bretwalda* as the *Anglo-Saxon Chronicle* would have it) was to hold some sort of recognized office which rotated around the rulers of the English kingdoms. Bede's account in the passage of Rædwald's position implies this most specifically, since its meaning, although much debated, seems to be that Æthelberht had only partial overlordship over Rædwald, who enjoyed military leadership over the East Angles while the former was alive, and assumed his overlordship after his death.[43]

The existence of such an overlordship, however, is very difficult to accept of the earliest kings in Bede's list. Virtually nothing is known of Ælle of Sussex or Ceawlin of Wessex. Given that they are supposed to have lived in the early days of the English settlement of England, however, it is utterly unlikely that they really ruled all the kingdoms south of the Humber.[44] The same must be true of Æthelberht of Kent, although in his case there are some indications that he had influence over the kingdom of Essex, at least to the extent of introducing Christianity to it.[45] Rædwald of the East Angles was clearly a ruler of some power, for it was at his court that Edwin of Deira took refuge when he was exiled by the Bernician Æthelfrith. Bede tells us that he refused to hand Edwin over to Æthelfrith and that he assisted Edwin to win his victory over his Bernician enemy at the Battle of the Idle in 616.[46] This, however, is very far from showing that he ruled over all the southern

[41] Bede, *Eccles. Hist.*, bk 2 ch. 5. For the obscurities of Bede's mention of Rædwald here, see below, n. 43.

[42] *ASC, s.a.* 827; one version of the chronicle has *brytenwealda* ('wide ruler'), but this appears to be a corruption.

[43] The classic discussions are Stenton (1970c) and John (1966c); for more recent discussion see, for example, Yorke (1981), Fanning (1991) and Dumville (1997). For Bede's words concerning Rædwald, see Wallace-Hadrill (1988), pp. 220–2 (cf. p. 59).

[44] *ASC, s.a.* 477, 485, 491 (Ælle); 556, 560, 568, 571 (E), 577, 584, 592, 593 (Ceawlin).

[45] Bede, *Eccles. Hist.*, bk 2 ch. 3.

[46] Bede, *Eccles. Hist.*, bk 2 ch. 12.

kingdoms. From Æthelberht onwards the case for the overlordship being a defined office is a little stronger, in so far as Bede describes the office descending to successive kings, which Ælle of Sussex and Ceawlin certainly were not. It is just conceivable that the office was passed from Kent to East Anglia and so to Northumbria. Some references in the *Ecclesiastical History* and Stephen's *Life of Wilfrid* support the idea of overlordship to the extent that they show that the kings named by Bede – Edwin, Oswald and Oswiu – had some influence south of the Humber. We have already noted the activity of Edwin's bishop Paulinus in Lindsey as a possible indication of the king's power there, and for Edwin we could add the fact that he was able to deploy an army successfully in Wessex, and the fact that the king of Wessex thought it worthwhile to dispatch an assassin to take his life.[47] For Oswald, we can note that he was present as godfather at the baptism of King Cynegils of Wessex, a role which may have implied political dominance, but need not have done. And we can note also an account in Bede in which Oswald acted jointly with Cynegils to grant land at Dorchester on Thames (Oxfordshire) to the missionary Birinus, which could be held to demonstrate the role of an overlord in validating land-grants by his sub-kings, one of whom according to this interpretation Cynegils would have been.[48] None of this need prove, however, more than that when the kings saw an opportunity they exerted their power and influence over other kingdoms. Where kings acted as godparents or simply provided missionaries, we are by no means authorized to assume that this necessarily implied that they had political authority over the kingdoms involved. Nevertheless, Bede thought the list of seven kings worth giving and for that reason we must take what he says seriously. But of what was he thinking? If there was a real office of overlord, how had it arisen and to what did it correspond? It has been argued that it was an ancient office going back to the earliest period of English activity in Britain, even that it somehow derived from the office of the 'proud tyrant' referred to by Gildas and identified with Vortigern.[49] But there is no evidence for such a contention and it is best disregarded.

It could be that overlordship involved something similar to the high-kingship of Ireland, and that the overlord had rights to levy payments of tribute on the kingdoms recognized as subservient.[50] There is a hint to this effect in the account in Stephen's *Life of Wilfrid* of how King Wulfhere of the Mercians (657–74) 'roused all the southern nations against our kingdom

[47] Bede, *Eccles. Hist.*, bk 2 chs. 16, 9.

[48] Bede, *Eccles. Hist.*, bk 3 ch. 7; on this, see John (1966c), p. 16.

[49] Higham (1995). [50] Charles-Edwards (1989), pp. 28–33.

[of Northumbria], intent not merely on fighting but on compelling them to pay tribute in a slavish spirit'. King Ecgfrith of Northumbria (670–85) then defeated Wulfhere, laid his kingdom 'under tribute' and, after his death, 'ruled in peace over a wider area (*plenius aliquod spatium pacifice imperauit*)'.[51] This well illustrates the difficulty of using narrative sources in discussing a subject of this kind. For Stephen simply does not give us enough information to form a judgment. Mercia might previously have been paying tribute to Northumbria, perhaps still in place from the short period after the Battle of the *Winwaed* when King Oswiu ruled northern Mercia directly. If so, Wulfhere was presumably in revolt against this.[52] On the other hand, Wulfhere's failed attempt to impose tribute on Northumbria may have itself have been the cause of the imposition of a Northumbrian tribute on his own kingdom which was a novelty and a punishment specifically for his aggression. We have no way of telling. Nor are we any clearer about the extent of the 'wider area', and the terminology used is hopelessly vague.

Evidence in favour of the idea that the Northumbrian kings laid the southern kingdoms under tribute may possibly be found in a curious document known as the Tribal Hidage, which is a list assessing in terms of hides the major midland and southern English kingdoms, together with a number of lesser units which may or may not have been kingdoms in any real sense.[53] Hides were units of assessment for tax and renders, and it is reasonable to suppose that the list might have been used as a tribute assessment document. There has been much debate about the identity of some of the territorial units listed in it, and about its date and provenance. Since it begins with Mercia and contains most detail about midland areas, it has usually been interpreted as a tribute list relating to the kingdom of Mercia, possibly in the time of King Wulfhere, possibly in the time of either King Æthelbald (716–57) or King Offa (757–96). There are, however, two reasons why it might have been a Northumbrian tribute list. The first is that it opens with a very high number of hides assigned to the core area of Mercia around the River Trent. If the document is indeed a tribute list, the implication of this is that Mercia was being subjected to a very heavy tribute, seemingly a punitive one. It is hard to understand why this should have been the case if Mercia were the beneficiary of the tributes listed, but easy to understand if the list were Northumbrian made at a time when the kings of Northumbria had defeated Mercia and were imposing a penal assessment on it. The second

[51] Stephen, *Life Wilf.*, ch. 20. [52] See below, n. 58.

[53] See, for the text and manuscripts, Dumville (1989b); for identification of the peoples, Wendy Davies and Vierck (1974). See also *Encyclopaedia ASE*, entry '*Tribal Hidage*' and references therein.

reason is the inclusion in the list of the *Elmed sætna*, that is people of Elmet, supposed to have been a British kingdom in the area of Leeds which was absorbed into Northumbria by King Edwin, but which is not known ever to have been subject to Mercia.[54] It has even been argued that the text was drawn up in the reign of that king, and that the inclusion of Elmet as separate from Northumbria reflected the circumstances of its fairly recent conquest by Edwin.[55] It should be noted, however, that the text of the Tribal Hidage has been preserved in manuscripts of the early eleventh century and later. It is evidently corrupt and the versions vary one against another. We cannot even be sure that it is a tribute list. It may have been a purely antiquarian exercise. If it is indeed a Northumbrian list of the seventh century, it is very hard to see why so many minor units in the midlands should have been included – units which Bede and our other sources never allude to at all.

The most the evidence presented above seems to authorize is the conclusion that at some periods of the seventh century the Northumbrian kings dominated England south of the Humber in some sense, possibly involving tribute. What is equally certain is that the kings of Mercia were vying with them for dominance, often successfully. As Bede emphasizes, Penda of Mercia was a threat to the very existence of Northumbria down to his death at the Battle of the *Winwaed* in 655. In alliance with Cædwalla, ruler of the British kingdom of Gwynedd in North Wales, he was responsible for the death of Edwin at the Battle of Hatfield, and he subsequently struck deeply into Northumbria, so that he even besieged Bamburgh, threatening to over-run the kingdom (map 5).[56] If Nennius is correct (and in this instance he probably is), King Oswiu was attacked even in *Iudeu* (*Guidi* – probably Stirling) in the Firth of Forth area, and escaped only by buying Penda off: 'Then Oswiu delivered all the riches that he had in the city into the hand of Penda, and Penda distributed them to the kings of the British, that is the "Distribution of Iudeu".'[57] King Oswiu retaliated, however, and, after the defeat of Penda at the *Winwaed*, which Nennius calls 'Gaius's Field' and mistakenly puts before rather than after the 'Distribution of Iudeu', Mercia was laid under direct Northumbrian control, although the southern part was granted to Penda's son Peada who had married into the Northumbrian royal family. After three years, however, 'the leaders (*duces*) of the Mercian race, Immin, Eafa, and Eadberht, rebelled against King Oswiu and set up as their king Wulfhere,

[54] See below, pp. 85–7. [55] Higham (1995), pp. 74–111.
[56] Bede, *Eccles. Hist.*, bk 2 ch. 20, bk. 3 chs. 9, 16–17, 24. [57] Nennius, *Hist. Britons*, para. 65.

Penda's young son, whom they had kept concealed; and having driven out the princes (*principibus*) of the foreign king, they boldly recovered their lands (*fines*) and liberty at the same time'.[58]

After the Battle of the *Winwaed* then, Northumbrian control of Mercia was real, but intermittent. Sometime after the coup described in this passage, Wulfhere, as we have seen, attempted to subjugate Northumbria but his kingdom was itself subjected to tribute. Mercia was soon in the ascendant again, however, with the defeat of King Ecgfrith of Northumbria at the Battle of the Trent in 679, when his brother Ælfwini was killed. The result of the battle was seemingly to wrest Lindsey from Northumbrian control.[59] When Bede finished his *Ecclesiastical History* in 731, he was able to state that the Mercian king Æthelbald had subject to him the southern kingdoms 'which reach right up to the Humber'.[60] Perhaps, it can be argued, he was the eighth of the overlords referred to in the passage quoted above, and he was not noted as such by Bede because the Mercians were enemies of the Northumbrians, and not noted in the *Anglo-Saxon Chronicle*'s entry because of enmity between the Mercians and the West Saxons.[61]

Bede's claims for the overlordship of the Northumbrian kings went beyond southern England. He tells us that Edwin 'held under his sway the whole realm of Britain, not only English kingdoms but those ruled over by the British as well', that Oswald held 'under his sway all the peoples and kingdoms of Britain, divided among the speakers of the four different languages, British, Pictish, Irish, and English' and that Oswiu ruled 'almost the same territory' and overwhelmed and made 'tributary even the peoples (*gentes*) of the Picts and the Irish who inhabit the northern parts of Britain'.[62] Further information in Bede's *Ecclesiastical History* and in Stephen's *Life of Wilfrid* gives some substance to the claim that the Northumbrian kings ruled part at least of the Pictish lands to the north, at any rate until 685 when the defeat of King Ecgfrith at the Battle of *Nechtanesmere* restricted them to south of the Firth of Forth. Bede informs us that Abercorn was created as a bishopric for the Picts, and it is possible that the establishment of such a bishopric implied political power over them, although it is just possible that the Picts in question were resident in Northumbria south of the Firth of Forth since there is, as we have seen, evidence for possible Pictish activity there.[63] Stephen is more explicit, describing how the Picts revolted against

[58] Bede, *Eccles. Hist.*, bk 3 ch. 24. [59] See above, p. 35. [60] Bede, *Eccles. Hist.*, bk 5 ch. 23.
[61] See above, p. 37 nn. 41–2. [62] Bede, *Eccles. Hist.*, bk 2 chs. 5, 9, bk 3, ch. 6.
[63] Bede, *Eccles. Hist.*, bk 4 ch. 26 (24); Thomas (1981).

King Ecgfrith of Northumbria in 671–3: 'the bestial peoples (*populi bestiales*) of the Picts fiercely resented their being subject to the Saxons; indeed they began to stir up revolt. Swarms of them gathered from every cranny of the north.' Ecgfrith, however, 'quickly mustered a troop of cavalry and putting his trust in God, like Judas Maccabaeus' he attacked them: 'He filled two rivers with the slain and his men crossed dry-shod over the corpses to slay the fugitives. Thus the Picts were reduced to slavery, a condition in which they remained until Ecgfrith himself was slain.'[64] The clear implication is that Northumbrian domination of Pictland endured until 685. If the site of the Battle of *Nechtanesmere* is correctly identified with Dunnichen Hill east of Dundee, Ecgfrith was demonstrating his power in Pictland through a campaign in the interior worthy of Edward I, even though the upshot of it was a disaster for him.[65] Certainly the evidence provided by a number of Pictish sculptured stones shows considerable Northumbrian cultural influence over the Picts.[66] It is possible, however, that this Northumbrian influence post-dated 685, for Bede includes in his *Ecclesiastical History* a letter in which Ceolfrith, abbot of Monkwearmouth and Jarrow, replied to a request from King Nechtan of the Picts (706–24) to be sent information on the Northumbrian method of calculating the date of Easter, and also architects to build stone churches for him; and Bede notes that at the time he was writing his *Ecclesiastical History* (731) there was a treaty of peace with the Picts. It may well have been in that happier period of Pictish–Northumbrian relations that Northumbrian culture influenced Pictland.[67]

As regards the status of the British kingdoms of Strathclyde and Gwynedd vis-à-vis Northumbrian power, Bede gives no specific information, but he makes the generalized statement at the end of his *Ecclesiastical History* that at that time they had been brought 'partly' under the rule of the English. If we were to accept literally his statements about Edwin, Oswald and Oswiu cited above, this present observation would indicate that Northumbrian power over the British also had waned by 731, which is consistent with another of Bede's statements to the effect that after the Battle of *Nechtanesmere* in 685 'some part of the British (*Brettonum pars*) recovered their independence which they have enjoyed now for about forty-six years'. In the case of King Edwin's alleged power, one corroborating detail given by Bede

[64] Stephen, *Life Wilf.*, ch. 19.

[65] *Duin Nechtain* in *Annals Ulster*, *s.a.* 686 (685), and in *The Annals of Tigernach* (Stokes (1993)), *s.a.* 685. This may be Dunnichen, three miles south-east of Forfar; the name means 'fortress of Nechtan' and there was formerly a lake in the vicinity which could have been the 'mere' of Symeon's name and that of Nennius. See Wainwright (1948).

[66] See below, pp. 128–9.

[67] Bede, *Eccles. Hist.*, bk 5 chs. 21, 23; and below, pp. 126–9.

is that king's conquest of Anglesey and the Isle of Man from the British, the former presumably from Gwynedd, the British kingdom of North Wales. Our confidence in this is increased by the fact that Bede is able to give figures for the number of hides (or units) at which Anglesey and Man were assessed, suggesting that the Northumbrian kings had laid them under tribute and had 'hidated' them the better to levy that tribute. Conquest of the islands must have involved very considerable naval and military capabilities in the west. In the later middle ages, Anglesey was the core of North Wales, being the principal grain-producing area, and there is no reason to suppose that it was any less important in the seventh century. Another detail offered by Bede in support of his general claim that Edwin at least had overlordship over the British is his presentation of King Cædwalla of Gwynedd's attack on Edwin as a revolt (*rebellauit aduersus eum*). Although these details go a little way towards corroborating Bede's general claim, however, the picture to be derived from them is very fragmentary. It is not clear, for example, whether Cædwalla's revolt permanently ended Northumbrian power over Gwynedd and, if not, how Edwin's successors re-asserted it. Of Strathclyde, we have very few details. As we have seen, King Eadberht (737/8–58) is represented in the Northern Annals as warring against the kingdom of Strathclyde, and his conquest of the plain of Kyle was presumably at its expense. But it is quite unclear whether this laconically reported military activity represented a new departure or related in some way to the general overlordship of the British which Bede assigned to Edwin, Oswald and Oswiu. In short, such details as we have suggest no more than an intermittent Northumbrian overlordship over parts at least of northern Britain; and we have no way of deciding whether to accept Bede's more general claim for that overlordship as factual or merely a rhetorical flourish.[68]

Let us now turn from the outer limits of Northumbria to the interior of the kingdom. What can we perceive of its internal divisions? As we have seen, Bede is explicit about the existence up to the time of King Oswald (634–42) of the two constituent kingdoms of Deira and Bernicia, which he seems to imply comprised between them the whole of the kingdom of Northumbria. Neither Bede nor any other contemporary writer is specific about the frontier between them. Bede implies, however, that his own joint monastery of Monkwearmouth and Jarrow, its components respectively south of the Rivers

[68] For the texts, Bede, *Eccles. Hist.*, bk 2 chs. 5, 9, 20, bk 4 ch. 26 (24), bk 5 ch. 23; Bede, *Continuations*, *s.a.* 750; *Hist. Kings*, *s.a.* 756 (? 755, p. 40). On Anglesey, see R. R. Davies (1987), pp. 9–11. See also, Bede, *Eccles. Hist.*, bk 5 ch. 23, on the general situation of the British in 731 (for discussion, below, pp. 57–8).

Wear and Tyne, lay in Bernicia, since he states that Bishop Acca consecrated one of the abbots. Acca was bishop of Hexham, definitely in Bernicia, and the inference is that Monkwearmouth and Jarrow also lay within that see (map 6). On the other hand, the monastery of Gilling (Yorkshire North Riding) some way to the south of the River Tees presumably lay in Deira since it was founded as a compensation payment to the Deiran royal house for the killing of King Oswine of Deira (map 4). Despite the statement of Richard of Hexham in the twelfth century to the effect that the boundary between Deira and Bernicia was on the River Tyne, it seems likely, therefore, that it was on the River Tees, or at least in its region, since the position of Gilling would not rule out the possibility that the boundary followed the southern watershed of the river rather than its course.[69]

Diocesan boundaries are less useful than might have been expected in providing clues to internal divisions. Originally, Bede tells us, there was only one diocese for the Northumbrians, which had its see at Lindisfarne, the island monastery founded off the Northumbrian coast in 635. After the synod held at Whitby in 664, the see was moved to York, but the diocese remained undivided – it may have been the resistance of the then bishop, Wilfrid, to its division by Archbishop Theodore (668–90) which led to the former's exile. We can only presume that the boundaries of this see were those of the kingdom itself, although as we have seen this was not in fact the case for the southern boundary of the see of York as it appears in the later middle ages.[70] In the end, however, the Northumbrian see was divided, despite Wilfrid's opposition, and by the 680s we find sees at York, Lindisfarne, Hexham (the first bishop of which was Eata (678–81)), and Abercorn (the first and last bishop of which, Trumwine, was expelled in 685), with Whithorn established by 731. What seems to be a free-standing document contained in the *History of St Cuthbert* purports to describe the boundary of the 'land' of Lindisfarne (*Lindisfarnensis terre terminus*) as it was in the seventh century. The description, which is detailed enough to follow on the ground, is of a substantial block of land embracing the areas later known as Islandshire and Norhamshire; but it may represent the estates in the possession of the church of Lindisfarne rather than the extent of its diocese.[71] For Hexham, we have Richard of Hexham's statement that 'some people said' the boundaries

[69] Peter Hunter Blair (1949). For the texts, see Bede, *Eccles. Hist.*, bk 3 chs. 1, 24; Bede, *Hist. Abbots*, para. 20; and, for Richard of Hexham, Raine (1864–5), I, p. 20.

[70] Bede, *Eccles. Hist.*, bk 3 chs. 4, 26; Stephen, *Life Wilf.*, ch. 16 (where the reference to York as a metropolitan see is not factually correct). On the boundary of the see of York, see above, p. 27.

[71] *Hist. Cuth.*, para. 4 (on which see Johnson South (2002), pp. 79–80).

of the former see of Hexham (which had ceased to exist in the early ninth century) lay on the River Tees to the south, the River Alne to the north, and as far as Wetherall (Cumberland) and Carlisle to the west (maps 3, 6). As we have seen, Acca, bishop of Hexham, was the diocesan for Monkwearmouth and Jarrow, since Bede refers to him as officiating in that monastery, so we can corroborate this to that extent that the joint monastery lay between the Rivers Tees and Alne. It has been possible to reconstruct with some plausibility the boundary between Hexham and Lindisfarne on the west by reference to the distribution of church dedications and other sites associated with one or other of the two cathedral churches. It is just possible that this boundary also represents that between Bernicia on the east and a separate political unit in north-western Northumbria, the area known as Cumbria but presumably excluding the northern shores of the Solway Firth, since Bede says that Whithorn was in the kingdom of Bernicia.[72] As for Whithorn itself, we can speculate that its diocese was congruent with the twelfth-century Lordship of Galloway, but we really have nothing to go on. As we have seen, there is a possibility that the diocese of Abercorn lay south of the Firth of Forth, although this is very uncertain and it may equally have lain to the north.[73] Diocesan boundaries, with the possible exception of the putative boundary of the see of Hexham along the River Tees with Deira and possibly with Cumbria, cannot be shown to follow political frontiers in Northumbria. In any case, all three sees of Hexham (678–?821), Whithorn (*c.* 731–*c.* 803) and Abercorn (?–685) were quite short-lived so, if they did reflect any underlying political units, those units were not strongly enough defined to preserve the sees from extinction.

Although the divisions within the kingdom of Northumbria are not well defined, some evidence can be gathered to demonstrate concentrated areas of royal interest and power, that is the kingdom's heartlands. In the area of the former kingdom of Deira, between the Tees and the Humber, there may have been three such heartlands in close proximity, which may in fact have formed a single heartland (map 4). The first, in broadly what was later the East Riding of Yorkshire, included York which, although we shall need to consider how far it remained a royal as distinct from an ecclesiastical centre, was nevertheless an important place for the Northumbrian kings. Bede writes that it was where Edwin was baptized and built his first church in 627.[74] It was just to the east of York, somewhere on the River Derwent,

[72] See above, n. 29; Raine (1864–5), I, p. 20; Phythian-Adams (1996), pp. 66–75.
[73] See above, n. 63. [74] Bede, *Eccles. Hist.*, bk 2 ch. 14, and see below, p. 77

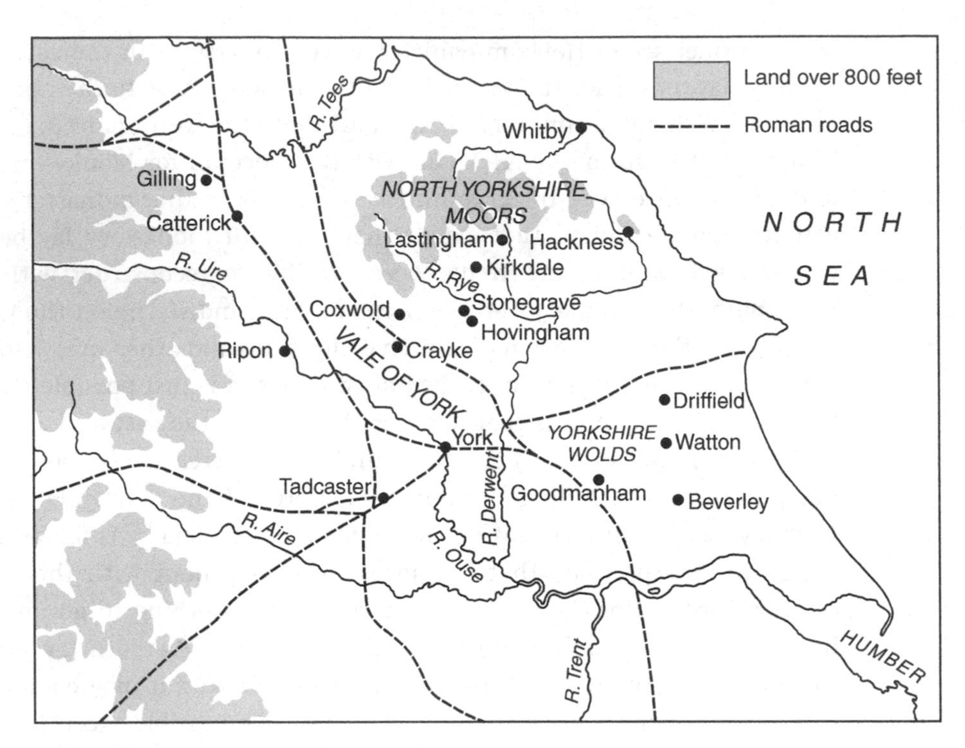

MAP 4 Map of the Deiran heartland

that Bede tells us there was a royal vill (*uilla regalis*) where Edwin was staying when an assassination attempt was made on him, and there too that his daughter was born.[75] Bede further tells us that on another occasion Edwin called a council to discuss whether or not he should adopt Christianity, and at this council the pagan priest Coifi decided to renounce his old religion and destroyed the temple at a place 'not far from York, to the east, over the River Derwent. Today it is called Goodmanham.'[76] Bede does not tell us where the council itself took place (possibly at York, possibly at the royal vill on the Derwent already mentioned, possibly at another royal vill further to the south) but it is clear from Bede's words that Coifi was the king's priest (*primus pontificum ipsius*) and it follows that Goodmanham, in the area of the Yorkshire Wolds, was a royal temple (map 4). Evidently the kings had lands and vills, perhaps palaces, in this area. The area also contained the early churches of Watton and Beverley (Yorkshire East Riding) which, although not in themselves royal, may have owed their existence

[75] Bede, *Eccles. Hist.*, bk 2 ch. 9. [76] Bede, *Eccles. Hist.*, bk 2 ch. 13.

to royal interest and influence.[77] In the same area, there are also significant numbers of inhumations in burial mounds, including a whole group of mounds at Driffield (Yorkshire East Riding), the place where Aldfrith, king of the Northumbrians, died in 705 on 14 December, when he was very likely preparing to celebrate Christmas at what was no doubt a royal vill there.[78] The picture thus derived from Bede and other written sources of a concentration of royal interest in the area of York and eastwards is graphically borne out by other evidence, principally archaeological but also textual. The greatest concentration of cemeteries associated with the earliest English settlers of Northumbria is found in the area of the Yorkshire Wolds and on the routeways leading away from York to the Humber and the sea, suggesting that this was an early core area of English settlement in this part of England.[79]

The second heartland was around the fringes of the North Yorkshire Moors, and included Whitby (Yorkshire North Riding), a royally founded monastery, at least two of the abbesses of which were princesses of the royal house, and where King Edwin himself was buried, after his body had been recovered from the battlefield of Hatfield.[80] In the south-east fringes of the Moors, all within the later Yorkshire North Riding lay Hackness, a monastery dependent on Whitby, while further east were two monasteries nestling in the skirts of the Moors: Lastingham, where the monk Cedd founded a monastery with lands donated by Œthelwald, sub-king of Deira; and Kirkdale, where inscriptions, ecclesiastical sculpture and archaeological remains bear witness to its status as a monastery in the pre-Viking period. Across the River Rye was Stonegrave, a monastery mentioned in a letter of Pope Paul I to King Eadberht, in which mention was also made of another monastery at Coxwold on the eastern fringes of the Moors. Nearby Hovingham is not documented as an early monastery, but its elaborate and sophisticated pre-Viking sculpture suggests strongly that it was one (map 4).[81]

[77] Bede, *Eccles. Hist.*, bk 5 ch. 3 (*Wetadun*), bk 5 ch. 6 (Beverley – *In Silua Derorum*). For the identification, see Cambridge and Morris (1989), pp. 9–10, and, further, John Blair (2001).

[78] Eagles (1979), pp. 427–8; *ASC* (D and E versions), *s.a.* 705. [79] Lucy (1999).

[80] *Encyclopaedia ASE*, entry 'Whitby'; for the texts, see Bede, *Eccles. Hist.*, bk 3 ch. 24, and *Life Greg.*, chs. 18–19.

[81] For relevant texts, see Bede, *Eccles. Hist.*, bk 3 ch. 23 (Lastingham), and bk 4 ch. 23 (21) (Hackness); and Haddan and Stubbs (1869–71), III, pp. 394–6, trans. *EHD I*, no. 184 (Stonegrave and Coxwold). For early ecclesiastical sculpture from Hackness, Hovingham, Kirkdale and Stonegrave, see Lang (2001). On archaeological excavations at Kirkdale, see Rahtz, Watts and Grenville (1996–7) and Rahtz, Watts, Okasha et al. (1997).

The third possible heartland was the Vale of York, and its western fringes, where Bede tells us of a number of places of importance to the kings. Ripon was the site of a monastery founded by the abbot of Melrose, Boisil, and numbering amongst its monks Cuthbert, future bishop of Lindisfarne. Alhfrith, who ruled Deira under Oswiu, expelled the monks of this community and gave the site to his protégé, Wilfrid, who proceeded to develop there the wealthy monastery of which he was abbot.[82] We have already mentioned Gilling, where King Oswiu founded a monastery following his killing of King Oswine.[83] Catterick (Yorkshire North Riding) is mentioned several times in the Northern Annals as a royal vill, which was the wedding place of two Northumbrian kings. Crayke (Yorkshire North Riding) was a monastery, seemingly founded on land granted to St Cuthbert, where the community of that saint found refuge during its long flight from Lindisfarne; and Tadcaster (Yorkshire West Riding) to the south-west of York makes an incidental appearance in Bede's work as another monastery (map 4).[84]

A Bernician heartland seems to be observable to the north (map 5). Bede tells us on several occasions of the presence of the Northumbrian kings at Bamburgh on the coast of Northumberland, and also at Lindisfarne, where one of them, Oswald, founded a monastery in conjunction with the Irish churchman, Aidan of Iona, in 635.[85] Just inland was Yeavering, sited in the valley of the River Glen on the edges of the Cheviot Hills. Here, Bede informs us, was another royal vill where in Edwin's time the missionary Paulinus spent thirty-six days catechizing and baptizing the people who came there. This vill, Bede goes on, was deserted by the succeeding kings in favour of a place called *Mælmin*, identified with nearby Milfield. Extensive archaeological evidence of sophisticated and impressive structures there corroborates Bede's account of Yeavering, and his account of Milfield is similarly corroborated by admittedly more restricted archaeological evidence.[86] In the same general area, studies from aerial photography of a site at Sprouston (Roxburghshire) have further emphasized the importance of this area by revealing halls which were probably built by people of high social status.[87]

[82] Stephen, *Life Wilf.*, ch. 8; Bede, *Eccles. Hist.*, bk 3 ch. 25; Anon, *Life Cuth.*, bk 2 ch. 2; and Bede, *Life Cuth.*, ch. 7.

[83] See above, p. 44; but cf. n. 26, where the possibility of identifying Gilling with a possession of Lindisfarne in Bernicia is noted.

[84] On Catterick, see *Hist. Kings*, *s.a.* 762, 769, 792 (pp. 42, 44, 54); on Crayke, see *Hist. Cuth.*, paras. 5, 10, 20; on Tadcaster, see Bede, *Eccles. Hist.*, bk 4 ch. 23.

[85] Bede, *Eccles. Hist.*, bk 3 ch. 3; for the date, see David Rollason (2000), pp. 20–1 n. 10.

[86] Bede, *Eccles. Hist.*, bk 2 ch. 14; see also below, pp. 82–5.

[87] See above, p. 18.

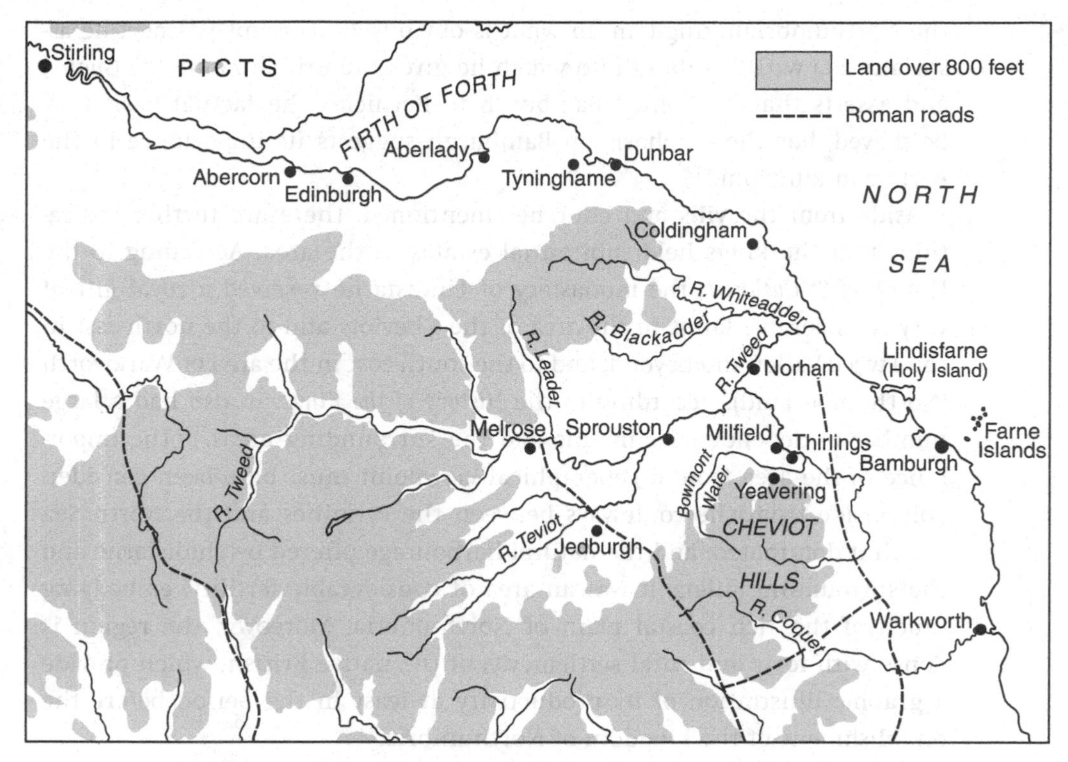

MAP 5 Map of the Bernician heartlands

Assuming the role of kings in founding monasteries was as important here as elsewhere, another indication of the area's importance to the kings is the site of Melrose on the Tweed, the monastery where Cuthbert first became a monk.[88] Nearby Jedburgh is mentioned as a possession of Lindisfarne in the *History of the Kings*, and the fact that it possessed an extremely fine sculptured panel, seemingly the end of a tomb or shrine, suggests that it was a monastery of some wealth.[89] Along the coast we come to Coldingham, where there was a monastery, of which Æbbe, another royal princess, was abbess. Bede tells us that this monastery was rich to the point that its inmates lived idle, debauched lives and were divinely punished for it.[90] It seems therefore that this area of northern Northumberland and south-east Scotland was a heartland of Bernician power. Within it, Bamburgh (Northumberland) may have had a particularly important role, for the early ninth-century British writer Nennius highlights its significance in the establishment of

[88] Anon., *Life Cuth.*, bk 2 ch. 3; Bede, *Life Cuth.*, ch. 6.
[89] *Hist. Kings*, *s.a.* 854 (p. 101 – *Geddewrd*); for the panel, Ralegh Radford (1956).
[90] Bede, *Eccles. Hist.*, bk 4 ch. 25 (23); on the site, see Alcock, Alcock and Foster (1986).

the Northumbrian kingdom. In what is obviously a corrupt passage, he associates Ida with Bamburgh (to which he gives the British name *Din Guaire*) and asserts that he 'joined Bamburgh to Bernicia': the factual basis may be flawed, but the emphasis on Bamburgh suggests its importance in the Bernician kingdom.[91]

Aside from the vills and churches mentioned, there are further indications that the kings held substantial estates in the area. According to the *History of St Cuthbert*, the monastery of Lindisfarne received a royal gift of very considerable lands in the area of the Cheviots and to the north-east in the Tweed Valley and beyond, and to the south-east in the area of Warkworth (Northumberland). According to the *History of the Kings*, it also had a large number of dependencies in this and the surrounding areas.[92] The importance of the area from a geographical viewpoint must have been considerable. Aside from the routeways between the Pennines and the North Sea which it dominated, and the natural harbourage offered by Lindisfarne and the surrounding inlets, it was an area of considerable fertility, embodying much of the rich coastal plain of Northumbria. Moreover, the region is dense with forts and rural settlements of the native British, which provide a graphic illustration of its productivity at least in the period before the establishment of the kingdom of Northumbria.[93]

A second Bernician heartland would seem to have been the area in which Bede himself lived, in the south of Bernicia (map 6). Monkwearmouth and Jarrow, respectively on the Rivers Wear and Tyne, were both royally endowed on a very substantial scale. Bede's own career and that of the monasteries' founder Benedict Biscop, who travelled widely and frequently to the continent in search of books and treasures, and the rich library which he established indicate the wealth of these establishments and the extent in this area of the royal resources available to endow them on such a scale.[94] Although we know virtually nothing of the early monasteries mentioned by Bede as existing at Gateshead and on or near the mouth of the River Tyne, it is very likely that they were of royal foundation or at least of royal endowment.[95]

Westwards, the valley of the Tyne forms an important natural routeway through the Pennines, leading via a gentle pass to the lowlands around

[91] Nennius, *Hist. Britons*, para. 61.
[92] *Hist. Cuth.*, paras. 3, 4, 7, 8; *Hist. Kings*, *s.a.* 854 (pp. 101–2). See Aird (1998b), maps 1.1–1.3.
[93] Jobey (1965). [94] See below, pp. 123, 140.
[95] Bede, *Eccles. Hist.*, bk 3 ch. 21, bk 5 ch. 6; Bede, *Life Cuth.*, ch. 3. For the identification of the monastery with South Shields, see Colgrave (1940), pp. 342–3. It may also have been Tynemouth (Northumberland), where Christian sculpture has been preserved, some of ninth-century date; see Cramp (1984), pp. 226–9, and Fairclough (1983).

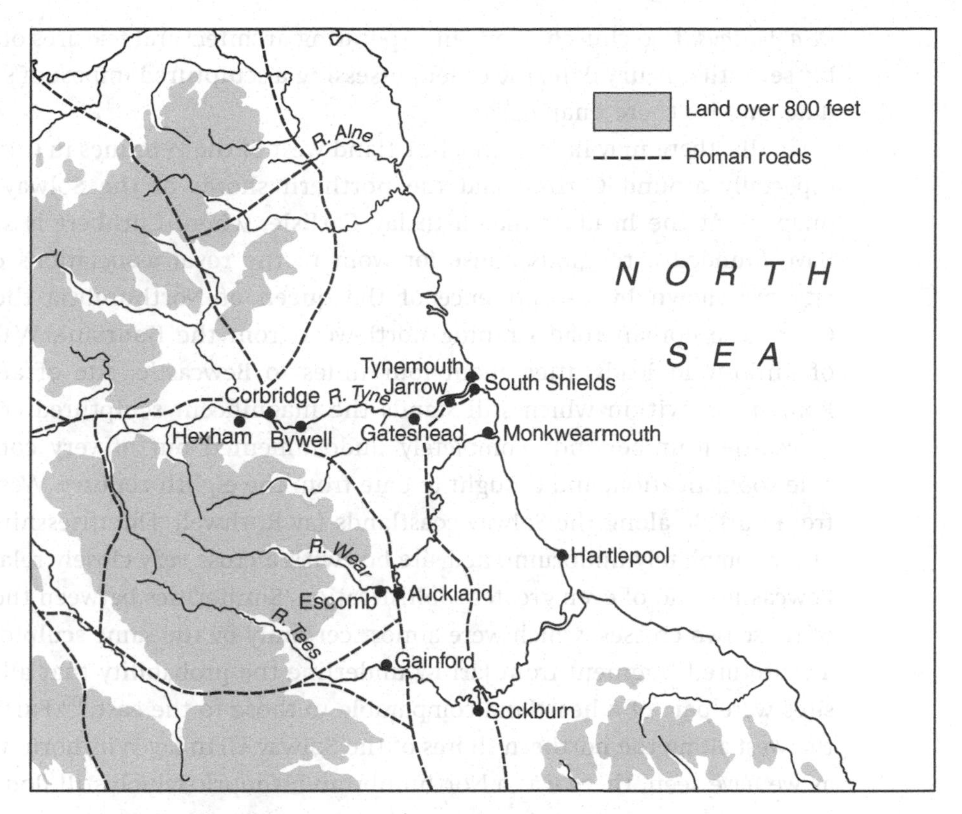

MAP 6 Map of the Tyne–Wear heartland

Carlisle. On this routeway, which had earlier been exploited by the Romans, lay the monastery of Hexham, founded by Wilfrid on land given to him by the queen of Northumbria, Æthelthryth, which was presumably part of her dowry from her husband King Ecgfrith. Stephen, in his *Life of Wilfrid*, describes the monastery for us and expatiates on the scale and magnificence of the church which Wilfrid built there.[96] Some four miles east of Hexham, the *History of the Kings* tells us, lay the church (*monasterium*) of Corbridge where a bishop of Mayo was consecrated in the eighth century, and where the existing church of St Andrew incorporates fabric and architectural features which have been dated to the seventh century.[97] The same source mentions the church of Bywell (*Bigewelle*), which may have been the monastic community which was in the early ninth century capable of producing the long Latin poem about its own inmates, Æthelwulf's

[96] Stephen, *Life Wilf.*, ch. 22.
[97] *Hist. Kings*, *s.a.* 786 (p. 51); Taylor and Taylor (1965–78), I, pp. 172–6.

De abbatibus. Two churches, one incorporating architectural features of possibly seventh-century date, the other possessing a sculptured impost of similar date, survive there (map 6).[98]

Finally, there may have been a heartland west of the Pennines in Cumbria, especially around Carlisle and the northern shores of the Solway Firth (map 3). At the head of that firth lay Carlisle, where Cuthbert is said to have founded a religious house for women; the royal associations of the city are shown by the presence of the queen of Northumbria there in 685.[99] The Roman road running north-west from the Hadrian's Wall fort of Birdoswald leads after about four miles to Bewcastle, site of another Roman fort within which still stands the magnificent sculptured cross of Bewcastle (Cumberland), completely undocumented but of very considerable sophistication, and thought to date from the eighth century. Westward from Carlisle along the Solway coastlands lay Ruthwell (Dumfriesshire), another completely undocumented site but with a cross very closely related to Bewcastle's and of even greater sophistication. Similarities between the style of these two crosses, which were almost certainly by the same sculptor, and a sculptured fragment from Jarrow underline the probability that all these sites were part of a heartland comparable to those to the east.[100] Further to the west along the northern shores of the Solway Firth lay Whithorn, where, as we have seen, there was a Northumbrian bishopric which still flourished in Bede's own day.[101]

Let us then take stock of the progress of our investigation. As regards frontiers, it appears that we are dealing with a kingdom with clearly recognizable, even linear, frontiers although in some areas, notably in the south, but also in the north-east, we can see reasons for regarding them as zones rather than as lines. Moreover, those frontiers seem always to have been more important than any internal frontiers. Northumbrian overlordship, however, transcended them in all directions, and we can see that overlordship at certain periods north of the Firth of Forth, south of the Humber, and possibly south-west into Wales, a situation which offers some comparison with the *regna* of Carolingian history, although it is perhaps closer to the sort of overlordship which the German monarchs exercised over the Slavs. Within Northumbria, the evidence enables us to glimpse a series

[98] See above, p. 17 n. 43; on the architecture, see Taylor and Taylor (1965–78), I, pp. 121–6; for the impost, Cramp (1984), p. 168.

[99] See above, n. 30.

[100] Bailey and Cramp (1988), pp. 61–72 (Bewcastle); on Ruthwell, see below, pp. 158–9; on Jarrow, see Cramp (1984), p. 115 (Jarrow no. 20).

[101] See above, n. 29.

of heartlands in Deira around the East Riding, the fringes of the North Yorkshire Moors and the Vale of York; in Bernicia, in the hinterland of Bamburgh and Lindisfarne, and in the valleys of the Rivers Tyne and Wear; and in Cumbria, around Carlisle and the northern shores of the Solway Firth. To the processes which operated on the territories we have been defining, in terms of ethnicity, culture and political development, we must now turn.

PART II

The creation of Northumbria

CHAPTER 3

The Northumbrians: origins of a people

At the beginning of the *Ecclesiastical History* Bede defines the peoples of Britain in terms of their languages: 'At the present time, there are five languages in Britain...These are the English, British, Irish, Pictish, as well as the Latin languages; through the study of the scriptures, Latin is in general use among them all.' The categorization of the peoples of Britain as fourfold – English, British, Irish and Pictish – is found elsewhere in the *Ecclesiastical History*. It is echoed in the document which formed one of Bishop Wilfrid's appeals to Rome and which refers to 'the northern part of Britain and Ireland and the islands, which are inhabited by the races of English (*Angli*) and British as well as Scots and Picts'. For Bede, the 'people of the Northumbrians' (*gens Nordanhymbrorum*) formed part of the English.[1]

There is strong evidence, both negative and positive, for the Englishness of the Northumbrian elite in Bede's time. The negative evidence consists, first, of Bede's summing up of the position of the British at the time he completed his *Ecclesiastical History* in 731:

Though, for the most part, the British oppose the English through their inbred hatred, and the whole state of the catholic church by their incorrect Easter and their evil customs, yet being opposed by the power of God and man alike, they

[1] Bede, *Eccles. Hist.*, bk 1 ch. 1, bk 3 ch. 6 (where Bede refers to four languages, that is omitting Latin); and Stephen, *Life Wilf.*, ch. 53 (pp. 114–15). Note that contemporary writers use the words *gens* (people), *natio* (nation), *populus* (people) in ways in which it is very hard for us to see how they were distinct or what any distinctions there may have been signified (see, for example, Putnam Fennell Jones (1929), under appropriate entries; for discussion, see Tugène (2001b), pp. 18–19).

cannot obtain what they want in either respect. For although they are partly their own masters, yet they have also been brought partly under the rule of the English.[2]

It is not clear what Bede meant by the British who are partly their own masters as against those who were partly under the rule of the English. Possibly he was thinking of the British kingdoms of Strathclyde to the north-west of Northumbria, of Gwynedd and the other Welsh kingdoms, and perhaps especially of Dumnonia in south-west Britain, the Britons of which he described as 'subject to the West Saxons'.[3] The important point, however, is that Bede did not envisage any British as being under full English rule. Within Northumbria, therefore, and presumably in the other English kingdoms also, Bede considered that the people who mattered viewed themselves as entirely English.

Secondly, recognizably British individuals and groups in Northumbria appear only very rarely in the writings of Bede and his contemporaries. Stephen, in his *Life of Wilfrid*, tells a story of how a woman brought her dead son to Wilfrid, who restored him to life, commanding that, when the child reached the age of seven, she should give him to Wilfrid for the service of God. The mother broke her promise and 'fled from her land' (*fugiens de terra sua*) with her son. But in vain – the bishop's prefect (*prefectus*) found the boy hiding 'amongst others of the British' (*sub aliis Bryttonum*) and brought him by force to serve in Wilfrid's church at Ripon. Stephen does not tell us where these British lived, nor does he make clear whether the woman was one of them, although the wording seems to imply that she was. At all events, neither her village nor the abode of the British who gave her sanctuary were distant enough to be beyond the power of Wilfrid's prefect. The British moreover were not high enough up the social scale as to be in a position to resist the prefect's demands.[4] British of perhaps even lower status appear in the account in the *History of St Cuthbert* of how Cuthbert was granted Cartmel (Cumberland) with the Britons dwelling on the land, presumably as slaves.[5]

In Bede's *Ecclesiastical History*, it is possible that Aidan's disciple Chad (*Ceadda*) was of British extraction to judge from his name which is probably

[2] Bede, *Eccles. Hist.*, bk 5 ch. 23 (for Northumbrian dominance of Gwynedd in King Edwin's time, see above, pp. 42–3).

[3] Bede, *Eccles. Hist.*, bk 5 ch. 18, probably referring to the letter of Aldhelm to King Geraint of Dumnonia (Lapidge and Herren (1979), pp. 155–60). On Dumnonia, see Pearce (1978).

[4] Stephen, *Life Wilf.*, ch. 18; for the possible identification of the woman's village with Tidover (Yorkshire West Riding), see G. J. R. Jones (1979), pp. 31 (fig. 2.5) and 33, and G. J. R. Jones (1995), pp. 22–6.

[5] *Hist. Cuth.*, para. 6. Although the grammar of the passage is corrupt, and Johnson South (2002), p. 49, interprets it to mean that the Britons in question helped to grant the land, the meaning here assumed seems to involve less emendation of the text.

British, although it is not clear that he came from Northumbria. The name of the Whitby cow-herd who received a miraculous gift of song, Cædmon, is certainly British, although part of the point of the story is that he was of low social class.[6] Aside from the possibility that the British bishops who consecrated Chad did so in Northumbria, the only British person to be identified in Bede's *Ecclesiastical History* as British and in English territory is in an account of one of the miracles of King Oswald of Northumbria. The person in question (*de natione Brettonum*) was passing near the place where the king had been killed in battle, when he noticed that

> a certain patch of ground was greener and more beautiful than the rest of the field. He very wisely conjectured that the only cause for the unusual greenness of that part must be that some man holier than the rest of the army had perished there. So he took some of the soil with him wrapped up in a cloth, thinking that it might prove useful, as was indeed to happen, as a cure for sick persons. He went on his way and came in the evening to a certain village, entering a house where the villagers (*uicani*) were enjoying a feast. He was received by the owners of the house and sat down to the feast with them, hanging up the cloth containing the dust he had brought on one of the wall-posts.

The upshot was that the feast engendered carelessness which led to a fire, which consumed all the house except the bag containing the soil and the post on which it hung – thus demonstrating the power and virtue of the soil and the sanctity of Oswald. As regards the British man, we are handicapped by not knowing for certain where the battle, called by Bede *Maserfelth*, took place. If it was at Oswestry in Shropshire, close to the Welsh Border, as was believed already in the twelfth century, the British person may have been an inhabitant of Wales passing through; and even if he were a resident this would probably have been of Mercia, in which kingdom Oswestry lay, rather than Northumbria. Even if *Maserfelth* was somewhere in Northumbria, however, and the British person was a resident of that kingdom, we should again be struck by the low social class assigned to him. He was clearly on a level with the villagers, the rustic country folk, and not with the elite. Of an identifiably British elite in Northumbria there is thus virtually no trace in our written sources, in which British names are notable for their almost complete absence.[7]

[6] Bede, *Eccles. Hist.*, bk 4 ch. 24 (22); on Chad, John McKinnell, pers. comm., and on Cædmon, Jackson (1953), p. 554.

[7] Bede, *Eccles. Hist.*, bk 3 ch. 10; on Oswestry see Stancliffe (1995b). In view of the paucity of references in our sources, it is hard to share the view that 'many people living in Northumbria in Cædmon's period are unambiguously named as British by our sources' (Moorhead (2001), p. 105).

The evidence of place-names, although hard to interpret because of the difficulty of establishing when the names were coined, suggests the presence of a British population in Northumbria. First, certain place-names contain elements referring to communities of British in Northumbria. These comprise place-names such as Walworth (County Durham) and Walton (Yorkshire West Riding), which have in them the Old English place-name element *Walh-*, that is Welshman or British person; and place-names which have in them the synonymous *Brettas-* (British), as in Bretton (Yorkshire West Riding), and *Cumbre-* (from *Cymry* or British), as in Cumberworth (Yorkshire West Riding). Names with the element *eccles*, derived from the British word for church, as in Ecclesall (Yorkshire West Riding), may also indicate the existence of British communities, since the church from which they were named was presumably one that served such a community. Elsewhere in England, names with the element *wickham* have been interpreted as indicating the existence of British communities, since the first element is supposed to be derived from the Latin *uicus* (settlement), and the distribution of these names suggests that the places to which they are attached may have reflected the Roman settlement pattern. Although such names do not occur in Northumbria, the element *wic* is found, as in Market Weighton (Yorkshire East Riding) and Barwick in Elmet (Yorkshire West Riding), and it too was presumably derived from *uicus* and indicates some relationship to the Roman settlement pattern. The occurrence of all these types of name thus helps to demonstrate the existence of communities of British in the Northumbrian heartlands, especially in the east, albeit – to judge from the distribution of the names – scattered and isolated ones.[8]

Other place-names, because they are formed from words or names derived from the British language, provide evidence for the currency of that language in at least some parts of Northumbria at some period. There is a number of such names in the Northumbrian heartlands: Yeavering is an example, so too is Auckland (County Durham), and so too is Catterick, which Bede uses in a form which, it has been argued, he must have heard spoken by British-speakers. The British name for the place where King Edwin had a royal vill, *Cambodunum*, was presumably in use in Bede's day (although it is no longer), and Bede seems to have learned this too by word of mouth, presumably from British-speakers. Bede knew Carlisle by its Roman name *Lugubalia*,

[8] Gelling (1997), pp. 67–74 and 87–105. For the place-name elements, see A. H. Smith (1956), I, pp. 50 (*Brettas*), 119–20 (*Cumbre*), 144–5 (*eccles*), II, pp. 242–4 (*walh*), 257–63 (*wic*, where the Roman associations are much less developed than in Gelling's work on *wickham* names). For Walworth, see Watts and Insley (2002), pp. 131–2.

itself derived from the British name *Luel* which he also used. The district of *Ahse* referred to in the Anonymous *Life of St Cuthbert* appears to derive from the Romano-British name for the fort of Great Chesters (*Aesica*), and that of Brougham (Cumberland) seems likewise to derive from the Romano-British name for the fort there (*Brocauum*). British names for rivers are also found, although much less so in the east than in the west, suggesting the continuance of a British population familiar with them, and a small concentration of British names in the North Yorkshire Moors may bear witness to a substantial British community there. Between the Tweed and the Forth generally, we find somewhat more numerous British names. These include names using British elements such as *aber* meaning 'river-mouth' as in Aberlady (Haddingtonshire), *pevr* meaning 'radiant' or 'beautiful' as in Peffer Mill (Midlothian), *pren* meaning 'tree' as in Pirnie (Roxburghshire), *cair* meaning 'fort' or 'manor', and *pol* meaning 'pool' or 'hole'. Certain British elements in place-names, namely *ucheldref* meaning 'high settlement' as in Ochiltree, and *pebyll* meaning 'tent' or 'temporary shelter' in Papple (Haddingtonshire), seem to relate to the practice of transhumance and suggest the presence not only of British language but also of British customs. So too does *mynydd* meaning 'mountain' but also 'upland pasture', as in Mindrum. It is nevertheless important to emphasize that in the lands south of the Forth and in the Northumbrian heartlands most villages and hamlets which do not have Viking names (which they must have received at a later period) have English names. Such names are dense even in the area north of the Tweed, and they include ones with elements regarded as early in date, such as *ingaham* meaning 'village of the people of' as in Tyninghame (Berwickshire). The dominance of English names throughout Northumbria east of the Pennines is extremely striking.[9]

This dominance of English names extends even to the remote settlements in the Cheviot Hills which are named in a passage in the *History of St Cuthbert* purporting to be a grant made to St Cuthbert in the late seventh century: 'Then the king [Ecgfrith] and all the English magnates gave to St Cuthbert all that land that lies near the Bowmont Water, with these vills: Sourhope (?),

[9] On Auckland, which in one form is identical to *Alcluith* (Dumbarton), see Watts and Insley (2002), pp. 9–10; on *Ahse*, Brougham, *Cambodunum*, Carlisle and Catterick, see Rivet and Smith (1979), pp. 242, 283–4, 302–4, 292–3, 402; on the North Yorkshire Moors, see Jackson (1953), pp. 238, 490, 680; on place-names in Scotland, see Nicolaisen (2001), pp. 61–108, and Geoffrey W. S. Barrow (1998), pp. 55–68. On rivers, see Kenneth Cameron (1977), pp. 33–46 and map; for mapping of rivers and British names, see A. H. Smith (1956), end-pocket map 2. Useful maps and summaries are to be found in Peter G. B. McNeill and MacQueen (1996), pp. 50–1, 61. See also Faull (1977), p. 45 (map 7b: Celtic place-names in Yorkshire).

Staerough, Old Graden, Pawston, Clifton, Shereburgh (?), Colewell, Halterburn, Thornington (?), Shotton, Kirk Yetholm and Mindrum.'[10] The only one of these tiny, rural settlements to have a name which is certainly derived from the British language is Mindrum, the first element of which is the British *mynydd*, noted above. Even the Bowmont, although a river-name which might be expected to be British in comparison with others, may be of English origin, meaning 'winding eel-river'.[11]

It does not follow that place-names necessarily reflect the ethnic composition of the settlements that bore them; the English names, for example, could have been coined by a not very numerous English ruling elite without reference to the language spoken by the inhabitants of the settlements to which they were given. In the face of such widespread English names, however, it is hard to avoid the conclusion that the population in all the lands south of the Firth of Forth came to regard itself as predominantly English and was principally English-speaking. This conclusion is reinforced by our Latin sources, which leave us in no real doubt that the vernacular language of those parts of Northumbria which Bede knew was Old English, the language of the English incomers, virtually uninfluenced (as the overall development of Old English suggests) by the language of the native British. As Bede lay dying, he broke into a poem in Old English:

For þam neodfere   nenig wyrþeð
þances snottra,   þonne him þearf sy,
to gehiggenne   ær his heonengange,
hwæt his gaste   godes oððe yfeles,
æfter deaðe heonen   demed wurðe.

Facing that enforced journey, no man can be
More prudent than he has good call to be,
If he consider, before his going hence,
What for his spirit of good or of evil
After his day of death shall be determined.[12]

Bede then turned in his last hours to translating St John's Gospel into Old English. Similarly, when Cædmon miraculously broke into song, it

[10] *Hist. Cuth.*, para. 3. The identifications are those in Johnson South (2002), p. 43.

[11] For Mindrum, see Mawer (1920). The fact that its second element is Gaelic *druim* meaning 'ridge' presumably reflects Scottish influence of a later period when Mindrum lay on the border of the kingdom of Scotland. For the Bowmont, see Ekwall (1928), pp. 45–6.

[12] Ed. and trans. Colgrave and Mynors (1969), p. 583 (trans. repr. McClure and Collins (1994), p. 301).

was – despite his apparently British name – in Old English that he did so. His song is one of a small group of Old English texts which can be assigned to Northumbria on grounds of the Northumbrian dialect in which they are composed. If we turn to the later middle ages, we find that a northern dialect of English, closely related to Lowland Scots, was the vernacular throughout the areas north of the Humber frontier, right up to the ancient northern frontiers of Northumbria on the Firth of Forth, and beyond. It was presumably the successor of Northumbrian Old English.[13]

The evidence of place-names west of the Pennines is more complex. There occur in Cumbria (that is the later counties of Cumberland, Westmorland and northern Lancashire), in addition to British river-names, five or six clusters of British settlement-names. One of these is around Carlisle which itself retained a version of its Romano-British name (*Luguuallium*), fused with the British element *cair*; and a second is found around the Hadrian's Wall fort of *Aesica* (Great Chesters), which may have given its name (as we have seen) to the district of *Ahse*. The evidence from this area, however, is complicated by the problem of whether the other, quite numerous British settlement-names to be found there represent survivals from the pre-Northumbrian period or whether they are new introductions of the tenth and eleventh centuries when the kings of the Cumbrians and of Strathclyde ruled Cumbria. The latter has been argued on semantic grounds, the former principally on the grounds that the British name clusters are mostly in out-of-the-way parts of Cumbria such as Tyndale, and it is therefore more likely that they are the result of survival in the face of English conquest than of the later conquest by Strathclyde British who would presumably have been in a position to impose their names on more central places. In either case, Cumbria looks to have been very English-dominated in the pre-Viking period. If the names are pre-Viking, they show only that British presence was peripheral while English presence was central; if they are late, they have no relevance to the period of the origins of Northumbria and the place-name evidence from Cumbria similarly points to English dominance. There is nothing in the archaeological record to contradict this. Although distinctively pagan English graves have only been found in the valley of the River Eden in the east of the area, their absence from more westerly areas could have resulted from

[13] On Cædmon, see Bede, *Eccles. Hist.*, bk 4 ch. 24 (22); for the Old English text of Cædmon's hymn, see A. H. Smith and Swanton (1978), pp. 38–41. On Northumbrian dialect, see below, p. 117. The most detailed mapping of late medieval English, which unfortunately deals only cursorily with southern Scotland, is Samuels, Laing and Williamson (1986). For Lowland Scots, see Murison (1979), pp. 2–3. For the general development of English, see Wakelin (1988), pp. 45–84.

the fact that the English only penetrated into those areas after they had ceased to place in their burials the grave-goods which allow archaeologists to recognize them.[14]

To the north-west, we come to the area north of the Solway Firth, as far as the Northumbrian episcopal see at Whithorn and beyond. Place-names with British elements are widely distributed in this area; but so too are names with English elements in their make-up, even as far west as Whithorn itself and the Rhinns of Galloway (map 3). Indeed, it has been possible in the case of Galloway and the area of Carrick to the north of it to build up a picture of a patchwork of English and British settlement, and to perceive the English as organized into three units with British in the intervening lands. The Old English place-name element *botl* meaning 'house' or 'palace' is important in this respect, for example in the name Buittle (Kircudbrightshire) which seems to be the centre of a concentration of English settlement, and Aisbutil, meaning 'frontier palace'. The existence around these places of an English administrative structure which levied tribute payments from the British settlements is hinted at by names such as Shirmers (embodying Old English *scir-(ge)-maere*, meaning 'shire boundary') and possibly Penninghame (Wigtownshire), which may embody the Old English element *pening*, meaning penny. The name would then mean 'farm on which a penny geld is payable' and thus be evidence of the imposition of an English taxation system. In Galloway and Carrick, we appear therefore to have a less thorough effacement of the British than to the east, but none the less English dominance.[15] The inference from the place-names that Old English was the predominant if not the dominant language is reinforced by the evidence of English culture in the region to be discussed in the next chapter. But we should here note that the craftsmen who erected the great carved and inscribed stone cross at Ruthwell well to the west of Carlisle, probably in the eighth century, inscribed on it a version of the poem *The Dream of the Rood* in Old English.[16]

In the light of linguistic evidence, then, it seems that Bede was right to regard Northumbria in his own time as essentially English. In short, that part of the Roman Empire south of Hadrian's Wall and native areas to the north of it, both inhabited by British, had been welded into a kingdom which was regarded as English, inhabited by the 'people of the Northumbrians', as part

[14] On place-names, see Phythian-Adams (1996), pp. 77–87, arguing against Jackson (1963a), pp. 74–7, who generally treated the names as late formations. See also below, p. 250. For the text, see Anon., *Life Cuth.*, bk 4 ch. 5; for the burials, see O'Sullivan (1996).

[15] Daphne Brooke (1991); on Penninghame, see Hough (2001). [16] See below, pp. 158–9.

of the wider lands in which lived the 'people of the English'. The question of how this had come about is controversial, and scholars' search for an answer has involved deploying highly complex arguments derived from often limited and ambiguous evidence of many types, textual, archaeological, topographical, semantic and so on. It seems best, therefore, first to set out the possible models or hypotheses which, in the light of what we know of Roman Britain and of developments elsewhere in western Europe after the end of the Roman Empire in the west, can be offered to explain the genesis of Northumbria; then to examine how far the available evidence can be marshalled to support one or more of them. Although, in reality, the evolution of Northumbria may have been a combination of one or more of these models, and although different parts of the kingdom almost certainly developed differently, the exercise of setting out the models and marshalling the evidence relating to them remains a valuable one, if only in showing the extent and limits of our understanding. The models in question, which we shall call Models 1–3, are the following:

Model 1 involves a controlled cession of government to barbarian (English) mercenaries or federates, who then proceeded to create the kingdom of Northumbria. In other words, the Roman rulers of Britain saw it as advantageous to hand over political power to the English, and did so peacefully and by agreement. We could envisage the Roman rulers in question as being the imperial authorities of the Roman Empire, or as being what are often called 'sub-Roman' authorities. In the latter case, with the end of direct Roman rule by the Empire, a sort of *de facto* Roman organization would have continued for some time, and it was the Roman or British leaders of this who would have effected the controlled cession of power to the English. Either way, Model 1 involves a direct and more or less peaceful transition from Roman or 'sub-Roman' to Northumbrian, with a change in the ruling elite but not in the make-up and organization of the population as a whole.

Model 2 similarly involves a more or less peaceful transition, but in this case from distinctively native British kingdoms to Northumbrian rulers, rather than directly or indirectly from the Roman Empire. According to this model, such kingdoms would have existed north of Hadrian's Wall during and after the Roman period, organized on native British lines and perhaps focused on British power-centres, such as hill-forts or fortified settlements (*oppida*). Similar British kingdoms would have emerged south of Hadrian's Wall after the end of Roman rule, or perhaps those to the north would have extended southwards. Such kingdoms might have been focused on previous Roman power-centres; or they may have eschewed even the relics of Roman organization. If the latter, they may have represented re-assertion of native British organization, and presumably culture, amongst people who had barely accepted Romanization in all the long centuries of Roman rule in Britain. The rulers of these kingdoms would then have handed over power to incoming English who, as in Model 1, created Northumbria with themselves as the ruling elite and the basis of native society unchanged.

Model 3 involves conquest of the area of Northumbria by incoming English with consequent destruction or degradation of the native British population and removal of its organizational structures. According to this model, there was no peaceful transition. Northumbria was a creation of the incoming English and owed little to either a Roman or a British past.

According to Model 3, then, Northumbria was English because the British had been exterminated, expelled or at best degraded by incoming English. Models 1 and 2 are based on the view that the Roman Empire in the area of Northumbria or its sub-Roman or British successors came to be dominated, for whatever reason, by an English ruling elite, who were eventually able to impose their language and culture on the British population, without necessarily altering in any fundamental way its organization and underlying social structure. According to these two models, language change to Old English was a cultural process, involving deliberate choice on the part of the British population to adopt English. In short, Models 1–2 envisage Northumbria as founded on British population, culture and political organization; and as in effect a British kingdom, the language and culture of which had been transformed by an English ruling elite. All three models can be evaluated both in terms of their plausibility in the wider context of the later Roman Empire and what succeeded to it elsewhere in Europe, and in terms of the specific evidence from Northumbria bearing on them.

The plausibility of Model 1 lies in the wider context of the role of barbarians in the later Roman Empire in the west, where the Roman authorities, especially in the fifth century, settled barbarians within the Roman Empire as federates (that is groups of soldiers retaining their own organization and command), with responsibility for assisting the Roman authorities in warfare.[17] As Roman rule disintegrated in western Europe, it was a real possibility that power would have passed into the hands of the commanders of these barbarian contingents, who had the military capabilities to exercise authority in place of their former masters. The Roman emperors in Constantinople may have favoured this process because they viewed western Europe as a nest of usurpers (as indeed it had been) and as a potent source of trouble. Handing it over piecemeal to local barbarian chieftains may have seemed a not unattractive alternative to continuing to garrison it with troops in Roman pay with the attendant risk of a usurper emerging from their midst. Moreover, the Roman authorities may have had the means to achieve this constitutionally in the institution of *hospitalitas*, which originated in arrangements for the billeting of soldiers. In the late Roman period, it involved the transfer of lands to barbarians garrisoned in the provinces for their maintenance, or possibly the transfer of tax revenues directly to them by the Roman authorities. If this last occurred, it is easy to see how barbarian military leaders were able to set themselves up as the successors to the Roman provincial governors with the same financial infrastructure as their predecessors.[18]

Could something of this sort have happened in Northumbria? The Roman imperial province of *Britannia inferior*, which extended from the north into the midlands, was almost certainly ruled from York, which was after its foundation (probably in AD 71) the headquarters of a Roman legion throughout the Roman period, and was manifestly an important place. At some point in the late Roman period, the fortifications of its legionary fortress were reinforced on a grand scale with a line of polygonal towers facing the River Ouse. The court of the emperor Septimius Severus resided at York for over two years during his military campaigns in Britain (208–11). At York, too, the soldiers raised up as emperor Constantine (306), who then seized power over the whole Roman Empire and began the process of making it officially Christian. At York too in the last days of Roman rule was stationed the duke of the Britains, who was almost certainly responsible for the defence of Britain. Nor was York just a military centre. Around the legionary fortress, particularly in the area to the east of it towards the River Foss, there had

[17] Nicasie (1998), pp. 87–8, and Southern and Dixon (1996), pp. 46–52, 69–72. Mercenaries serving in the Roman armies themselves appear to have been less important: they were not overwhelmingly numerous in the Roman armies, at any rate before the late fourth century (Nicasie (1998), pp. 97–116).

[18] Goffart (1981) and, for *hospitalitas*, Goffart (1980).

grown up a civilian area of some pretension, where there were temples and presumably also civilian housing. To the south of it, on the other bank of the River Ouse, lay the civilian town (*colonia*), which was also of some wealth and grandeur, as is shown by the elegant tombstones recovered from the cemeteries which surrounded it as was normal for Roman towns (map 7).[19] York thus provides evidence that Roman rule south of Hadrian's Wall was strong enough to have been capable of overseeing the sort of hand-over to barbarian federates which this model envisages. Carlisle was another important place for the Romans, the site of two legionary fortresses; and Roman power and interest extended north of Hadrian's Wall, for Roman roads led across what is now northern Northumberland and southern Scotland, especially to the so-called outpost forts, which were probably occupied until at least the early fourth century. Beyond them, even the native tribes may have been subject to Roman-influenced organization, which was focused on tribal centres (*loca*).[20] A plausible context for Model 1 thus existed in terms of Roman organization.

There is, however, a serious chronological difficulty in proposing a direct hand-over from Roman authorities based in York and perhaps Carlisle to English incomers in that both written and archaeological sources suggest a substantial interval of time between the end of Roman rule and the arrival of English incomers. As regards written sources, a passage in the work of the Byzantine chronicler Zosimus suggests that direct Roman rule was withdrawn from Britain in 410, but the earliest presence of 'Saxons' in Britain was (according to the *Life of St Germanus* by Constantius) in 429. A 'Saxon' take-over of Britain is assigned to 441 by the so-called *Gallic Chronicle* of 452, and to 449 by Bede in his *Ecclesiastical History*. (The only reference to 'Saxons' in Britain before the 410 withdrawal of Roman rule is a reference to the year 408 in another passage in the *Gallic Chronicle*, the dates of which are, however, very suspect indeed because of its inconsistent use of different dating systems.) The fragmentary and unreliable character of the record these sources provide, the confusions of dating and events they embody and the unlikelihood that their authors were well informed about Britain might lead us to set them aside altogether.[21] The picture they give, however, is supported by archaeological evidence, to the extent that we cannot establish the presence

[19] See summary of scholarship in *Sources*, pp. 35–44. On the *colonia*, see the summary by Ottaway (1993), pp. 64–95; see also Brinklow (1984) and Ottaway (1984).

[20] On *loca*, see Mann (1978); on the outpost forts, see Frank Graham (1983) and, for their abandonment, Breeze and Dobson (2000), pp. 241–3.

[21] For general assessments, see Salway (1982), pp. 446–501; for discussion and more recent references, see Snyder (1998), pp. 29–49.

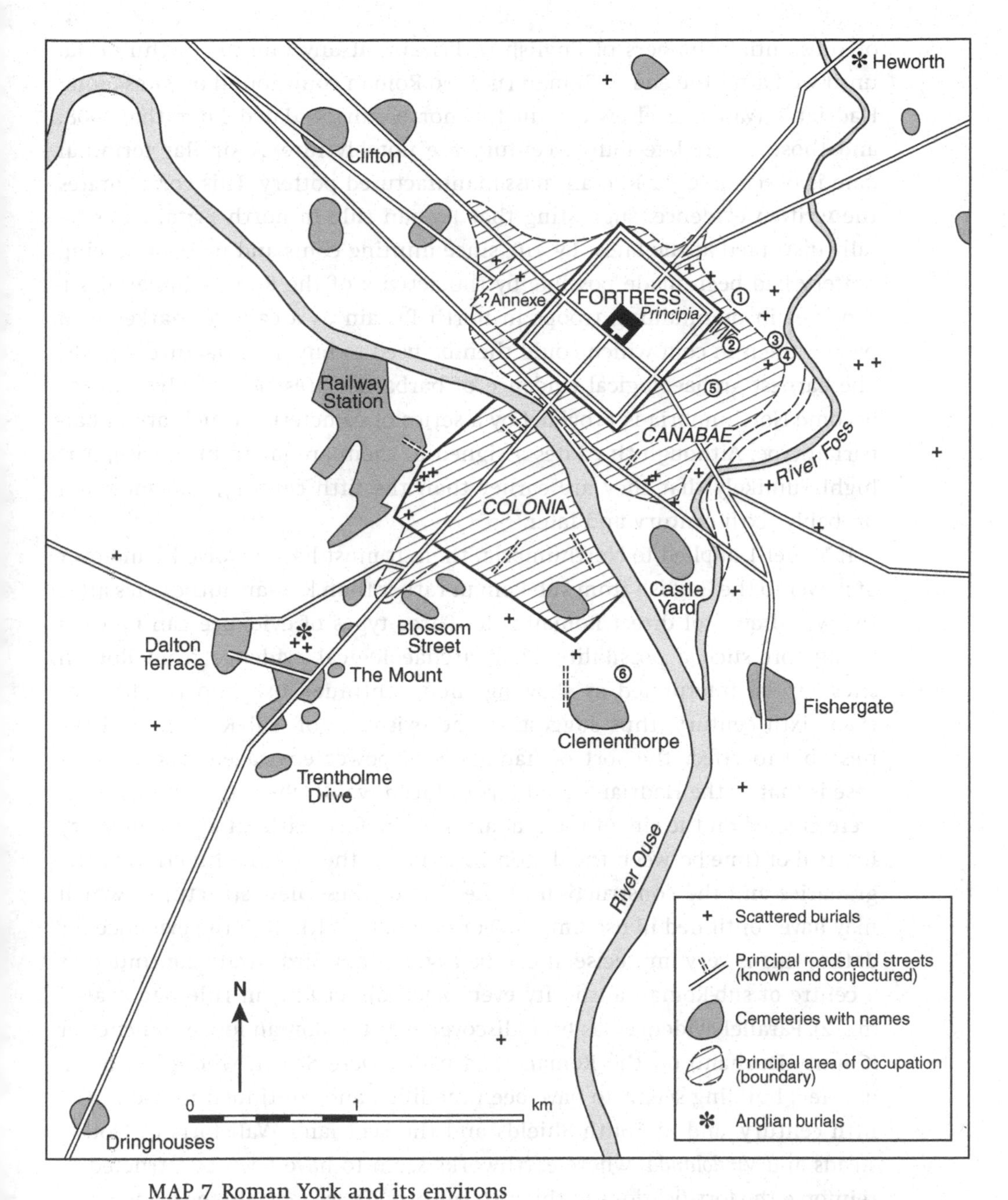

MAP 7 Roman York and its environs
Published in Ottaway (1993), fig. 1, where more details of the sites marked can be found. Note especially the location of the fortress, *colonia* and *canabae* (civilian area), and also the English ('Anglian') burials at Heworth and The Mount.

of substantial numbers of English in Britain, at any rate in Northumbria, until well after the end of Roman rule. No Roman coins found on sites along Hadrian's Wall and elsewhere in the north can be dated later than 408, and those of the late fourth century are notably rare. A similar terminal date is observable for Roman mass-manufactured pottery. This corroborates the written evidence, suggesting that Roman rule in north Britain was fatally disrupted at the latest by 410, since minting coins and mass-producing pottery had been made possible by the activity of the Roman imperial authorities in maintaining troops in north Britain as a captive market, and paying them in coin which could then be used to buy manufactured goods. The earliest archaeological evidence of barbarian presence in what was to become Northumbria is provided by a series of cemeteries which are of barbarian type. Although the dates assigned to them are far from precise, it is highly unlikely that they are earlier than the fifth century, and most are probably sixth century and later.[22]

If Model 1 applied to Northumbria, then, it must have involved hand-over of power to the English from sub-Roman rather than Roman authorities after the withdrawal of direct Roman rule. Three types of evidence can be used to support such a possibility. First, archaeological evidence from Roman sites can be interpreted as showing their continued use into the fifth or even sixth century, thus suggesting the existence of 'sub-Roman' authorities able to effect the sort of hand-over of power envisaged. The clearest case is that of the Hadrian's Wall fort of Birdoswald, where two timber halls were erected on the site of the granaries of the fort, without apparently any interval of time between the discontinuation of the original function of the granaries and the construction of the first of these new structures, which may have continued in use until 520 or even later, although the evidence for dating this is very imprecise. It can be argued that Birdoswald continued as a centre of sub-Roman authority even when direct Roman rule had ceased (fig. 2). Parallel evidence has been discovered at the Roman fort of Binchester (County Durham) on the Roman road called Dere Street, where the headquarters building seems to have been modified and continued in use in the fifth century, and at South Shields and the Hadrian's Wall forts of Housesteads and *Vindolanda*, where earthworks seem to have been constructed to reinforce the fortifications in the fifth century. There may also be evidence of

[22] On coins, see Brickstock (2000); on pottery, see Evans (2000); for cemeteries, see below, pp. 76–7. Welsby (1982), p. 164, has concluded that we should not expect to find barbarians in Britain at the end of Roman rule because 'there does not appear to have been any policy of settling barbarian groups within Britain' before *c.* 410.

FIGURE 2 Birdoswald (Cumberland), reconstruction of the hall and service buildings of the second phase of post-Roman timber buildings (drawn by Kate Wilson)
The hall, which measures 23 × 6.8 m, was built partly over the foundations of the granary and partly oversailing them to cover the main street (*uia principalis*) of the fort. At this time, the gateway at the end of the street (visible beyond the hall) had been blocked, and it has been suggested that the hall was deliberately sited in this way so that the now disused gateway could provide an imposing backdrop (Wilmott (2001), pp. 121–2).

ILLUSTRATION 4 York, north-west side of the fortifications, the 'Anglian' tower (tower 19)
The rubble-built wall on the left is the core of the Roman wall of the legionary fortress, from which the ashlar facing stones have been robbed at some period. The 'Anglian' tower is the small, rectangular building with the round-arched opening. Its construction is quite different from that of the Roman wall, both in the stone used and in the masonry, which is laid in coursed rubble rather than the ashlar facing with a rubble core of the Roman wall. The opening with its voussoirs has, in conjunction with another paired one on the opposite side, been compared with the church towers of early Northumbrian churches, such as Monkwearmouth.

fifth-century construction activity at York, where the rubble-built, so-called 'Anglian' tower repairing a breach in the fortress wall may be of 'sub-Roman' date (ill. 4).[23]

Secondly, there is the archaeological evidence of the distribution of the earliest 'barbarian' graves in Yorkshire, that is those which involve cremation. At York such graves have been excavated in the Roman cemeteries at The Mount outside the walls of the Roman civilian settlement (*colonia*), and at Heworth less than a mile from the Roman fortress (map 7). The largest of

[23] On Birdoswald, Wilmott (2001), pp. 121–4; on Binchester, Ferris and Jones (2000), and on this and other sites Snyder (1998), pp. 168–73 and Snyder (1996), pp. 45–7. For the 'Anglian' tower, see Buckland (1984), and, for summary of other scholarship, Ottaway (1993), pp. 109–11.

the cemeteries at Sancton (Sancton I, Yorkshire East Riding) is strategically placed relative to the Roman roads leading south to the Roman ferry-crossing of the Humber at Brough on Humber (*Petuaria*, Yorkshire East Riding), and north to the Roman fort of Malton. The cemeteries at Londesborough (Yorkshire East Riding) and elsewhere show the same sort of relationship to Roman lines of communication. It is possible that these cemeteries were those of federates in the pay of sub-Roman authorities, whom they would eventually supplant, although the type of objects found in such cremations do not themselves provide evidence for this.[24]

Thirdly, there is evidence of the use in the kingdom of Northumbria of Roman sites and buildings, evidence which is consistent with the English having assumed power in a negotiated way from 'sub-Roman' authorities, and thus continuing to use Roman power-centres. The excavations under York Minster, occasioned by the need to consolidate that building, are potentially important in this respect. The Minster is sited more or less in the middle of the Roman legionary fortress. The excavations under the south transept revealed, as would have been expected, the remains of the Roman headquarters building (*principia*) and in particular the great basilican cross-hall on the north side of the courtyard of that building (fig. 3). The unexpected conclusion arising from the excavators' report, however, was that this cross-hall had not been abandoned at the end of the Roman period, but had continued in use until the ninth century. The cross-hall was a massive stone structure, lined with impressive columns, and it would have provided an excellent palace for the new kings of Deira and then of Northumbria (fig. 4). What more likely than that, when the Roman authorities or their 'sub-Roman' successors ceded power to the barbarian federates and mercenaries in north Britain, one group of those barbarians – the future kings of Deira – should have continued to be based in the York legionary fortress, and should have converted the cross-hall into a royal hall which they and their successors continued to use until the coming of the Vikings in the ninth century?[25]

Something similar may have happened at a Northumbrian royal vill which Bede calls *Ad Murum* ('On the Wall'), a site not now identifiable but which was possibly a Hadrian's Wall fort which had passed into the hands of the Northumbrian kings. Comparable may have been *Cambodunum*, an

[24] On the cemeteries in general, see Eagles (1979), pp. 43–6, 240–1, and, for a gazetteer, pp. 421–54. See also Lucy (1999). On York, see Edward James and Heywood (1995), p. 9, and on Sancton, Myres and Southern (1973) and Timby (1993).

[25] Phillips, Heywood and Carver (1995), pp. 64–8.

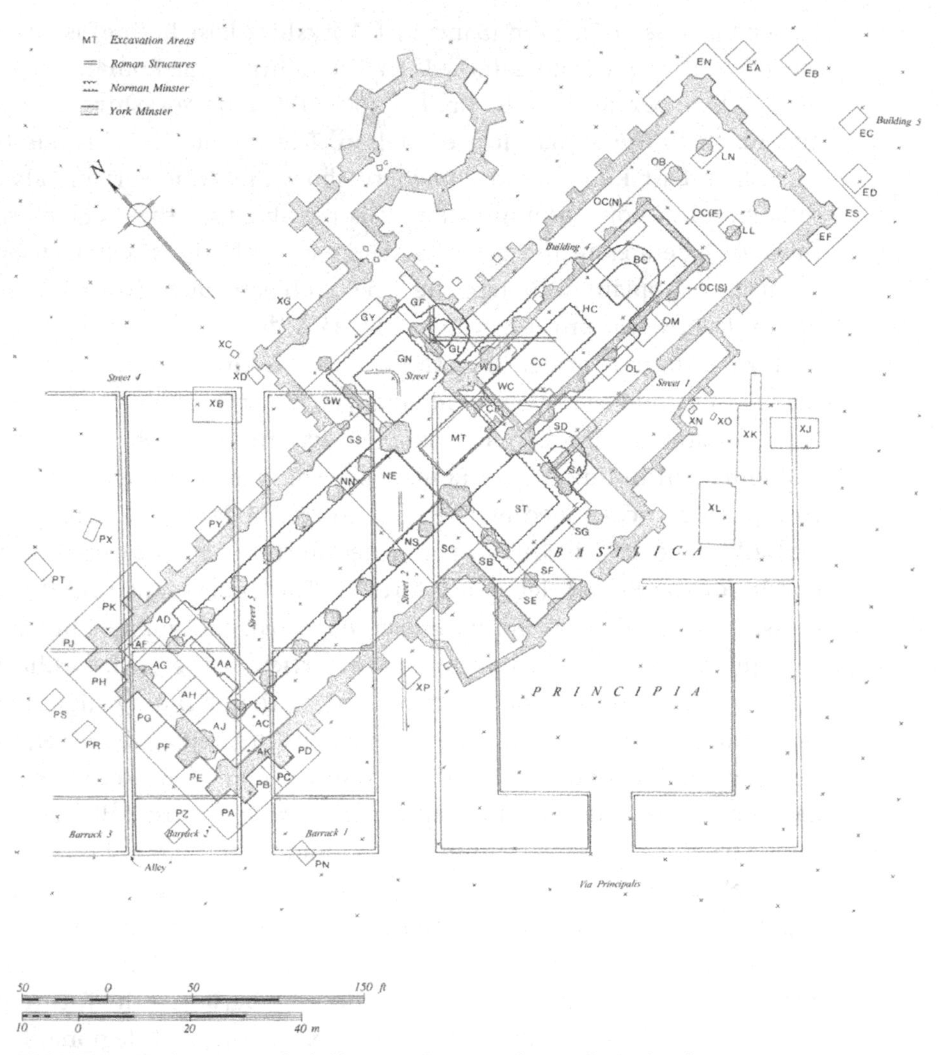

FIGURE 3 York, Roman buildings and streets beneath York Minster

unidentified Roman fort in the list of Roman sites known as the *Antonine Itinerary*. Here, Bede tells us, there was in King Edwin's day another royal vill. The persistence as a district-name of the name *Ahse*, possibly derived from that of the Hadrian's Wall fort of *Aesica*, may be explained by the existence of yet another royal vill in a former Roman fort, from which the district was governed. The discovery of fifth- and sixth-century pottery and brooches from within the Roman fort at Corbridge (Northumberland), and

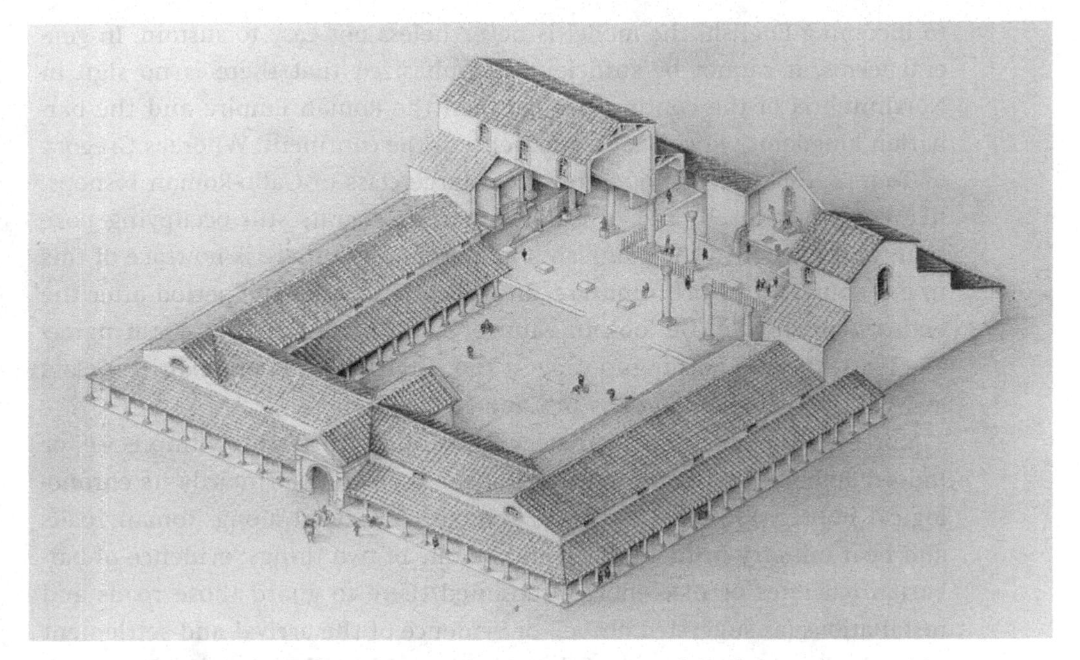

FIGURE 4 York, reconstruction of the *principia* of the Roman fortress
The cross-hall (basilica) is shown in cut-away section at the top right of the drawing.

of the sixth-century burials and artefacts from the fort at Binchester, could be interpreted to mean that these forts also continued in use as centres of Northumbrian power. Remains of barbarian-type buildings and burials of late fifth- and early sixth-century date excavated at the Roman town of Catterick suggest that the site was used by incoming English continuously until it emerged later as an important Northumbrian royal centre. Even from the Northumbrian royal fortress of Bamburgh there is evidence in the shape of archaeological remains, never fully published, which suggest that it had originally been the site of a Roman signal beacon, north of Hadrian's Wall and garrisoned by barbarians. Perhaps, pursuing our model, those barbarians had eventually assumed control, and Bamburgh as a Northumbrian royal centre developed as a result.[26]

Although the evidence surveyed above is consistent with Model 1 in so far as it involves a peaceful hand-over of power from sub-Roman authorities

[26] For *Ad Murum*, see Bede, *Eccles. Hist.*, bk 3 chs. 21–2; for *Cambodunum*, see Bede, *Eccles. Hist.*, bk 2 ch. 14, and below, pp. 86–7 nn. 47–8. On *Aesica*, see above, p. 61; for Catterick, see Cramp (1999), pp. 4–5, and P. R. Wilson, Cardwell, Cramp et al. (1996), pp. 50–4; for Bamburgh, see Hope-Taylor (1977), pp. 301–2.

to incoming English, the model is nevertheless not easy to sustain. In general terms, it cannot be sufficiently emphasized that there is no sign in Northumbria of the continuities between the Roman Empire and the barbarian kingdoms which are so apparent on the continent. Whereas Gregory of Tours's *History of the Franks* reveals a large class of Gallo-Roman bishops, like Gregory himself, and of Gallo-Roman aristocrats still occupying positions of influence in the Frankish kingdom of Gaul, there is no trace of this in our sources for Northumbria.[27] In Bede's account of the period after the end of Roman rule, we look in vain for Latin names or even Latin name-elements; there is nothing to suggest the survival of a Romano-British class as with the Gallo-Roman class in Gaul.

Moreover, the archaeological evidence is often frustratingly imprecise, or indeed ambiguous. For the cemeteries, the problem is primarily its chronological imprecision. The barbarian cemeteries located along Roman roads and near military installations could be one of two things: evidence of barbarian federates or mercenaries stationed there to guard those roads and installations (as suggested above); or evidence of the arrival and settlement of the barbarian conquerors of those roads and installations. Was Sancton I, for example, the cemetery of those who were protecting York and Malton from barbarian attacks along the Roman road from Brough on Humber? Or was it the cemetery of those who had used the Roman road as their route of invasion? Everything hinges on the date of these cemeteries. The earlier they are (i.e. the more firmly in the period when we can plausibly postulate the existence of sub-Roman rule) the more likely the first hypothesis is to be correct; the later they are (and thus the further into the post-Roman period) the more likely the latter is to reflect the truth. The only means of dating them is the typology of the cremation urns which they contain (that is the way in which the styles and designs of them developed) in comparison with urns from cemeteries from the continental homelands of the English. This is a very imprecise process, since dating chronologies on the continent as well as in England are often founded on the record of historical texts which are themselves problematic to interpret. It has been argued that the earliest 'barbarian' cemeteries such as Sancton I began in the late fourth century; but the weaknesses of the comparative dating methods applied make this seem dubious, and the earliest phases in such cemeteries are more likely to be no earlier than the mid-fifth century,

[27] See the frequency of Latin personal names in the full and clear index to Thorpe (1974); for example, Cato, Cautinus, Ecdicius, Hospicius, Injuriosus, Lupus, Modestus, Romanus.

with the majority of the cemetery evidence relating to the sixth century and later.[28]

The evidence for York is similarly subject to ambiguity. In the first place, doubt has now been cast on the notion that the cross-hall of the *principia* at York continued in use in the kingdom of Northumbria. This notion rests on a very small quantity of pottery of 'Anglian' date found on the site, and this, it is now argued, could as easily have been dropped accidentally on a ruined site as have been actively used in a building which continued fully in use.[29] In any case, the overall archaeological and historical evidence from York is little in favour of the idea that there was a smooth transition from Roman authorities to barbarian rulers there. Bede's description of what happened after the conversion to Christianity of King Edwin is instructive:

> He was baptized at York on Easter Day, 12 April [627], in the church of St Peter the Apostle, which he had hastily built of wood while he was a catechumen and under instruction before he received baptism. He established an episcopal see for Paulinus, his instructor and bishop, in the same city. Very soon after his baptism, he set about building a greater and more magnificent church of stone, under the instructions of Paulinus, in the midst of which the chapel which he had first built was to be enclosed.[30]

Assuming, as is very likely, that the church referred to here was near the site of the present York Minster, it is striking that the church which Edwin built was in the heart of the Roman fortress. This could be seen as supporting the idea that the king used the cross-hall for his residence, with the church built in its courtyard. A passage in the eighth-century *Life of Gregory the Great* can be interpreted as describing just such a setting. It refers to an event apparently shortly after Edwin's conversion involving the bishop Paulinus who had been principally responsible for converting him.

> When the king was hurrying to the church for the instruction of those who were still bound not only to paganism but also to unlawful marriages, he [Paulinus] hurried with him from the hall where they had previously been encouraging them

[28] On general issues of dating, see the illuminating comments of John Morris (1974). For the early dating of Northumbrian cemeteries, see Myres and Southern (1973), Myres (1969), pp. 74–5, and Myres (1977), pp. 121–3. On the first of these, see Kydd (1976). For more recent assessments, see Eagles (1979), pp. 82–142, and Timby (1993) who, while agreeing that Sancton I was probably in use during the fifth, sixth and possibly seventh centuries, notes that 'there is as yet no acceptable absolute chronology' (pp. 311–12). The most recently excavated inhumation cemetery, West Heslerton (Yorkshire East Riding), is assigned a date-range of *c.* 475–*c.* 650 (Haughton and Powlesland (1999), p. 81).

[29] Carver (1995), pp. 187–90.

[30] Bede, *Eccles. Hist.*, bk 2 ch. 14.

[the catachumens] to amend their lives in both respects, when a crow set up a hoarse croaking from an unpropitious quarter of the sky. Thereupon the whole of the royal company, who were still in the *platea populi*, heard the bird and turned towards it, halting in amazement as if they believed that the 'new song' in the church was not to be 'praise unto God', but something false and useless. Then, while God looked down from his heaven and guided everything, the reverend bishop said to one of his youths, 'Shoot the bird down quickly with an arrow.' This was speedily done and then the bishop told him that the arrow from the bird was to be kept until the instruction of the catechumens was finished and then brought into the hall.

This story, the primary purpose of which was presumably to demonstrate the superiority of Paulinus's religion over paganism, is not explicity located in York. But if York was in fact the place where the incident was considered to have taken place, it is possible to argue that the hall (*aula*) was the cross-hall of the *principia*, and the *platea populi* was the courtyard in front of it, in which stood the church (*ecclesia*), which was the one referred to by Bede above.[31] Aside from the difficulty of whether York is meant at all, however, let us note how little support either this or the previous text really give to the idea that there was at York a relatively peaceful hand-over of power by the Roman authorities. Bede makes it clear that the church built by Edwin was a new one, made in the first instance hastily of wood. There is no suggestion of a Roman church being reused, as for example there is in Bede's account of Christ Church, Canterbury. We should not of course expect to find such a church in the fortress, but there must have been one in the civilian part of York for there was in the late Roman period a bishop of York who attended the church council of Arles in 314.[32] The most economical interpretation of the evidence set out above is that there was no real continuity between Roman York and Edwin's York, even via a postulated 'sub-Roman' phase, and York does not provide evidence for Model 1. Nor is that model required to explain why Edwin should have wished to use York. Pope Gregory the Great had explicitly instructed Augustine, the leader of his mission to England, that York should be the site of a metropolitan see; and Edwin's mentor Paulinus was a member of that mission.[33]

The archaeological evidence from other Roman sites is in general too fragmentary and sporadic to be anything but equivocal. What does it mean that

[31] *Sources*, pp. 127–9 (A.2.1 – *Life Greg.*, ch. 15 (pp. 96–9)).

[32] On Canterbury, see Bede, *Eccles. Hist.*, bk 1 ch. 33: 'a church which, as he was informed, had been built in ancient times by the hands of Roman believers'. For the bishop of York, see below, p. 111.

[33] Bede, *Eccles. Hist.*, bk 1 ch. 29.

the only very loosely dated post-Roman structures at Birdoswald are on the site of the granaries? Was this the most convenient place on the fort for some 'sub-Roman' official or a barbarian mercenary warlord to whom the Roman authorities had ceded power to build a new headquarters? Or are we seeing simply re-use of a convenient site without any real continuity with its former Roman use? Are the buildings and other remains at Catterick early enough to indicate any sort of hand-over by Roman authorities (dating evidence has recently been revised to the sixth rather than the fifth century)?[34] It is in this connection instructive that certain of the finds of barbarian material in Roman forts are of burials within them, as at Binchester where the English burials have been recovered from the headquarters building itself. Is this really continuity of usage or some sort of quite unconnected expediency? Burials in such situations would have been unthinkable in the Roman period, when the dead were placed in extramural cemeteries and thus strictly segregated from the living. As for *Cambodunum* and *Ad Murum*, it seems likely that Northumbrian kings, in common with other barbarian kings, wanted to be thought of as imitators of the Roman Empire just as, Bede tells us, King Edwin had a standard which Bede (and probably Edwin too) thought was a typically Roman standard called a *tufa*.[35] Whether they were right or not in this belief, the aspiration was clear enough; and it was precisely such an aspiration which might have led an ambitious Northumbrian king to re-use a Roman fort long after any connection with Roman organization had been lost.

The evidence for a 'sub-Roman' phase succeeding Roman Britain proper is itself open to serious question at least in the area which was to become Northumbria. The evidence we have surveyed proves little more than that there was continuing activity of some sort at certain Roman sites. There is little to suggest that this was 'sub-Roman' activity involving, for example, continued use of Roman buildings for the same sort of purposes for which they were constructed. There is also a notable scarcity of evidence for continuing Roman culture in the fifth century and beyond, which might be an indicator of 'sub-Roman' activity. Nor is there very convincing evidence for the survival in Northumbria of Roman units of government which might have been preserved by 'sub-Roman' authorities, handed over to the Northumbrian rulers and so preserved at least in attenuated form. Attempts have been made to equate the Roman *ciuitas* of the Brigantes with Bernicia, but considerable violence to the testimony of Roman authors is required

[34] P. R. Wilson, Cardwell, Cramp et al. (1996), p. 51.
[35] Bede, *Eccles. Hist.*, bk 2 ch. 16; see Bruce-Mitford (1974), pp. 7–17.

to achieve this. It has also been suggested that the Roman *ciuitas* of the *Caruetii* in north-west England had an after-life in a shadowy Cumbrian kingdom, as reflected in the continued existence of the Roman centre of Carlisle (*Luguuallium*), known as *Lugubalia* or *Luel* in Northumbrian sources. This is possible, but it cannot be proved. Carlisle is a very strategic site on a number of routeways, and its importance in the Northumbrian kingdom need not have derived from any continuity via a 'sub-Roman' phase. The story told by Bede of St Cuthbert being shown the Roman walls and a Roman fountain at Carlisle by the 'citizens' (*ciues*) need indicate no more than that Roman remains were still visible in Carlisle, and were an object of interest then as now. The fullest archaeological evidence – that from Birdoswald – suggests discontinuity of function since it envisages the conversion of granaries into residential or meeting halls, even if there was no interval of time between Roman use of the granaries as such and what succeeded it.[36]

The evidence for a 'sub-Roman' phase north of Hadrian's Wall is even more nugatory. The existence of such a phase has been argued for the British kingdom of Strathclyde in what is now south-east Scotland on the grounds that the genealogy of its kings contains names, notably a certain Paternus 'of the Red Robe', which have been held to show that its rulers adopted Latin names and Roman costume, and that the kingdom was therefore the product of Roman influence or even Roman political intervention in creating it as a buffer state.[37] The evidence is late; the inference seems strained in the extreme; and it is in no way supported by other evidence of activity which can definitely be identified as 'sub-Roman' in Strathclyde or other areas. A hoard of Roman silver from the British fort of Traprain Law near Dunbar might indicate that the lord of the fort favoured Roman silver and was therefore behaving in a 'sub-Roman' way; but it could as well have been the result of plunder or of tribute payment by the Roman authorities, and in any case it appears to date to the latter part of the fourth century and is thus too early to cast light on the end of Roman Britain.[38] It is very hard, therefore, to see how Model 1 can apply at all north of Hadrian's Wall.

The idea behind Model 2, which postulates a transition by more or less peaceful means from distinctively native British kingdoms to Northumbrian rulers, rather than directly or indirectly from the Roman Empire, has been

[36] On the Brigantes, see Dark (1994), p. 74; cf. Rivet and Smith (1979), entry 'Brigantes'; and on Carlisle, see Phythian-Adams (1996), pp. 62–4, and, for archaeological evidence of post-Roman activity there, McCarthy (2002), pp. 134–9; for the texts, Anon., *Life Cuth.*, bk 4 ch. 8, and Bede, *Life Cuth.*, ch. 27. On Birdoswald, see above, pp. 70, 71.

[37] Peter Hunter Blair (1947), pp. 28–9; cf. Dumville (1989a), p. 216.

[38] Armit (1997), pp. 116–17; for the treasure, see Curle (1923).

influenced by archaeological research on south-west and west Britain. This has revealed the re-use in the post-Roman period within Roman Britain of native British fortresses, such as South Cadbury or Cadbury Congresbury in Somerset, or Dinas Powys and Castell Degannwy in Wales. This can be interpreted as indicating the resurgence of British political organization, presumably British kingdoms, following the end of Roman rule (the British population in effect 'going native') within the former Roman Empire.[39]

The evidence which can be adduced for the applicability of this model to Northumbria is wide-ranging, though very problematic to evaluate. There is some evidence to suggest that Deira and Bernicia, the two original constituent kingdoms of Northumbria, were British kingdoms taken over as going concerns by an incoming English elite as Model 2 postulates. In the case of Deira, all we really have to go on is the name, which is certainly not an English one, and may be British although its meaning is unclear. The name Bernicia seems to be British, even if it does not refer to the British tribe of the Brigantes (as was once supposed) but means something like 'land of mountain passes'. There is, however, rather more to say about this kingdom, or at least about the parts of it north of Hadrian's Wall. First, it is very striking that several of the important centres in it appear, to judge from the earliest forms of their names, to have been of British origin. Bamburgh had a British name, *Din Guaire*, in Nennius's account and may well have been of British origin. Coldingham, site of a Northumbrian monastery, is called by Bede *Coludi urbs*. Since Colud is a British name, this suggests that Coldingham too was originally a British centre taken over by the Bernicians, most likely the hill-fort on St Abb's Head, rather than the site of the modern village of Coldingham. King Ecgfrith's prefect imprisoned Bishop Wilfrid at a royal centre in Dunbar, which also has a British name (*Dinbaer*) and was almost certainly in origin a British fort defending the promontory on which modern Dunbar stands. Possibly it was associated with the remains excavated on the nearby site of Old Dunbar (Doon Hill) of a chieftain's residence of post-Roman date, built in a way strongly suggestive of British styles, and then replaced by a residence of more Northumbrian style, suggesting just the sort of take-over we are postulating. Another of the king's prefects imprisoned Bishop Wilfrid at *Inbroninis*, another place with a British name, although we do not know where it was.[40]

[39] Alcock (1971), pp. 209–29; on South Cadbury, see Alcock, Stevenson and Musson (1995).

[40] For the names Deira and Bernicia, see Bede, *Eccles. Hist.*, bk 3 ch. 1, and, for comment, Jackson (1953), pp. 419–20, 701–5. For the sites referred to see Alcock (1989) and, on Doon Hill specifically, David M. Wilson and Hurst (1966) and Hope-Taylor (1983). See also Snyder (1996), p. 46; and, for

Further south within Bernicia, the most striking evidence in support of Model 2 is provided by two juxtaposed archaeological sites at Yeavering. The first is a massive hill-fort crowning the steep outlier of the Cheviot Hills called Yeavering Bell, a major fortified centre or *oppidum* as Roman writers might have called it, with considerable evidence of occupation within it in the form of hut-circles (ill. 5). This fort, which is of pre-Roman Iron Age date (early first century AD), was very probably a political centre of some importance for the surrounding region. Since we find a high density of rural settlements of broadly Roman date distributed around it, that importance is likely to have continued through the Roman period. The second site is a complex of timber buildings, discovered by aerial photography and then excavated, on a low gravel ridge above the River Glen at the foot of Yeavering Bell (fig. 5). Beginning as a religious site and a cemetery, this second site developed in various stages into a substantial complex of halls and other buildings, and was almost certainly the place referred to by Bede in connection with the preaching of Paulinus:

> So great is said to have been the fervour of the faith of the Northumbrians and their longing for the washing of salvation, that once when Paulinus came to the king and queen in their royal vill (*uillam regiam*) at Yeavering (*Adgefrin*), he spent thirty-six days there occupied in the task of catechizing and baptizing.[41]

Bede goes on to describe Paulinus baptizing in the River Glen 'which was close at hand', which tends to confirm the identification. As we might expect in a period when we have virtually no coin evidence to help us, the dating of the phases of the site as revealed archaeologically is difficult to establish; but it does look as if the earliest belong to the period before there was any possibility of Bernicia having existed and therefore relate to some British rather than Northumbrian activity. Moreover, the name-form used in Bede's account is of very considerable interest. The Latin prefix *Ad-* should be disregarded in that it appears to be a translation of the standard Old English prefix *Æt-* meaning 'at the' (as in the Old English version of Bede, *Ætgefrin*), but in the context of place-names having little precise

the view that the earlier hall may in fact be prehistoric, Ian Mervin Smith (1991), p. 267. For St Abb's Head, see Alcock, Alcock and Foster (1986). For the texts, see Nennius, *Hist. Britons*, para. 61 (Bamburgh); Bede, *Eccles. Hist.*, bk 4 ch. 19 (17) (Coldingham); Stephen, *Life Wilf.*, ch. 38 (Dunbar) and ch. 36 (*Inbroninis*; for a speculative identification of this place with Lindisfarne, see G. J. R. Jones (1990)). Lindisfarne had a British name, *Metcaud*, although we cannot be sure that the English ever used it (Nennius, *Hist. Britons*, para. 63 (*Metcaud*)). There is no evidence for Deira being a British kingdom centred on York and under the rule of King Coel Hen, the ancestor of several of the lines of northern British kings recorded in the Welsh geneaologies (Peter Hunter Blair (1947), pp. 45–8).

[41] Bede, *Eccles. Hist.*, bk 2 ch. 14.

ILLUSTRATION 5 Yeavering Bell (Northumberland) from the air, looking north
The rampart of the hill-fort, which consists of a massive wall of tumbled rocks, is clearly visible, with indentations indicating the sites of dwellings within it. The palace site lies beyond, between the two woods at the foot of the Bell. The farthest wood occupies a steep site dropping down to the River Glen beyond.

significance. Thus the name of Yeavering was for Bede *Gefrin*, which evolved over the centuries into *Yever* in 1242, *Yeure* in 1329 and so to Yeavering. This shows that although at first glance Yeavering might appear to be an English name in its modern form (with an *-ing* ending as in Sonning on the River Thames), it is in fact a British one, derived from British *gafr* meaning 'goat', probably in combination with British *bryn*, meaning hill,

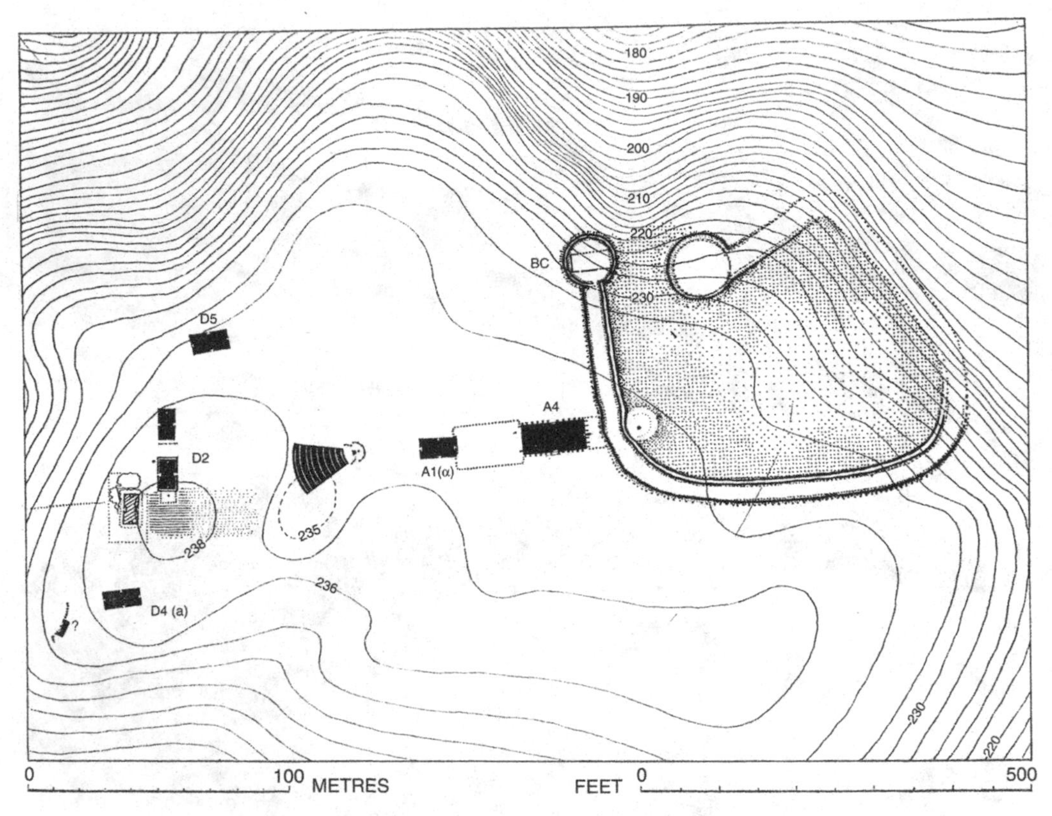

FIGURE 5 Yeavering (Northumberland), palace site, post-Roman Phase IIIc buildings and earthworks (from Hope-Taylor (1977), fig. 77)
This phase of the palace-site's development has been tentatively assigned to the reign of Edwin, with evidence for its destruction by fire being associated with the invasion and attacks of King Cædwalla of Gwynedd. On the right, the earthwork known as the Great Enclosure served either as a defence or as a cattle corral; and building A4 is a grand timber hall on a large scale. The quadrant building has been interpreted as a sort of amphitheatre, perhaps for the king to address his people; building D2 has been interpreted as a temple converted into a Christian church (Hope-Taylor (1977), pp. 277–8).

Thus the name in Bede's time meant 'hill of the goats'. So *Gefrin* cannot have referred in origin to the gravel ridge at the base of Yeavering Bell; it must have referred to Yeavering Bell itself and must have been the name not of the site on the gravel ridge but of the hill-fort. In short, Edwin's royal vill, where Paulinus preached, had a name derived from a British political centre, and seems very likely to have originated as a British site. Indeed some of the building techniques observable in the timber buildings at Yeavering have been interpreted as British in character, as has the probably fortified enclosure at the east end of the site, the bulbous mounds at the entrance to which are characteristic of forts in British areas. The royal vill which Bede says succeeded to Yeavering, that is Milfield (*Mælmin*), can similarly be interpreted as British in character, for it too has a British name and appears to have remains notably similar to those of Yeavering.[42]

The evidence set out here for sites in the heartlands of Bernicia has been used to argue, in line with Model 2, that the formation of Bernicia consisted of the peaceful take-over by an incoming English elite of a British kingdom, of which Yeavering was one of the centres, or perhaps the principal centre. According to this interpretation, the chronology and the character of Yeavering make it impossible to believe that the Northumbrians could have established themselves there in the interior of Northumbria so early in any other way. The incoming Anglo-Saxons initially acted as warlords for the native British and assumed political power because it was in everyone's interests for this to happen.[43] These arguments have some plausibility, but are weakened by the lack of evidence for the British kingdoms which are supposed to have lain behind Bernicia and Deira.

Evidence can be adduced for the existence of other British kingdoms. The most unequivocal reference, which is in the section on northern history in Nennius's *History of the Britons*, refers to a kingdom called Elmet: according to this King Edwin (616–33) 'occupied Elmet and expelled Ceretic (*Certic*), king of that country'. This account finds some corroboration in a passing reference in Bede's *Ecclesiastical History*, according to which Hereric, the father of St Hild of Whitby, himself a member of the Deiran royal family, was killed at the court of King Ceretic (*Cerdic*), and presumably this was the same king mentioned by Nennius. The two events may even have been linked – maybe Edwin invaded Elmet because Hereric had been poisoned by his hosts there; or maybe Edwin himself had Hereric poisoned because he was a dissident member of the Deiran royal house in exile, and then invaded Elmet to punish Ceretic for harbouring such an exile.[44]

[42] For the name Yeavering, see Hope-Taylor (1977), pp. 15–16, and, for the excavated features, pp. 205–39; see also entry 'Yeavering' in Ekwall (1960) and Mawer (1920) ('clearly a Celtic name'). For Milfield, see Hope-Taylor (1977), pp. 13, 276–7, and for a more up-to-date plan and evidence for the characteristic 'barbarian' houses known as Grubenhäuser on the site, see Gates and O'Brien (1988). The possibility that the buildings at Yeavering were in fact part of a hybrid English–British tradition is discussed by Simon James, Marshall and Millett (1984).

[43] Hope-Taylor (1977), pp. 276–324. A similar argument for Sprouston, another site discovered by aerial photography nearby on the River Tweed, is proposed by Ian Mervin Smith (1991), p. 285.

[44] Nennius, *Hist. Britons*, para. 63 ('occupauit Elmet, et expulit Certic, regem illius regionis'), and Bede, *Eccles. Hist.*, bk 4 ch. 23 (21).

Elmet and Ceretic are British names, so it is certainly possible that we are dealing with a post-Roman British kingdom, and one which, if we can so interpret an inscription erected in North Wales to a certain Aliortus of Elmet, had connections with the North Welsh kingdom of Gwynedd.[45] It is possible too that the inhabitants of this kingdom, or at least their descendants, were the *Elmed sætna* who figure in the list of peoples called the Tribal Hidage.[46] A possible clue to the location of Elmet is provided by Bede in the following passage:

In *Cambodunum* where there was also a royal dwelling (*uilla regia*), he [Paulinus] built a church which was afterwards burnt down, together with the whole of the buildings, by the heathen who slew King Edwin. In its place, later kings built a dwelling for themselves in the region (*regio*) known as *Loidis*. The altar escaped from the fire because it was of stone, and is still preserved in the monastery of the most reverend abbot and priest Thrythwulf, which is in the forest of Elmet.[47]

The region of *Loidis* has a British name, related to modern Leeds, as well as to the nearby place called Ledsham (Yorkshire West Riding). *Cambodunum* can be interpreted as a lost fort in this area. Assuming that the altar was heavy and so unlikely to have been transported far, we can suppose that Thrythwulf's otherwise unknown monastery was nearby, thus locating Elmet in the vicinity of Leeds. This is confirmed by the survival of place-names containing the name Elmet in the same area: Sherburn in Elmet, Barwick in Elmet and so on. It is not certain, however, that these names really preserve the memory of a kingdom of Elmet. In the passage quoted above, Bede refers to Elmet as the name of a forest; he actually makes no mention of its being the name of a kingdom. So the place-names could refer to a lost Forest of Elmet; the association of the name Elmet with the putative kingdom is solely due to Nennius. Identifying other kings of this kingdom is a speculative business. There is a king called Gwallawg who appears earlier in the same paragraph from the *History of the Britons* already quoted; he is conceivably to be associated with Elmet because the British bard Taliesin speaks of him attacking nearby York – but then the same poem credits him with battles over a much wider geographical area. A warrior called Madog of Elmet (*Elfed*) appears in *The Gododdin*, but the association with the Elmet we are discussing is not certain, nor is there any indication that this person was a king. In short, Nennius may have made an association between Ceretic

[45] G. J. R. Jones (1975), pp. 3–4. The inscription is Nash-Williams (1950), no. 87 (p. 88).
[46] See below, pp. 39–40; see also G. J. R. Jones (1975), pp. 13–14.
[47] Bede, *Eccles. Hist.*, bk 2 ch. 14.

and Elmet which was well founded; or he may have confused a forest name as that of a kingdom and created an imaginary kingdom of Elmet. Even if we accept Nennius's testimony, we have no way of ascertaining how far beyond the region of *Loidis* the putative kingdom extended. Brave attempts have been made to use linear earthworks and the putative boundaries of the supposedly British kingdom of Craven to the west to reconstruct them; but any certainty is impossible – the earthworks are largely undated, and the only evidence for the existence of a British kingdom of Craven is provided by later documents. These refer to a district of unknown origin, which may not have been a kingdom at all, although it admittedly had a British name.[48]

To the north-west of *Loidis* scholars have located a post-Roman British kingdom called Rheged, the existence of which is even more shadowy. The name is not mentioned by Bede or any other English writer. Even Nennius makes no mention of it, although he does refer to a king called Urien (the name is British), a figure who is usually identified with the hero of many of the poems of Taliesin, describing in eulogistic style a series of campaigns fought by Urien, who is there referred to as 'lord of Rheged'. On the strength of this, one of the later Welsh genealogies of the 'Men of the North' in which Urien's name appears is assumed to have been the genealogy of the putative kingdom of Rheged. Neither the poems nor the genealogy can be regarded as sober factual sources, and it is hard to know what to make of their testimony. Nor are we on any firmer ground as regards the extent of the kingdom. Taliesin's poems point us towards Wensleydale (Yorkshire West Riding), identified by modern scholars with *Gwaith Gwenystrad* where Urien is said to have won a victory, and Catterick, identified with *Catraeth* of which Urien is said to have been lord. Place-names supposedly incorporating the name Rheged have also been pressed into service: the claim that Dunragit in Galloway means 'The Fort of Rheged' has led to the view that the kingdom extended right along the north coast of the Solway Firth, even though the site of Dunragit (admittedly a fortified one) has yielded no corroborating evidence. Rochdale (Lancashire) may also incorporate the name, but this is highly doubtful; the name may equally be English meaning 'the valley of the Roch', and even if it is British it may mean simply 'district or river opposite to the forest'. Because of its strategic position and its Roman origins, Carlisle is sometimes considered to have been the centre of Rheged. It was evidently

[48] On Elmet in general, see Faull and Moorhouse (1981), I, pp. 157–63 and 171–8; on *Cambodunum*, see Rivet and Smith (1979), pp. 292–3; on the forest names, see Hind (1980). For the poems, see Taliesin, *Poems*,'The Battles of Gwallawg' (Pennar (1988), pp. 107–12), and Aneirin, *Gododdin*, B.22. On Craven, see P. N. Wood (1996), where interesting parallels with other British kingdoms are drawn.

an inhabited and important place when St Cuthbert visited it in the late seventh century to see the queen, but there is no proof that it was in any way a political centre.[49]

Speculative as all this is, however, there is one point of slightly greater solidity in Nennius's statement that the first wife of King Oswiu of the Northumbrians was 'Rieinmellt, daughter of Royth, son of Rhun', this last being identified by Nennius as 'son of Urien', king of Rheged. The fact of the marriage finds some corroboration in the occurrence of the name Rieinmellt (*Raegnmaeld*), a highly unusual one in an English context, in the list of queens who were remembered in the ninth-century *Liber Vitae* ('Book of Life') usually assigned to the church of Lindisfarne. If there really was a marriage between Oswiu and a British princess, and if that princess really was a daughter of the ruling house of Rheged, we may here have evidence of one of the ways in which a British kingdom was taken over as a going concern by the Northumbrian kings. Indeed, it has been argued that the account of Oswiu's two marriages, first to Rieinmellt, then to the Deiran princess Eanflæd, 'lies at the very heart' of Nennius's *History of the Britons* and emphasizes Oswiu's role in uniting Rheged to Northumbria. It has been further argued that a line of sub-kings ruled the former kingdom of Rheged as a Northumbrian sub-kingdom, retaining a strong British identity. These kings may, it is argued, have belonged to a subsidiary line of the Bernician royal house, specifically the five generations of descent from King Ecgfrith of Northumbria which the genealogies preserved by Nennius represent as culminating in an otherwise unknown Oslaf. The argument is an alluring one, but the evidence is slim – for all we know, Rieinmellt might have been enslaved and forcibly married to Oswiu following the destruction of the kingdom of Rheged. Equally problematic is Nennius's statement that King Edwin of the Northumbrians was baptized by Rhun son of Urien, a statement which was clumsily glossed in the text at some unknown period to bring it into line with Bede's unequivocal account of the baptism of Edwin by Paulinus who had come from Kent. Attempts have been made to reconcile Nennius's account and Bede's but, if there was any reality behind Nennius's account, it is very difficult to be sure what it was.[50]

More shadowy still is the kingdom of the Gododdin, supposedly centred on Edinburgh and thus in the area into which Northumbria ultimately

[49] On Aneirin and Taliesin, see Nennius, *Hist. Britons*, para. 62, and Taliesin, *Poems*, 'The Battle of Wensleydale'; for the genealogy, see Miller (1975), p. 265. On Dunragit, see Alcock (1987), pp. 236, 238, and on Rochdale, Ekwall (1960), entry 'Rochdale'.

[50] Nennius, *Hist. Britons*, para. 57; see below, p. 121. For the *Liber Vitae*, see Gerchow (1988), p. 304.

expanded northwards to the Firth of Forth. The only early mention of this kingdom is in the long poem named after its people: *The Gododdin* attributed to the British bard Aneirin, which through a series of eulogies of fallen warriors tells the story of a battle at a place called *Catraeth*, identified with Catterick. The participants, who are represented as having been drawn from various areas of Britain as well as from the people of the Gododdin, set out from *Din Eidyn*, a place identified with Edinburgh. This latter piece of information is what has led scholars to regard Edinburgh as the centre of the kingdom of the Gododdin, the name of which is apparently derived from the Latin name *Votadini*, assigned by the first-century Roman geographer Ptolemy to a British tribe located between the Firth of Forth and either the River Wear or the River Tyne. Northumbria's absorption of the area north to the Firth of Forth is utterly obscure, never being mentioned explicitly by any writer. All we have to go on are two words under the year 638 in the *Annals of Ulster*, a point at which they incorporate other Northern Annals perhaps from the monastery of Iona: *obsesio Etin* ('siege of Edinburgh'), without any reference to the identity of either besiegers or besieged. It is on this slender basis that modern accounts of how King Oswald of the Northumbrians besieged Edinburgh in 638, destroying the kingdom of the Gododdin and absorbing its territory, have been based.[51]

Model 2 remains a plausible way of envisaging the development of Northumbria, but the paucity of evidence relating to the British kingdoms which are central to it makes it difficult to establish it with any degree of certainty. That there were several British kingdoms of some sort in the period after the end of Roman Britain is likely but not provable, except perhaps in the case of Elmet, and certainly in the case of Strathclyde which remained a neighbour of Northumbria.[52] There was presumably some sort of relationship between certain of these kingdoms and the expanding Northumbria which accounts for the take-over of apparently British centres; but we cannot prove that this was by the sort of peaceful, agreed means that Model 2 postulates. It might as easily have been the result of violent conquest.

We may make more progress towards evaluating Models 1 and 2, however, if we turn our attention to one of their general implications: that, whether the transition from Roman Britain to Northumbria was from sub-Roman or British political structures, that transition was sufficiently peaceful as to permit the survival of the basic agrarian organization of the native British

[51] Kirby (1974a), pp. 1, 7; Jackson (1959); on the name, see Rivet and Smith (1979), pp. 508–9. For the sources, see Aneirin, *Gododdin*, and *Annals Ulster*, *s.a.* 638.

[52] Kirby (1962), p. 91; see also above, p. 85 n. 44.

population, and by inference of that population itself. The principal proof adduced by scholars to establish this is that of the organization of the land in Northumbria as it is revealed in much later sources: *Domesday Book*, for the lands south of the Tees; two tenth- and eleventh-century records of the lands of the church of York; the *History of St Cuthbert* of the eleventh century and *Boldon Book* (the survey of the lands of the bishop of Durham made in 1183) for lands associated with the church of Durham; and later manorial records.[53] Let us look for a moment at the entry from *Boldon Book* which concerns West Auckland (County Durham), which begins:

> In West Auckland there are 18 villeins who hold 21 bovates, and pay from each bovate 5s, and provide from each bovate in the autumn 2 men a week for mowing and raking the whole meadow, and they prepare the hay and lead it away and at that time they have subsistence once, they lead corn for 2 days, and they render 22 hens, 180 eggs and 1 cow for metreth (*uaccam de meteride*) and they cart loads between Tyne and Tees...Elstan the dreng held 4 bovates, and pays 10s, and does 4 obligatory days in the autumn with all his men except his own household, and ploughs and harrows 2 acres and will go on missions for the bishop between Tyne and Tees.[54]

The services listed here, which are characteristic of those noted in the other documents mentioned above, are notably light in comparison with services recorded elsewhere in England. They do not, for example, involve large amounts of weekly work on the lord's demesne, being generally restricted to harvest duties. They also include services which are arguably appropriate to men higher up the social scale than villeins (dependent peasants). Carrying and messenger services seem more in line with freemen than with villeins, and are reminiscent, for example, of those imposed on that intermediate social class, the radknights of eleventh-century southern England. In short, it has been argued that the men who appear in these documents were not in origin villeins, but rather free subjects of the kings who had been depressed to the level of dependants, while still retaining indications of their former more elevated position.

The presumed significance of this emerges more clearly if we examine the end of this entry which reads:

[53] *Domesday Book*; Robertson (1956), nos. LIV (statement by Oswald, archbishop of York, regarding church lands in Northumbria) and LXXXIV (types of tenure among church lands in Northumbria); *Hist. Cuth.*; and *Boldon Book.*

[54] *Boldon Book*, p. 37.

> All the villeins of Aucklandshire, that is North (Bishop) Auckland and West Auckland and Escomb and Newton Cap, provide 1 rope at the Great Chases of the bishop for each bovate and make the hall of the bishop in the forest 60ft. in length and in width within the posts 16 ft with butchering facilities and a store-house and a chamber and a privy. Moreover they make a chapel 40ft. in length and 15ft. in width, and they have 2s as a favour and they make their part of the enclosure around the lodges and on the bishop's departure a full barrel of ale or half if he should remain away. And they look after the hawk eyries in the bailiwick of Ralph the Crafty and they make 18 booths at St Cuthbert's Fair. Moreover all the villeins and leaseholders go on the roe-hunt on the summons of the bishop and to the working of the mills of Aucklandshire.

This passage, which is also characteristic of our documents, shows that West Auckland was part of a wider area: Aucklandshire, which contained a number of vills. *Domesday Book* records similar structures in Yorkshire: estates consisting of a series of subordinate vills ('berewicks' or 'sokelands') dependent on a main vill. Such estates, sometimes termed 'multiple estates' by modern scholars, were clearly cognate to Aucklandshire and the other 'shires' described by *Boldon Book*. Given the evidence for the relative freedom of the inhabitants of these estates, provided not only by the sources considered here but also by later manorial records, it has been possible to argue that 'multiple estates' or shires were in origin administrative units in the hands of the kings, and that the people who appear as villeins in the eleventh century and later were the descendants of the free inhabitants.[55]

The full significance of this for the present discussion emerges if we ask, what were the origins of these arrangements? It has been argued that they lay in the British past, and that essentially for three reasons. First, certain of the estate centres have British names (Auckland is a case in point) or are known to have been important places before the emergence of Northumbria (Aldborough in Yorkshire West Riding, for example, which was a Roman town), and this has been held to suggest that the estates attached to them were therefore of pre-Northumbrian origin.[56] Secondly, some of the dues referred to in the documents could be British in origin. Thus *metreth*, found in the Auckland entry, appears to be a British word; and another due found in the documents, a cattle tax called 'cornage and hornage', can be compared with cattle taxes in medieval Wales. In this

[55] The crucial work is Jolliffe (1926), developed in a series of studies by G. J. R. Jones, for example, G. J. R. Jones (1979). On Aucklandshire, see Roberts (1977), pp. 13–18.

[56] On Auckland, see above, p. 61 n. 9; and, on Aldborough, G. J. R. Jones (1979), pp. 29–32.

connection, it is urged that the specialized duties imposed on the inhabitants of the estates (building the bishop's hall in the example quoted above) reflect duties imposed on the king's tenants in the medieval Welsh laws, and are therefore of British origin, having survived in parallel in Northumbria and in Wales. Thirdly, 'multiple estates' are held to have once occurred very widely throughout England and southern Scotland when those areas were in British hands, and this is held to be only explicable if they were in origin the common base of British organization, preserved at first universally, then altered to the manorial system in the south and the midlands, but preserved to a large extent in Northumbria. Moreover, it is argued that this British origin of multiple estates is corroborated by parallels between them and estates documented in the high medieval laws of Wales.[57]

If these arguments are correct, we must envisage that British influence on Northumbria was deep at the level of the organization of the land. But are the arguments correct? An answer to this question depends to a large extent on how we are prepared to evaluate the evidence presented. There certainly were similarities between Northumbrian estates as represented in eleventh-century and later documents and those of Wales, and of other parts of England at an earlier date. But it is not certain that these similarities were so specific as to justify the hypothesis that all had a common origin in the pre-English, British past. Nor is it clear that the services, such as the construction of the bishop's hall for the great hunt (in origin, so the argument goes, the king's hall) were really characteristic of that British past. In the case of the construction of a hall for the great hunt, for example, it is just as arguable that it was a newly imposed service arising out of the development of hunting and hunting forests in post-Conquest England. It does not follow moreover that the use of estate centres known to have been important in Roman times proves that the estates attached to them had retained their framework of organization unchanged. The absence of week-work for the villeins may prove not that they were the descendants of free citizens but merely that demesne working, with its attendant demands for week-work, had become more important in the south and midlands than it had in Northumbria. Setting aside the difficulty of imagining continuity over such long and turbulent periods as the arguments examined here require, these concerns suggest that the evidence of agrarian organization

[57] Jolliffe (1926), pp. 40–1; G. J. R. Jones (1979), pp. 9–18; and Geoffrey W. S. Barrow (1973b).

adduced in favour of Models 1 and 2 is equivocal, or at least not of a type which compels its acceptance.[58]

Finally, then, we turn to Model 3 which postulates violent conquest of the area of Northumbria by incoming English with consequent destruction or degradation of the existing British population and removal of their organizational structures. Unlike Models 1 and 2, this model can claim no plausibility by reference to developments elsewhere in the Roman Empire where scholars have increasingly interpreted the process by which the provinces of that Empire became barbarian kingdoms as one of seamless transformation. In particular, it is argued, the nature of the Roman frontiers, with Roman economic and political activity extending well beyond them, had led to the erosion of distinctions between those dwelling on either side of them, so that the resulting difficulty of distinguishing between Romans and barbarians made the transformation all the easier.[59] Northumbria, however, may have been different from other areas of the Roman Empire, and we should therefore not dismiss Model 3 out of hand.

The evidence can be marshalled as follows. First, we have the evidence of language as reviewed above. The Englishness of Northumbria in this respect could, as we noted, be the result of British assuming English language under the influence of a dominant English-speaking elite; but the widespread and often near-complete character of the change to English is at least not inconsistent with Model 3, although the place-name evidence does require us to qualify this to the extent that some of the British population survived in Northumbria at some social level.[60]

Secondly, the archaeological evidence. The archaeology of the small number of rural settlement sites which have been excavated in Northumbria has been held to have provided some evidence of possible continuity of British population: in particular West Heslerton in the Yorkshire Wolds where the fact that the settlement appears to have been focused on a Roman shrine has been held to demonstrate its British origins; and the area of Driffield at the eastern extremity of the Vale of Pickering, where the evidence of the continuing importance of iron-working sites through the Roman and post-Roman periods is held to demonstrate continuity of settlement. This evidence,

[58] For judicious criticism of scholarship relating to multiple estates and exploration of their complexity and diversity, see Hadley (2000), pp. 94–164; Kapelle (1979), pp. 50–85; and Dark (1994), pp. 148–51. On forests, see the suggestive comments of Cox (1905), p. 97.

[59] Whittaker (1994), pp. 233–9; see also Isaac (1992), pp. 372–418.

[60] See above, pp. 60–1, 63–4.

however, is ambiguous and limited. So too is that from the excavations at the deserted medieval village of Wharram Percy in the Yorkshire Wolds; the site seems certainly to have continued in use from the Roman period onwards, but with such radical re-organization, and with a possible interruption in use following the Roman period, that it is hard to see that any real continuity is involved. At Cottam in the Vale of York evidence for two successive layouts of the village and its fields has been found; the first (Cottam A) may be, but need not be, of Romano-British date and, as at Wharram Percy, what succeeded it (Cottam B) was an entirely different layout suggestive on the face of it of discontinuity rather than continuity.[61]

The significance of the cremation cemeteries found in York and the East Riding of Yorkshire, and of the larger number of mostly somewhat later inhumation cemeteries and isolated burials, distributed not only in the East Riding but also more widely in Northumbria, is equally uncertain. These cemeteries, which mostly date from the fifth to the seventh centuries, are barbarian in character and appear to provide evidence for a dominant English population. Cremation was not at this period a rite practised by the British of Roman Britain, who were in any case presumably Christian, so the practice of cremation suggests the arrival of barbarians who were largely uninfluenced by Roman culture. Indeed, the urns that they used can be closely paralleled in some cases with urns in the areas across the North Sea from which the English are thought to have come, notably Funen in Denmark (fig. 6). The transition visible in some of the cemeteries from cremation to inhumation was part of a Europe-wide trend and need in no way have been influenced by the sort of contact with the British population postulated by Models 1 and 2. The grave-goods with which the bodies were generally buried are generally of barbarian type: weapons, pottery, jewellery and so on. There are admittedly some objects, the style and design of which it is suggested indicate that they had been made by British craftsmen; but they could have been in the graves because they were booty or maybe purchases (fig. 7). Moreover, details of certain of the burial rites appear to be in line with barbarian practices rather than with British: burial alive of a female at the cemetery of Sewerby, for example, is directly in line with early medieval English law if she were an adulteress (fig. 8). When we add that there

[61] On West Heslerton, see Powlesland (1999); no full report has yet been published. On Driffield, see Loveluck (1996); on Cottam, see Richards (1999) and Richards (2000a); on Wharram Percy, see Beresford and Hurst (1991), pp. 71–2, and for discussion Rahtz (1988). Currently, no convincingly detailed basis has been presented for claiming that the early medieval settlement pattern emerged in the late Roman period and is therefore itself evidence of continuity.

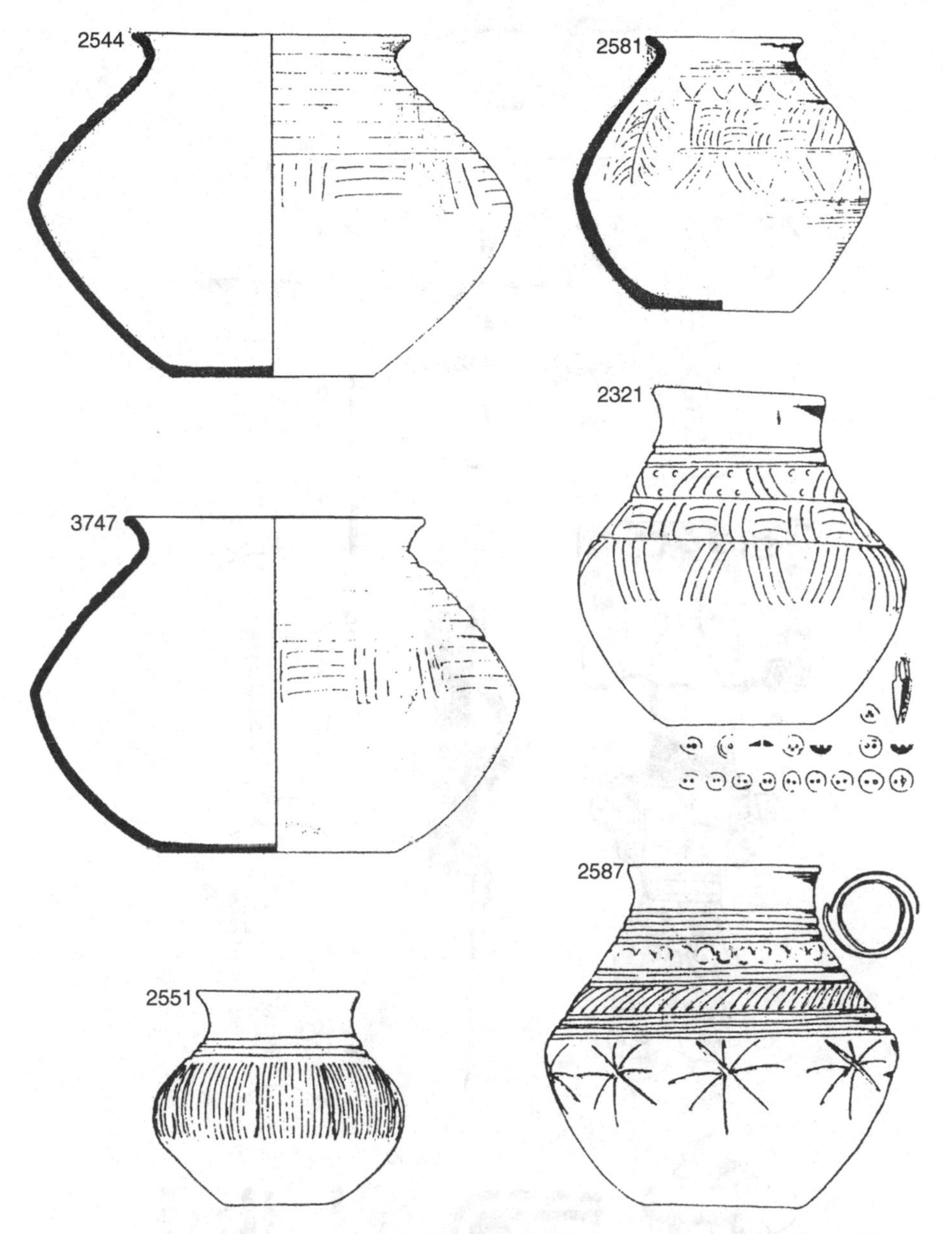

FIGURE 6 Sancton I cemetery (Yorkshire East Riding), cremation urns (from Myres and Southern (1973), fig. 26)
Note the incised and stamped designs on the urns, which are not wheel-thrown. No. 2321 particularly resembles urns from a cemetery in Funen (Denmark), and is thus evidence of contacts across the North Sea. See Myres and Southern (1973), and compare Myres (1969), pp. 75–6.

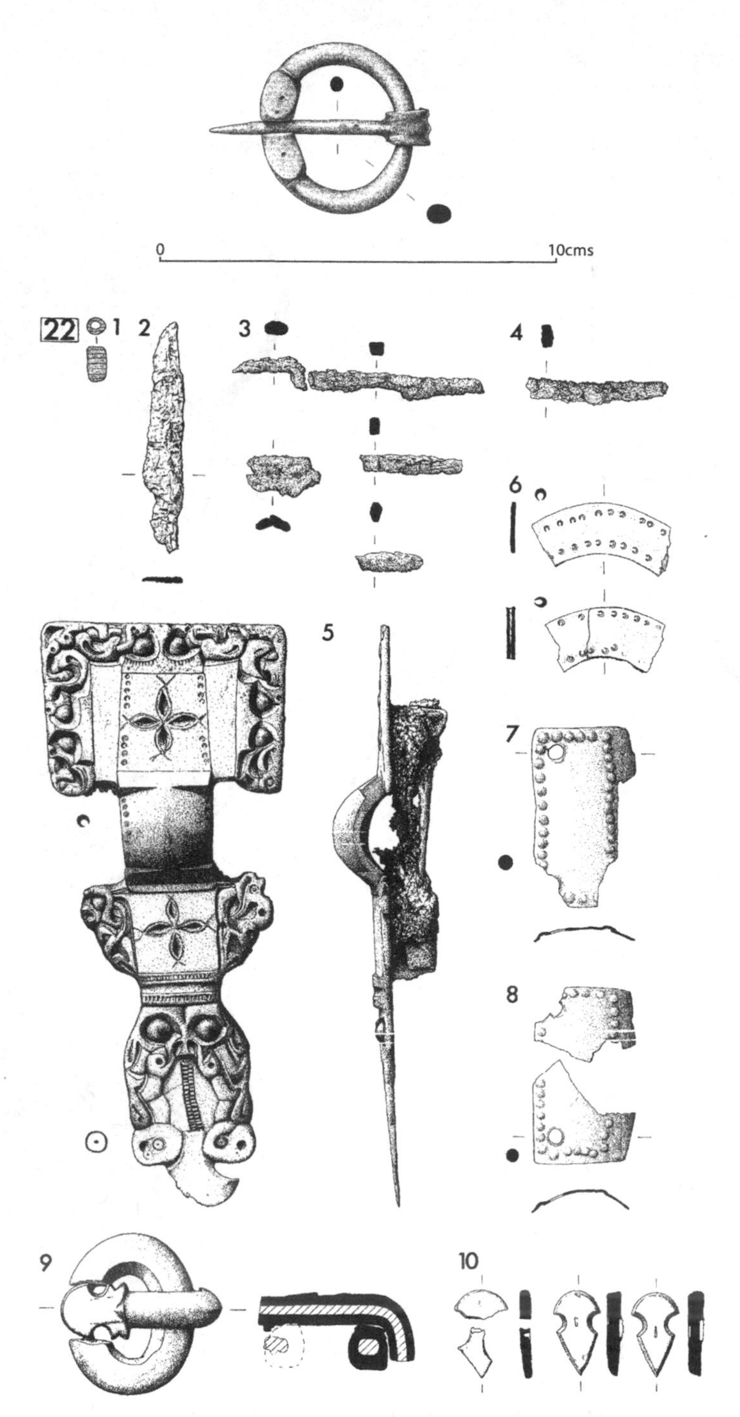

FIGURE 7 Norton-on-Tees (County Durham), grave-goods from the cemetery
The objects from Grave 22, that of a female, include a cruciform brooch (no. 5) of typically barbarian type, an annular brooch (no. 6), fragments of a wrist clasp (nos. 7–8) and a 'shield-on-tongue' buckle (no. 9). An object in British style is the penannular brooch from Grave 65. See Sherlock and Welch (1992), pp. 133, 167.

FIGURE 8 Sewerby (Yorkshire East Riding), reconstruction of the 'live burial' (G41) (drawing by David A. Walsh, from Hirst (1985), frontispiece)
The reconstruction represents Susan Hirst's interpretation of the burial as evidence of judicial killing for adultery resulting from rape; for views critical of this, see below, n. 62.

are no cemeteries at all which are *prima facie* British to set alongside these apparently barbarian cemeteries, our conclusion must be that the evidence presented here is strongly in favour of Model 3 and provides no support for Models 1 and 2.[62]

[62] Lucy (1999) and Thomas (1981), chs. 4–9. For British objects in cemeteries, see Faull (1977), pp. 4–5. On Sewerby, see Hawkes and Wells (1975); and discussion by Lucy (2000a), pp. 71–2, and Hirst (1993).

Four objections, however, can be raised to this conclusion. First, that the cemeteries we have been considering have been identified as English precisely because of the use of the rites of cremation and then of inhumation with grave-goods. British graves, it can be argued, may have been numerous but they would have been of inhumations without grave-goods. Such graves, lacking the evidence of dating and cultural affinity which grave-goods or cremation urns provide, may not have been discovered at all by archaeologists, or if discovered may not have been identified as graves of this period. The British were therefore far more numerous in Northumbria than excavation of cemeteries has allowed us to perceive. The discovery at Yeavering of such 'unfurnished' graves has given some substance to this objection since these have been interpreted as British graves.[63] Such unfurnished graves elsewhere are, however, impossible to date precisely, since dating relies almost exclusively on the dating of grave-goods according to their development in terms of style. They could therefore as well be English graves but after the conversion of the English to Christianity when the use of grave-goods declined and eventually came to an end.

Secondly, that the social status reflected in the English graves indicates that they are those of an English elite, ruling over a British Northumbria as in Models 1 and 2. This objection has been urged most cogently for Bernicia, where it has been argued that the inhumations with grave-goods contain objects of notably higher status than contemporary graves in southern England and therefore point rather to the presence of an elite than to an in-depth settlement of English as in Model 3; and that their scarcity (as opposed to their relative frequency in Deira) is also consistent with their being the graves of such an elite.[64] This objection, which in any case does not really apply to Deira, is itself open to being undermined. The higher status apparent in the Bernician graves is not especially clear-cut; what higher-status grave-goods there are need not in any case reflect on the actual social status of those buried, but rather on the wealth and upwardly mobile ambitions of those who survived them; and the relative scarcity of burials in Bernicia may simply reflect the fact that English settlement there was later, when grave-goods were already going out of fashion – so as a result we have not found or not identified more numerous English cemeteries and burials there, even if they existed.[65]

Thirdly, that the burial rites reflected in the cremations and inhumations with grave-goods may appear English, or at least 'barbarian', but were in

[63] Hope-Taylor (1977), pp. 70–8. [64] Alcock (1981).

[65] Miket (1980); on the use of grave-goods in general, see Härke (1990) and Geake (1997).

fact deeply influenced by a British population surviving on the sort of terms postulated by Models 1 and 2. Burials in which the bodies had their heads in a general northerly direction or were buried in a crouched position are found as a proportion of graves in Northumbrian cemeteries, and these positionings can be interpreted as British rather than English practices, on the grounds that they are found in burials of the Roman period at any rate in Yorkshire, whereas they are not generally characteristic of English burials elsewhere in England.[66] This objection too can be undermined, however: such positionings are found elsewhere in England in English contexts, and we cannot be sure therefore that what we see in Northumbria is British influence rather than a local variation on an English custom; nor is it easy to accept that the evidence of pre-Roman graves is germane to British custom as it might have existed in the period of the origins of Northumbria.[67]

The fourth objection is the opposite of the third. It derives from the view that burial customs reflect the aspirations of the living rather than the status and cultural affinities of the dead. In other words, those who arranged the burials were making a statement about their position in the world which may not have had any direct connection with their real position or that of the deceased.[68] Once this argument, which has an inherent plausibility, is accepted, the way is open to argue along the same lines as we have seen done in the case of language. The British, still very important in Northumbria as in Models 1 and 2, wished to align themselves with the dominant English elite. They therefore hastened to adopt English customs, of which burial rites are the most obvious to us in the archaeological record, and so effectively reconstructed their cultural identity as English. Thus what the apparently English cemeteries and burials actually show us is diametrically opposed to what they appear to do: they are evidence for British survival albeit under an English elite.[69] This objection is engagingly bold and contrary; but it suffers from the drawback that it requires what is in effect a leap of faith to accept that the evidence shows us the opposite of what it appears to, and is also unprovable.

If these objections are discounted, there is therefore nothing in the evidence we have which absolutely requires us to discard Model 3 in favour of the more peaceful processes of Models 1 and 2, with their emphasis on co-existence between British and English and the ultimate transformation

[66] Faull (1977), pp. 5–8.
[67] Sherlock and Welch (1992), pp. 27–30, and Lucy (2000a), pp. 171–2.
[68] Lucy (1999), pp. 22–3, and Lucy (2000b). See also Lucy (1998), pp. 17–21 and 104–6.
[69] For this view, see Powlesland (1997).

of British population into English. Moreover, an important objection to accepting Models 1 and 2 is the pronounced emphasis in our written sources on English–British hostility. Assessing the overall political situation in 731 when he finished his *Ecclesiastical History*, Bede noted that 'for the most part, the British oppose the English through their inbred hatred'. Earlier he had praised the pagan king of Northumbria, Æthelfrith (d. 616), for ravaging the British 'more extensively than any other English ruler'. He continued: 'No ruler or king had subjected more land to the English race or settled it, having first either exterminated (*exterminauit*) or conquered the natives.' Bede's willingness to lavish such praise on a pagan king underlines the hatred of the British which this passage embodies. The British were the prey; their lands were the spoils; their removal justified extermination where necessary. British retaliation came from the king of Gwynedd, Cædwalla, who, after defeating and killing King Edwin at the Battle of Hatfield in 633, proceeded to mete out similar treatment to the Northumbrians: 'With bestial cruelty he put all to death by torture and for a long time raged through all their land, meaning to wipe out the whole English nation from the land of Britain.' A year later Cædwalla was back to kill the two kings who had succeeded Edwin, and 'after this he occupied the Northumbrian kingdoms for a whole year, not ruling them like a victorious king but ravaging them like a savage tyrant, tearing them to pieces with fearful bloodshed'.[70]

There seems no question here of a picture of English–British relations founded on peaceful transition from British to English rule, no fusing of peoples, no agreed take-over. What Bede is representing is a scenario all too familiar to modern observers, the forcible removal of one people to make way for another. Stephen, too, in his *Life of Wilfrid*, evokes a picture of violent expropriation: Wilfrid, he tells us, had received between 671 and 8 for his monastery at Ripon 'holy places' taken from the British clergy (*clerus Bryttannus*) who had fled from 'the point of the hostile sword wielded by the hand of our people' (*aciem gladii hostilis manu gentis nostre*).[71] The same hostility, no doubt reflecting that of the Northumbrians, is evident in the report of the papal legates to northern England in 786. Castigating the evil practices of Northumbrian Christians, this document makes the following accusation: 'You wear your clothes in the manner of the gentiles, whom by God's favour your fathers expelled from the land by force of arms.' It is not clear what exactly is being referred to since strictly speaking the British

[70] Bede, *Eccles. Hist.*, bk 5 ch. 23, bk 1 ch. 34, bk 2 ch. 20, bk 3 ch. 1.
[71] Stephen, *Life Wilf.*, ch. 17.

were not 'gentiles', that is pagans; but the most natural interpretation of the passage seems to be that Alcuin was thinking of a violent expulsion of the British by the founding fathers of Northumbria.[72]

The British sources reciprocate this hostility by painting a picture of bitter conflict between the native British and the incoming English. The poems of Taliesin and *The Gododdin* attributed to Aneirin have, as we have seen, been used as if they constituted factual records, bearing for example on the date and political context of the Battle of *Catraeth* in *The Gododdin*, or of the Battle of *Gwaith Gwenystrad* in the poems of Taliesin.[73] Such an approach seems misguided. The events which provide the narrative framework of *The Gododdin* and the Taliesin poems might be based on actual history (as the Old English poem *The Battle of Maldon* was), but they might equally well be fictional or semi-fictional, like the Old English epic *Beowulf* or the Arthurian romances. In any case, they are primarily literature and not history. Together with the genealogies of the British, they evoked a vision of an age when the 'Men of the North' (*Gwyr y Gogledd*) battled heroically but with tragic futility against the English (*Lloegrwys*) who were depriving them of their birth-right, the rule of all Britain. The Welsh believed themselves descended from these 'Men of the North' through the (probably legendary) northern British ruler Cunedda who had, according to tradition, migrated with his eight sons from a place in the north called Manau Gododdin to found the various kingdoms of Wales.[74] For the medieval Welsh, then, Taliesin's poems and *The Gododdin* were part of their nostalgic view of an heroic past they believed was theirs, like Geoffrey of Monmouth's legendary *The History of the Kings of Britain*. As an aspect of what we might view as their national consciousness its evidence is clear: hostility to the English and a tradition of warfare against them was fundamental to it. A stanza from the A-version of *The Gododdin* reads:

> Warriors mustered. They met
> together. With a single intention they attacked.
> Their lives were short. Their friends' grieving for them was long.
> They killed seven times their number of men of England.
> In combat they made wives widows,
> many a mother with her tear on her eyelids.[75]

[72] For the text, Dümmler (1895), no. 3, ch. xix (p. 27): 'Vestimenta etiam uestra more gentilium, quos Deo opitulante, patres uestri de orbe armis expulerunt, induitis' (trans. *EHD I*, no. 191). For comment, suggesting that the Picts may have been the people referred to, see Wormald (1991), p. 33 and n. 19.

[73] See above, pp. 87, 89; and further Alcock (1987), pp. 253–4, and Koch (1997), pp. xiii–xxxiv.

[74] *Hist. Britons*, para. 62; see Miller (1976–8). [75] Aneirin, *Gododdin*, stanza 56.

The theme is identical in Taliesin's poetic tribute to Urien of Rheged:

> You cause havoc
> when you advance;
> before dawn
> houses aflame
> before the lord of Erechwydd
> (fairest Erechwydd with her
> most generous men).
> The English are without protection
> because of the most courageous stock
> you are the best.[76]

The appeal of this in thirteenth-century Wales, the age of Llewelyn of Gwynedd's wars against the English and Edward I's savage conquest of Wales, is easily comprehensible. But how far did it go back? Both Taliesin and Aneirin are mentioned by Nennius in a passage which that writer appears to synchronize with the time of King Ida of the Northumbrians (who was active – Bede believed – between 547 and 559), so it could be assumed that the poetry dates from the period of the very origins of Northumbria.[77] Even if that is so (and we only have Nennius's uncorroborated synchronization to go on), it is likely that the texts have been significantly modified over the intervening centuries. In the case of *The Gododdin*, the process of change is certain, for the thirteenth-century manuscript in which it is preserved contains two versions, designated by scholars A and B, the latter more archaic in language and containing some additional stanzas. Although scholars agree that the language in which the B-version is composed is more ancient than the thirteenth century, it need not on grounds of the spelling of its text be older than *c.* 1100.[78] But it has been argued that the B-version allows us to reconstruct two earlier versions: the first written before 638 presumably in the kingdom of the Gododdin, the second in the mid-seventh century in the kingdom of Strathclyde. In these early versions hostility between the British and the English was lacking, for neither version viewed the English as enemies but rather as allies of the British in what was a battle between two British kingdoms: Gododdin and Deira (here regarded as a British kingdom). Only after the text reached North Wales some time after the mid-seventh century was it modified to represent the English as the enemies of

[76] Taliesin, *Poems*, 'You Are the Best' (Pennar (1988), pp. 53–8).
[77] *Hist. Britons*, para. 62; Bede, *Eccles. Hist.*, bk 5 ch. 24.
[78] Lucidly explained by Jarman (1988), pp. xv–xvii.

the British. In its earlier phases, therefore, *The Gododdin* could be seen as support for Model 2, and hostility between British and English would have been a subsequent development. It is an intriguing idea, but the argument involves subtle and somewhat conjectural argumentation about what the original version might or might not have contained.[79]

It is clear, moreover, English–British hostility was not a later development in other British sources. Even if we were to reject the dating of Taliesin as sixth century and accept a dating for his poetry as we have it of not later than *c.* 1100, we should still find a persistent theme of English–British hostility in Gildas, who was certainly writing before 600. For him, the English incomers were 'the ferocious Saxons (name not to be spoken!), hated by man and God'; their coming led to revolt and violent conquest by them, so that 'the cities of our land are not populated even now as once they were; right to the present they are deserted, in ruins and unkempt'.[80] Gildas may have been referring to other parts of Britain, but Nennius, writing in the first half of the ninth century, was definitely referring to Northumbria when he described how the sons of Ida, king of the Bernicians, had war waged on them by King Urien of Rheged, King Rhydderch Hen of Strathclyde, and two persons called Gwallawg and Morcant, the former possibly king of Elmet, as we have seen. One of Ida's sons 'fought vigorously' against Urien, who besieged him on Lindisfarne. Nennius was evidently not envisaging anything like Model 2, any more than he was with his account of King Edwin's expulsion of King Ceretic from Elmet.[81] The prominence of English–British hostility in texts from both sides of that divide is thus very striking, and it seems clear that in the minds of the writers the area of Northumbria played a large role in its genesis.

If we were disposed to accept Models 1 or 2, we could argue that this ethnic fault-line was paradoxically the result of the transformation of British into English which those models postulate. The process of defining a new identity for the former British, now *parvenu* English, of Northumbria involved hostility to British outside Northumbria as an aid to that process, while some parallel process of self-definition amongst those British involved reciprocal hostility to the English. Bede's *Ecclesiastical History* suggests that such definitions of identity were indeed in process of construction in his time, for it lays emphasis on a series of origin legends which were no doubt

[79] Koch (1997), pp. lxxxix–cx, reviewed (with useful summary) by Padel (1998).

[80] Gildas, *Ruin*, paras. 23, 26.

[81] Nennius, *Hist. Britons*, para. 63; see Williams, Smyth and Kirby (1991), entry 'Riderch Hen'. On Edwin and Ceretic, see above, p. 85.

part of them. The British, Bede tells us, had 'sailed to Britain, so it is said, from the land of Armorica, and appropriated to themselves the southern part of it'. He then goes on to give us an origin legend for the Picts. As the British were taking possession of the island beginning from the south, the Pictish race came from Scythia and 'sailed out into the ocean in a few warships and were carried by the wind beyond the furthest bounds of Britain, reaching Ireland and landing on its northern shores'. The Irish race (*gens Scottorum*) refused them permission to settle but advised them to go east and settle in Britain, which they did in the north 'because the British had seized the southern regions'. The Picts then asked the Irish for wives, which were granted on condition that the Picts should 'elect their kings from the female royal line rather than the male'.[82] We need not regard these legends as necessarily based on fact. Their purpose was to define peoples who had no over-arching political organization, the British being divided in Bede's time into separate kingdoms such as those of Strathclyde and Gwynedd, the Picts into at least the northern and southern Picts, and possibly other units since Bede speaks of 'kingdoms of the Picts'. The English were certainly envisaged in this way, and so too were the Irish, who were divided into a large number of kingdoms (*tuatha*). The British did have a linguistic coherence, however, speaking the same language, a Celtic language of the so-called P-type, known to modern scholars as 'British' and an ancestor of medieval Welsh. We have virtually no texts in Pictish apart from a king-list, and we can only begin to reconstruct the Pictish language from this scanty source, from words found embedded in place-names, and from some very obscure inscriptions. It seems, however, that the Pictish language was simply a different dialect of P-Celtic, although an earlier pre-Celtic language may also have been used. Regarding themselves in Bede's day as a coherent 'people of the Picts' and distinct from the British may therefore have been a quite recent development for the people north of the Forth–Clyde line.[83]

After providing origin-legends for the British and the Picts, Bede goes on to give an origin-legend for the Irish who lived in Dalriada, the kingdom centred on Argyll and the Western Isles. 'These came from Ireland under their leader Reuda, and won lands among the Picts either by friendly treaty or by the sword. These they still possess. They are still called Dalriadans (*Dalreudini*) after this leader, *Dal* in their language signifying part.' The

[82] Bede, *Eccles. Hist.*, bk 1 ch. 1.

[83] For the texts, see Bede, *Eccles. Hist.*, bk 3 ch. 4 and bk 5 ch. 21. On the Irish, see O Croínín (1995), p. 111, and also Binchy (1970), lectures I and III. For the British language, see Jackson (1953), for the Pictish language, Jackson (1955) and Nicolaisen (2001), pp. 192–204.

implication of the use of *Dal* is presumably that Reuda had obtained part of north Britain for his people.[84] Bede then completes the picture with an origin-legend for the English. He repeats the story told by Gildas in *The Ruin of Britain* to the effect that under the rule of a 'proud tyrant', whom Bede identifies as Vortigern, the British invited three ship-loads of Anglo-Saxons as mercenaries, but these were joined by others, and the result was a revolt which effectively destroyed Roman Britain. The English, he writes:

came from three very powerful Germanic tribes, the Saxons, *Angli*, and Jutes. The people of Kent and the inhabitants of the Isle of Wight are of Jutish origin and also that people (*gens*) opposite the Isle of Wight, that part of the kingdom of Wessex which is still today called the nation (*natio*) of the Jutes. From the Saxon country, that is the district now known as Old Saxony, came the East Saxons, the South Saxons, and the West Saxons. Besides this, from the country of the *Angli*, that is, the land between the kingdoms of the Jutes and the Saxons, which is called *Angulus*, came the East *Angli*, the Middle *Angli*, the Mercians, and all the Northumbrian race (*progenies*) – that is those peoples (*gentes*) who dwell north of the river Humber – as well as the other peoples (*populi*) of the *Angli*. *Angulus* is said to have remained deserted from that day to this.[85]

Some of the subsidiary peoples named in this passage had their own origin myths. Those of the West Saxons and the South Saxons constitute the early annals in the *Anglo-Saxon Chronicle*, which are now widely recognized as being mythological rather than factual. For example, the annal for 495 which must reflect the West Saxon origin legend, complete with eponymous place-name: 'Here two chieftains, Cerdic and Cynric his son, came to Britain with five ships at the place which is called Cerdic's Shore.'[86] The origin myth of the people of Kent, to which Bede refers, was clearly bound up with the story of Vortigern inviting the English leaders Hengest and Oisc to Britain, a story which Nennius also tied closely to the kingdom of Kent, presumably using information derived from its people.[87]

The origin myth of the Northumbrians may be contained in a story given by Nennius of how Vortigern also invited to Britain the English leaders Octha and Ebissa with forty ships and they 'came and occupied many districts beyond the Frenessican Sea, as far as the borders of the Picts'. Although

[84] Bede, *Eccles. Hist.*, bk 1 ch. 1. On Dalriada, see Marjorie Ogilvie Anderson (1973) and Bannerman (1974); on the reality of the migration, see Cummins (1995), pp. 50–6, and Lane and Campbell (2000), pp. 32–4, and references therein.

[85] Gildas, *Ruin*, paras. 22–4; Bede, *Eccles. Hist.*, bk 1 ch. 15.

[86] See, for example, Sims-Williams (1980); and see also Howe (1989); for the text, *ASC*, *s.a.* 495.

[87] Nennius, *Hist. Britons*, paras. 36–8, 43–6; Bede, *Eccles. Hist.*, bk 1 ch. 15, bk 2 ch. 5.

the name Frenessican Sea is obscure, the reference to 'the borders of the Picts' presumably indicates that Northumbria is the area being referred to. Scholars have tried to interpret the passage as an actual description of the English settlement of Northumbria, but it seems obvious that, if the passage has any value, it is as an origin myth of the Northumbrian people, to which we should no more attribute factual reliability than we should to the story of the *Angli*, Saxons and Jutes coming in three ships. It may be reflected in Gildas's more general observation that the English (*Saxones*) came to the 'east side' of Britain 'to beat back the peoples of the north' (i.e. the Picts and the Irish).[88]

These accounts in Bede are evidence for a situation in which concepts of 'peoples' were in process of formation in early England, and it seems likely that what we are seeing in our written sources is a reflection of the process by which the ruling families and their elite followers defined and consolidated their positions by recording or fabricating the origin myths of their peoples. That this was a recent development for the English in general and the Northumbrians in particular in Bede's time may be indicated by uncertainty and variation in our sources in the use of the terms 'English' and 'Northumbrian'. *Angli* in the writings of Bede and his contemporaries is often translated 'Angles', but in fact it is simply the Latin word for 'English'. In the passage quoted above, however, Bede is using it to mean only one of the ethnic groups which made up the English people, of which the other two were the Jutes and the Saxons. But in a preceding passage he referred to the 'Saxons or the English (*Angli*)' as if the names were interchangeable. In the title of the *Ecclesiastical History*, which should be describing the history of the English (*Angli*), Saxons and Jutes in common, and elsewhere in the *Ecclesiastical History*, Bede simply uses 'the English people' (*gens Anglorum*) or 'the English' (*Angli*) to refer to all the inhabitants of Britain who were not British, Picts or Irish.[89] This inconsistency may be the result of novelty. It is possible that the word 'English' had only recently been used for all the inhabitants of Britain who were not British, Picts or Irish; and that the development of this usage had been influenced by the papacy. Gregory the Great appears to have been the first writer to refer to the Germanic inhabitants of Britain under the generic name 'English'. The usage may further have been influenced by the story which Bede and the anonymous Whitby

[88] Nennius, *Hist. Britons*, para. 38, and Gildas, *Ruin*, para. 23. If by 'east side', however, Gildas meant the south-east and East Anglia, the passage may not relate to Northumbria at all. For discussion, see Peter Hunter Blair (1947), pp. 13–17.

[89] Putnam Fennell Jones (1929), under appropriate entries. For discussion, see Brooks (2000), pp. 6–7.

biographer of Gregory the Great told about that pope seeing English slave-boys in the market at Rome: on being told that they were English (*Angli*), he declared that 'they have the face of angels (*angeli*), and such men should be fellow-heirs of the angels in heaven'. The pun only worked for *Angli*, not for *Saxones*, and it is conceivable that this spurious etymology for the name, with its associations with missionary activity, influenced the choice of *Angli* as the generic name for the Germanic inhabitants of Britain. Bede himself may have played a part in establishing that choice, in deliberately seeking in his *Ecclesiastical History* to create a concept of English nationality, perhaps as part of his support for an over-arching English church under the jurisdiction of the see of Canterbury. There may even have been a deliberate change in Bede's usage, from 'Saxons', when the English he was writing about were pagans, to *Angli* when they became Christian.[90]

Similar uncertainty and variation of terminology in contemporary writers is apparent in the use of *Northanhymbri* (Northumbrians), alongside other apparently synonymous names, *Aquilonales* (people of the North), *Septentrionales Angli* (northern English), *Transhymbrana gens* (people across the Humber), *Ultrahumbrenses* (people from beyond the Humber) and *Humbrenses* (people of the Humber). All this suggests the concept of a people in a state of genesis. That the term 'Northumbrian' was new and required explanation may be suggested by Bede's repeated and, to our mind, tautologous explanations of it as 'those who dwell to the north of the Humber'.[91] As an explanation for the emphasis on English–British hostility in our sources, however, the notion that the formation of ethnic identities was recent and likely therefore to have produced tensions is inadequate. Bede, for example, gives no such account of hostility between the English and the Picts who, despite having killed King Ecgfrith of the Northumbrians in 685 and a Northumbrian prefect in 698, are described as being on terms of peace and friendship with the Northumbrians. The Irish are treated with similar friendliness, and Bede even regrets the attack made on them by King Ecgfrith in 684.[92]

Were we to accept Model 3 we should not need to look further than it for an explanation for English–British hostility. On the British side it would have arisen from the way in which the invading English had treated them;

[90] See Tugène (2001b), pp. 87–8, and Brooks (2000), pp. 7, 16–17, 18, 22. On the importance of Canterbury in creating a sense of Englishness, see Wormald (1983), pp. 122–9. For the texts, see Letter to Augustine, quoted in Bede, *Eccles. Hist.*, bk 1 ch. 29; and *Life Greg.*, ch. 9 (p. 91); the same story is found in Bede, *Eccles. Hist.*, bk 2 ch. 1.

[91] For references and discussion, Peter Hunter Blair (1948), pp. 99–104, and Tugène (2001b), p. 63; on Bede's use of 'Northumbrians', see above, p. 25 n. 11.

[92] Bede, *Eccles. Hist.*, bk 4 ch. 26 (24), bk 5 ch. 24.

whilst fear of British reprisals would then have lain at the root of hostility on the English side. In accepting the reality of hostility between the two peoples, we need not accept that the violent displacement or degradation of the British population postulated by Model 3 necessarily occurred shortly after the end of Roman rule, which may well have been followed by relations more like those postulated by Model 1 or 2, although we have seen that that is difficult to prove. The events which made such friendly relations inconceivable to Bede and his contemporaries may have occurred relatively recently. The evidence suggests that Northumbrian expansionary warfare against the British occurred from the late sixth century onwards. Nennius's account of the siege of Lindisfarne by the confederation of British kings may admittedly not be a factual account; but, if it is, it relates to the end of the sixth century. Whatever our doubts about Nennius, Bede's account of the attacks on the British made by King Æthelfrith of Northumbria (d. 616) must surely reflect a genuine tradition about his activities transmitted in Northumbria. The acquisition of lands from the British clergy used for the endowment of Wilfrid's monastery at Ripon may well have occurred shortly before their being granted, that is the years 671–8. The places in question in this endowment were Ribble, Yeadon, Dent and Catlow, which, if they have been correctly identified, lay west of the Pennine watershed and thus represent a westward Northumbrian advance at the expense of the British.[93] Northumbrian expansion along the northern shores of the Solway Firth may have happened around the same time to judge from the appearance in the early eighth century of the first English bishop of Whithorn, which was then established as a Northumbrian see. The establishment of a Northumbrian bishopric at Abercorn on the Firth of Forth, the only known bishop of which was Trumwine (681–5), may have been part of the same round of expropriations of the British. Warfare against them may have continued, for King Eadberht of Northumbria is recorded in a laconic entry in the Northern Annals to have conquered the Plain of Kyle in 750, and he is later recorded to have attacked the kingdom of Strathclyde in alliance with the Picts.[94]

The question of the origin of the Northumbrians is thus a highly complex one. There is some evidence to suggest that there was a peaceful transition from British to English in Northumbria so that the kingdom of Northumbria

[93] Stephen, *Life Wilf.*, ch. 17; on the identifications, see Colgrave (1927), n. to ch. 17, citing H. M. Chadwick (pers. comm.). On Dent and Yeadon, see G. J. R. Jones (1995), pp. 29–36, who considers Yeadon to have been in the West Riding.

[94] See above, pp. 32, 43. The reluctance of the British to accept the 'Roman' calculation of the date of Easter favoured by Bede may have played a part in that writer's hostility to them (Charles-Edwards (1983)); but it cannot provide the entire explanation for the reciprocal hostility discussed above.

essentially remained a British area with a population of British extraction which had assumed an English identity and acquired an English language. But the evidence is not so strong that we need feel compelled to accept it; and, since that process would have effaced much of the evidence for that population's original Britishness, it is not easy to prove or disprove it.[95] As we have seen, it is not unreasonable to accept that the population of Northumbria was predominantly composed of English incomers who had killed, displaced or degraded the native British inhabitants, making Northumbria English by those means. The most prominent feature of the period, however, is the extent to which, by whatever process, Northumbria was English at least by Bede's time, and the hostility which is manifested in the texts between those who claimed that English identity and their British neighbours. How far that hostility was consolidated by cultural differences between British and English, and how far culture played a part in the creation of Northumbria, is the subject of the next chapter.

[95] See the most recent survey of research by Loveluck (2002).

CHAPTER 4

Culture and identity in pre-Viking Northumbria

We saw in the previous chapter that reaching any firm conclusion about the ethnic origins of a people is hampered by our difficulties in assessing whether or not a particular ethnic group (in our case, the British) might not have deliberately assumed the culture and identity of another ethnic group (in our case, the English). Whatever importance such a possibility may have, it reminds us of the potential of culture in defining – or at least characterizing – a political unit, either a complete state or a sub-group within a state. Modern governments have been acutely aware of this, not least in their policies towards education and the maintenance, or suppression, of the languages of minority groups. How then can we assess the culture of the kingdom of Northumbria? We have seen that by the time of Bede our evidence suggests that English was the predominant language of the kingdom, and that the Northumbrians could be characterized as English.[1] Aside from language, however, we need to consider: how far the culture of Northumbria was distinctively English and distinct from British culture; how far it was created by influences from outside Northumbria, and was thus the result of Northumbria's relations, friendly and unfriendly, with its neighbours in the British Isles; and, above all, how far it was shaped by Northumbria's conversion to Christianity, with all that meant for the opening up of relations with the wider world of western Christendom.

Let us begin with the culture of Northumbria in the period of its origins. The obscurity of the history of the British in the immediately post-Roman

[1] See above, pp. 57–64.

period extends to their culture. The British south of Hadrian's Wall had been under the authority of the Roman Empire and in contact with its culture for over four centuries; and to some extent so too had those north of Hadrian's Wall, partly because Roman rule extended to them periodically, partly because they must always have been within the orbit of the Roman Empire, as the discovery of the great Roman treasure at the British hill-fort of Traprain Law, enigmatic as it is, shows. The extent to which, or the level at which, the British had absorbed Roman culture is less clear. The northern part of Roman Britain was a heavily militarized zone. There was a reasonably wealthy city at York, capable of producing funerary monuments and inscriptions of a fairly high level; by 314 there was a bishop, so there must have been some literary activity surrounding the Christian church; there must presumably have been schools for the children and young people of York. But neither at York nor anywhere else in northern Britain is there any indication of scholarly activity of an advanced or remarkable type. There is nothing to show that there were schools of the distinction of Gaul, for example; and we would by no means expect this given the militarized character of the province. We must assume that the native British population had been Christian in the Roman period, although in truth there is little trace of it, aside from a rather ambiguous inscription from York and the fact that in 314 the city had a bishop. The Christian church recently discovered in the courtyard of the commander's house at the fort of *Vindolanda* (Chesterholm) proves no more than that Christianity was practised by the commander, who was almost certainly a foreigner doing his stint in Britain.[2]

The immediately post-Roman period has yielded only exiguous evidence of any continuing Roman cultural tradition, or even of Christianity, amongst the British. Whereas Wales and south-west Britain had after the end of Roman rule a vigorous tradition of erecting memorial stones with Latin inscriptions, clearly drawing on Roman models, such monuments are very rare from north Britain. There are isolated examples at Yarrowkirk in the Scottish borders, at Kirkliston (the 'Cat Stane') near Edinburgh and from *Vindolanda*

[2] On York in general, see Ottaway (1993); on the *colonia*, see Ottaway (1984); on evidence for the bishop (Acts of the Council of Arles, 314), see Rivet and Smith (1979), pp. 49–50. For a summary of written evidence relating to Roman York, see *Sources*, pp. 84–124. On Gaul, see Riché (1976), and Nora Kershaw Chadwick (1955). The work of Gildas shows that a high level of classical rhetorical training was available in Britain, but there is no reason to suppose that this was in the north (Lapidge (1984)). On Christianity in Roman Britain in general, see Thomas (1981) and J. Wall (1965, 1966); on the York inscription, see *Sources*, p. 111 (R.5.4); on the putative church at *Vindolanda* and other possible churches, Robin Birley (2002), p. 439; and on typical Roman careers, see Anthony Richard Birley (1981).

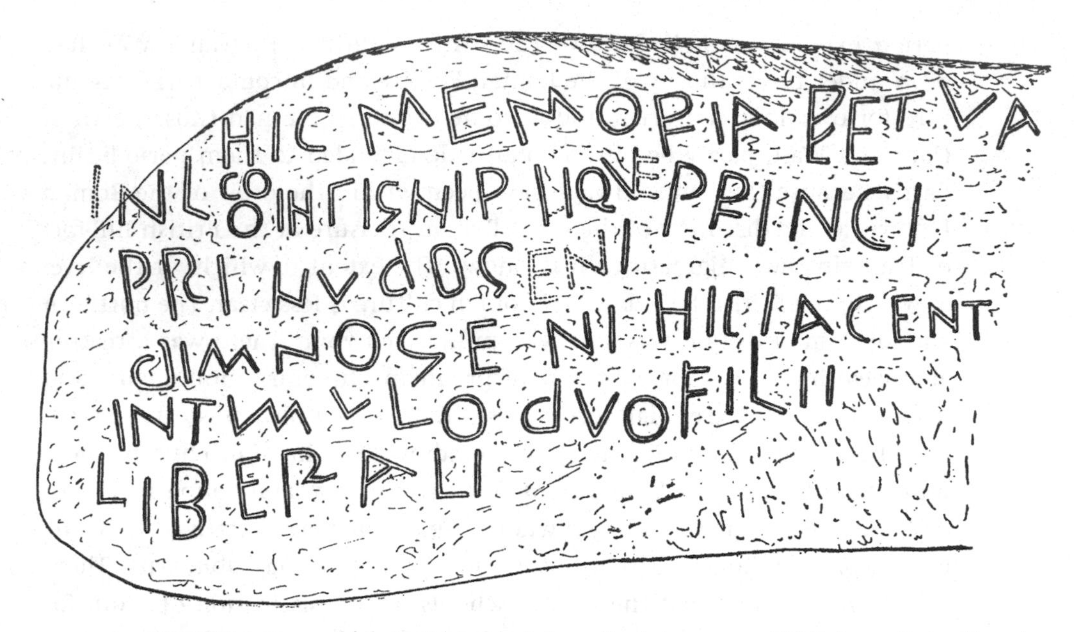

FIGURE 9 Yarrowkirk (Selkirkshire), inscribed stone near Whitehope Farm (drawing from Macalister (1945), no. 515).

(Northumberland). British names appear on those from *Vindolanda*, which is inscribed 'here lies Brigomaglos', and at Yarrowkirk, the inscription on which refers to a prince called Nudogenos, and possibly another called Dubnogenos (fig. 9). The inscriptions, however, are remarkable for the crudeness of their execution and the low level of their Latinity. Only at Whithorn and neighbouring Kirkmadrine in the extreme west of what was to become Northumbria are there memorial stones of post-Roman date which can compare with those of Wales (fig 10).[3]

We have little more to go on as regards native British culture. The poems of Taliesin in praise of King Urien of Rheged and *The Gododdin* were written in the British language and were presumably a product of British, oral, literary culture. But we have seen in the previous chapter the difficulties of accepting the often expressed belief that Taliesin was writing his poems at some point in the late sixth century in the British kingdom of Rheged in what was to become north-west Northumbria; and that Aneirin was producing *The Gododdin* in what was to become north-east Northumbria at about

[3] On Wales, see Nash-Williams (1950); on south-west Britain, see Okasha (1993); see also Thomas (1994). On north Britain, including Whithorn and Kirkmadrine, see Macalister (1945), nos. 498, 511, 514–20 (where the Yarrowkirk stone is wrongly located in 'Edinburgh Museum' and the county in which Yarrowkirk is situated is wrongly given as Stirlingshire).

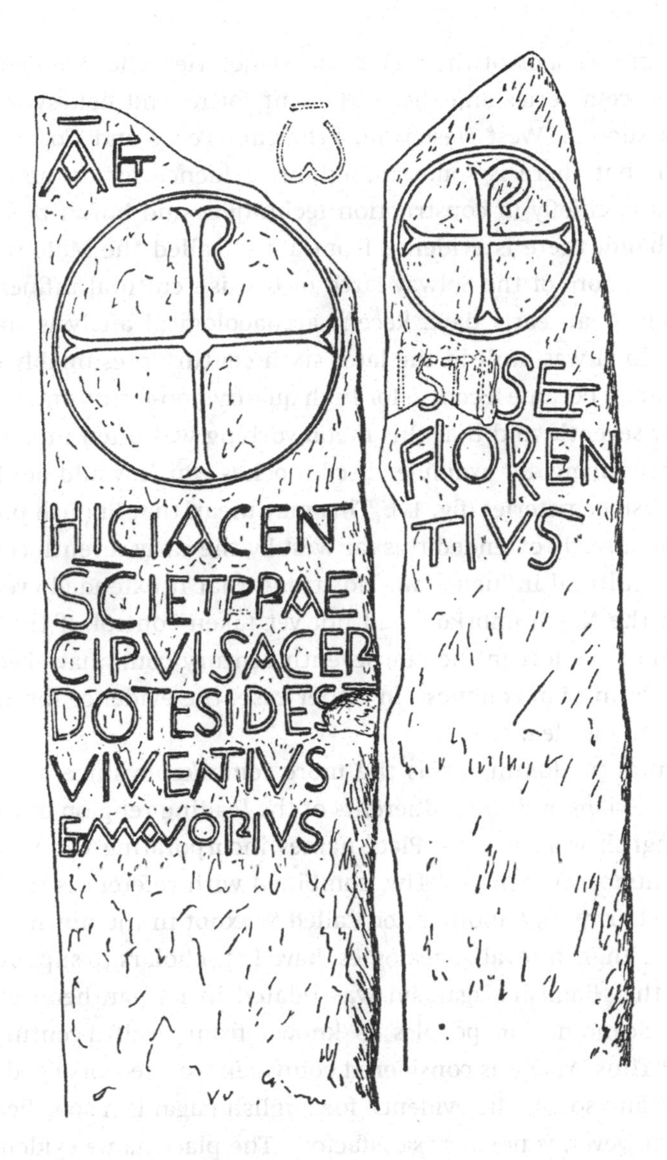

FIGURE 10 Kirkmadrine (Wigtownshire), inscribed stones (drawings from Macalister (1945), nos. 516–17)

the same time or a little later. As for the material culture of the British in the post-Roman period in the future Northumbria, we have virtually no evidence, apart from the few British – or British-style – objects (penannular brooches, for example) in apparently English graves which might have been obtained from anywhere.[4] If there was nevertheless a developed native British culture, it was displaced to a very large extent by English culture. The evidence for this has been surveyed already: language and place-names,

[4] See above, pp. 94, 96, 101–3.

the ritual and grave-goods of the cremation cemeteries, the grave-goods of the inhumation cemeteries and the settlement forms and field-systems revealed at sites such as West Heslerton, Wharram Percy and Cottam.[5] We have also seen that there is some ambiguous evidence for an admixture of British culture, chiefly in construction techniques and burial practices.[6] On the other hand, there is evidence from a fort called the Mote of Mark on the northern shore of the Solway Firth of English cultural influence on the British even at an early date. Recent archaeological analysis suggests that the fort, which was built in the later sixth century presumably by the British of that area, became a centre for high-quality non-ferrous metalworking in the early seventh, and that this metalworking was often in markedly English style, producing, for example, disc brooches which would not be out of place in English cemeteries (fig. 11). This may mean that English political domination had already extended this far west by the early seventh century; or that English cultural influence was so strong that it extended even into areas of which the Northumbrians had not yet taken control. If the latter, the destruction of the fort in the late seventh century could have been the result of a Northumbrian conquest. In either case, the evidence for English cultural hegemony is clear.[7]

English cultural predominance is the more remarkable since, while the British were Christians and thus adherents of the leading religion of western Europe, the English were pagans. Place-names incorporating the names of gods such as Thunor, Woden and Tiw, combined with references to Woden in Old English charms and another god called Seaxnot in the ninth-century collection of the English royal genealogies, have led scholars to suppose, not unreasonably, that English paganism was related in its pantheon of gods to that of the Scandinavian peoples as known from twelfth-century and later writings.[8] Thus Woden is considered equivalent to the Norse god Odin, Thunor to Thor and so on. The evidence for English paganism specifically in Northumbria, however, is not very satisfactory. The place-name evidence for the names of gods does not apply to Northumbria itself, the most northerly names being for places like Wednesbury (Staffs.) meaning 'the fortress of Woden'.[9] But the genealogies of the royal houses of Deira and Bernicia are traced back to Woden like their southern counterparts.[10] Beyond this, Bede

[5] See above, pp. 60–4, 93–9. [6] See above, pp. 84–5, 94, 96.
[7] Longley (2001); see also Laing and Longley (forthcoming). For illustrations, see Curle (1914).
[8] On English paganism generally, see Owen (1981), pp. 5–39, and, for the archaeological evidence, David Wilson (1992). For Scandinavian paganism see Turville-Petre (1964).
[9] Stenton (1941) and Gelling (1997). [10] Dumville (1976).

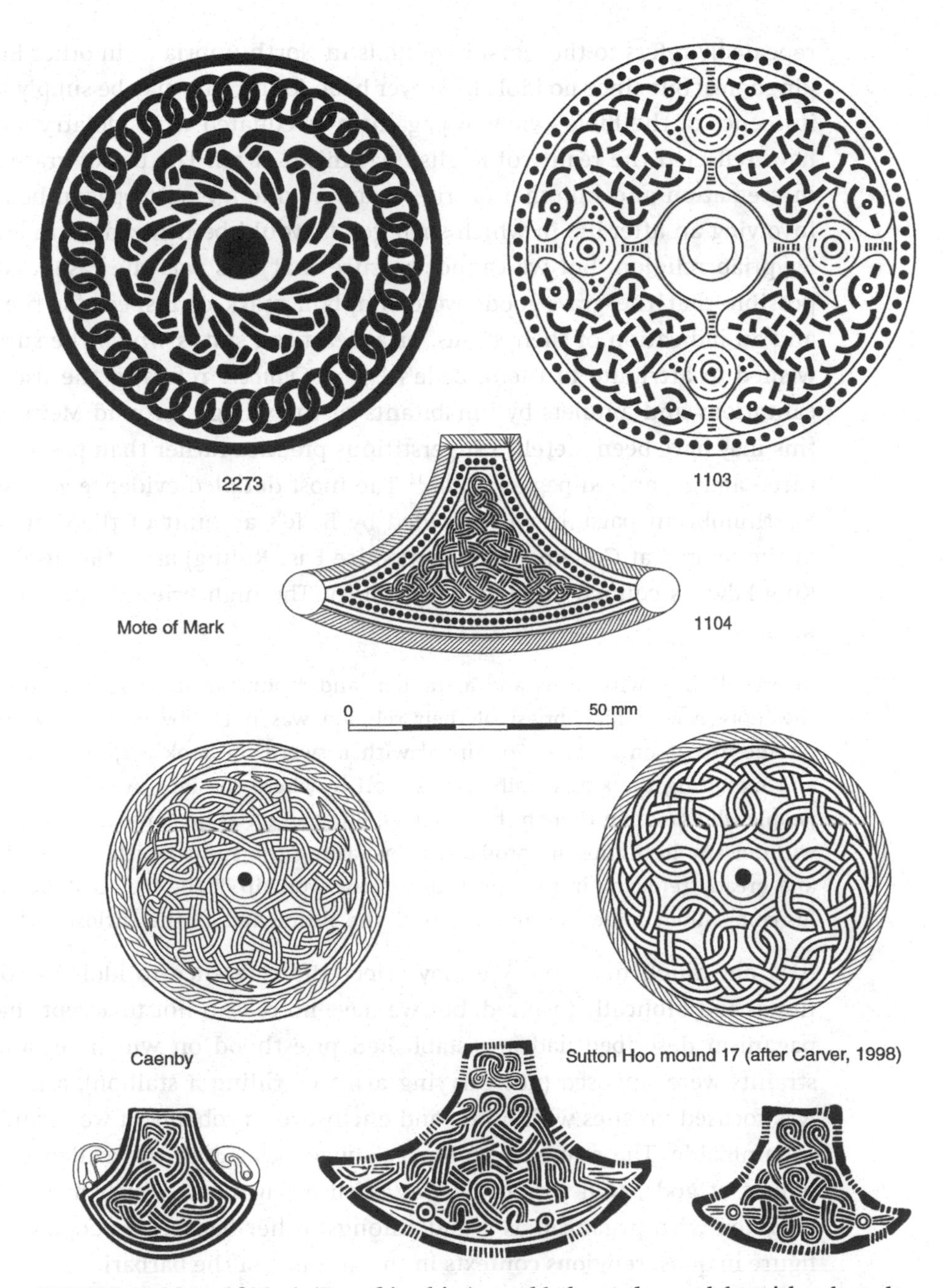

FIGURE 11 Mote of Mark (Dumfriesshire), axe blades and roundels, with selected comparanda from Caenby (Lincolnshire) and Sutton Hoo, mound 17 (Suffolk) (drawings by David Longley from Longley (2001); those of objects from Sutton Hoo after Carver (1998))

repeatedly refers to the worship of idols in Northumbria as in other English kingdoms; but since no idols have ever been found this may be simply a reminiscence of the Bible's view of paganism as equated with idolatry and not have reflected the reality of English paganism at all. The use of grave-goods in cremation and inhumation rites may provide a clue to pagan beliefs as involving an after-life in which such goods would be utilized, as in ancient Egyptian religion; but we cannot be sure that this was the case, and it is possible that the grave-goods were simply a mark of respect for the dead and an indication of their status, or at least the status which the survivors wished to attribute to them. Bede's *Life of Cuthbert* refers to the use of incantations and amulets by inhabitants of the villages around Melrose, but this may have been merely a superstitious practice rather than part of structured and organized pagan ritual.[11] The most detailed evidence we have for Northumbrian paganism is provided by Bede's account of the destruction of the temple at Goodmanham (Yorkshire East Riding) after the decision by King Edwin's council to adopt Christianity. The high priest, Coifi, asked the king

> to provide him with arms and a stallion; and mounting it he set out to destroy the idols. Now a high priest of their religion was not allowed to carry arms or to ride except on a mare. So, girded with a sword, he took a spear in his hand and mounting the king's stallion he set off to where the idols were. The common people who saw him thought he was mad. But as soon as he approached the shrine, without any hesitation he profaned it by casting the spear which he held into it; and greatly rejoicing in the knowledge of the worship of the true God, he ordered his companions to destroy and set fire to the shrine and all the enclosures.[12]

It is not much to go on. We may reject the reference to idols as conventional and biblically inspired, but we have no reason not to accept that the paganism described had an established priesthood on which certain constraints were imposed (not carrying arms or riding a stallion), and that it was focused on sites with altars and enclosures, probably of wood and thus inflammable. The reference to horses suggests, as does the evidence of the names of gods, that we are dealing with a paganism which had traits in common with paganism as found amongst other barbarian peoples. Horses figure in quasi-religious contexts in the account of the barbarians east of the Rhine written in the first century AD by the Roman historian Tacitus; and it is notable that horse sacrifice, which is widely known from barbarian burials, may have been a feature of the burial of the Frankish chieftain Childeric

[11] Bede, *Life Cuth.*, ch. 9. [12] Bede, *Eccles. Hist.*, bk 2 ch. 13.

in the late fifth century at Tournai in Belgium, at least to judge from the pits of apparently sacrificed horses found near his grave.[13] Interpreted thus, the paganism of the English was a common cultural heritage with the English elsewhere in England and with barbarians anywhere in Europe. It almost certainly distinguished them from the Christian British, and it is likely that it differed substantially from British paganism, assuming that some of the British had continued to be pagan in the post-Roman period.[14]

If Northumbrian paganism was not distinctive in an English or more generally a barbarian context, the same is true of the other aspects of English culture and language. Philologists identify a Northumbrian dialect in Old English, but this is based on very few texts: the Old English poem inscribed on the stone cross at Ruthwell in Dumfriesshire; the hymn supposedly composed by the Whitby cow-herd Cædmon and found as a contemporary appendix to Bede's *Ecclesiastical History* in its earliest manuscripts; the Old English inscription on the whale-bone casket known as the Franks Casket; and the so-called Leiden Riddle. Only the first two can be definitely associated with Northumbria, so this is a slender basis on which to found firm conclusions; but it is clear nevertheless that 'Northumbrian' did not diverge greatly from other dialects of Old English and was certainly not the sort of distinctive language which would have contributed to defining a people's identity.[15]

So much for the period of Northumbria's origins. After Northumbria's conversion to Christianity, the evidence for its culture is more extensive, although it principally concerns the secular elite and the church. As regards the former, it has been increasingly recognized that the upper class laity of early medieval Europe had some degree of literacy, and some kings in particular were literate in a sophisticated way. In the case of Northumbria, we know that Bede sent his *Ecclesiastical History* to King Ceolwulf of Northumbria for revision. And the great English scholar Aldhelm dedicated to King Aldfrith of Northumbria (686–705) discussions of biblical typology and Latin metrics, while works of literature could plausibly be attributed to him, even if incorrectly.[16] The overwhelming proportion of written evidence surviving from early medieval Europe was nevertheless generated by the church, and

[13] Mattingly (1948), ch. 10, and Edward James (1988), pp. 62–4.

[14] For a rather speculative argument that English shrines were influenced by 'Celtic' ones, see John Blair (1995). For British paganism, see Ross (1967) and, for evidence of pagan survival amongst the British (which does not specifically apply to Northumbria), see Thomas (1981), p. 266.

[15] Hogg (1992), pp. 422–3, and Wakelin (1988), pp. 49–51.

[16] McKitterick (1990); on Aldfrith, see Lapidge and Herren (1979), pp. 31–47, and Ireland (1999), pp. 48–56.

in the case of Northumbria this is almost 100 per cent. The only written material definitely attributable to lay people was a letter written by King Alhred of Northumbria (765–74) and his queen Osgifu to the English missionary on the continent, Lul.[17] Physical objects surviving from the post-conversion period are also overwhelmingly associated with the church. Once the practice of burying grave-goods ceased, we are admittedly dependent on chance survivals of objects that might have belonged to the laity, sometimes found by metal detectors, sometimes at excavated sites.[18] Such objects are generally of quite humble character, however, such as pins and other fastenings for clothing. The only really important high-status object to have been recovered from pre-Viking Northumbria is the superb helmet from Coppergate in York, which has strong ecclesiastical associations.[19] In general, indeed, what objects – and manuscripts – have survived have done so because they were connected with the church. This is not to say that distinguished objects owned by the laity did not exist, simply that they have in general not survived. We have, for example, no silver dishes such as the one King Oswald is said to have had divided amongst the poor as a gesture of humility and alms-giving;[20] but we do have, for example, some of the treasures from the shrine of St Cuthbert, the Lindisfarne Gospels and other manuscripts probably from Northumbrian churches.[21] Even allowing for deep bias in our evidence, it is nevertheless hard to escape the conclusion that conversion to Christianity was either the most formative process in the development of culture in Northumbria, or opened the way to processes which were.

In order to appreciate how these processes operated, we need to sketch briefly the course of the conversion of Northumbria and the subsequent consolidation of Christianity there.[22] According to Bede and the *Life of Gregory the Great*, two sources which may or may not be independent of each other, the first Christian king of Northumbria, Edwin (616–33), was converted by the Roman missionary Paulinus, who had come to Kent in 601 to join Augustine, sent to Kent four years earlier at the behest of Pope Gregory the Great. Paulinus came north to Northumbria with the Kentish princess Æthelburg when she married the then pagan Edwin, and the king was baptized in 627. The initial conversion of Northumbria was thus an extension of a mission dispatched from Rome. Paulinus's own work, however, was brought to an

[17] *EHD I*, no. 187. [18] See, for example, Leahy (2000).
[19] See below, pp. 205–7. [20] Bede, *Eccles. Hist.*, bk 3 ch. 6.
[21] On the Cuthbert treasures, see Battiscombe (1956); on the Lindisfarne Gospels, see below, pp. 148–51.
[22] The best accounts of the conversion of the English are Mayr-Harting (1991) and Fletcher (1997). Important studies include: James Campbell (1973), Bullough (1982) and Angenendt (1986).

end in 633 when King Edwin was killed by King Cædwalla of Gwynedd and Penda of Mercia at the Battle of Hatfield. His two successors, Osric who ruled Deira, and Æthelfrith's son Eanfrith who returned from exile in the north to rule Bernicia, reverted to paganism. These kings were killed in the following year by Cædwalla, who proceeded to rule Northumbria himself until he was defeated at the Battle of Heavenfield by Oswald, another of Æthelfrith's sons returned from exile amongst the Irish of Dalriada. According to Bede, he had become converted to Christianity there, and he now sought a bishop from the Irish monastery of Iona to convert his new kingdom. The first to be sent was, Bede tells us, not of suitable character; but he was replaced by Aidan, who founded the Northumbrian monastery of Lindisfarne.[23]

Bede makes no mention of any native British contribution to the conversion of Northumbria, either from within the kingdom or from outside it. On the contrary, one of his greatest objections to the British was precisely that they had done nothing to convert the English: 'to other unspeakable crimes, which Gildas their own historian describes in doleful words, was added this crime, that they never preached the faith to the Saxons or Angles who inhabited Britain with them'.[24] Now, it is possible to argue that Bede's prejudices led him to deny a British role when there really was one. But, although the fact of Bede's antipathy is not in doubt, such a view would require us to find either evidence of British missions from outside Northumbria evangelizing that kingdom, or evidence of surviving British Christianity within the area of Northumbria which might have affected or contributed to the conversion of Northumbria from within the kingdom. Of the latter, the only site which has yielded convincing evidence that it was a British church continuing to flourish in the kingdom of Northumbria is Whithorn. The evidence of archaeological excavation as well as of the inscribed stones from the site there leaves no doubt that it had had a long history as a church before it became a Northumbrian bishopric by the early eighth century at the latest. Moreover, according to Bede, the church was founded by a British churchman called Ninian at some indeterminate date. His relics were at Whithorn, and the existence of a Latin *Life* and a hymn devoted to him and dated to the eighth century shows that his cult was still flourishing when the church was a Northumbrian see. There was thus clear continuity between the British and the Northumbrian church.[25] But Whithorn was in

[23] Bede, *Eccles. Hist.*, bk 3 chs. 1–3. On Oswald, see Stancliffe (1995a).

[24] Bede, *Eccles. Hist.*, bk 1 ch. 22.

[25] Bede, *Eccles. Hist.*, bk 3 ch. 4. For the texts, see Levison (1940) and MacQueen (1990); for the site and the cult, see Peter Hill (1997).

the extreme west of Northumbria, in an area where we might reasonably expect greater evidence of British influence; and in any case the adoption of Whithorn as a Northumbrian church was long after the conversion of Northumbria, so that the British are unlikely to have effected that conversion – the Northumbrians who took over Whithorn were already Christian. Elsewhere in Northumbria we look in vain for English ecclesiastical foundations on the sites of known British churches. York had a bishop in the early fourth century and so presumably had a cathedral at that date; but as we have seen the church built there by King Edwin showed no continuity with it; and the Roman building discovered under St Mary Bishophill Junior in the *colonia* of York was apparently a secular structure associated with fish-processing and in no way the precursor of the present church.[26] The English churches of Abercorn and Melrose had names derived from the British language (*Aebburcurnig*, meaning 'horned confluence', and *Mailros*, meaning 'bare promontory'), but there is nothing to prove that these names were preserved because they were attached to British churches.[27] Hoddom may have been associated with the British churchman Kentigern; but the source for both him and his association with Hoddom is late and probably unreliable.[28] The only other evidence we can deploy is that provided by the distribution of the place-name 'Eccles' and its compounds (derived from Primitive Welsh *egles*, meaning a church); it has been argued that such place-names 'indicate centres of British Christian worship, each with its church', and that the names show that such centres persisted into the Northumbrian period. If so, it is striking how peripheral is their distribution in Northumbria, mostly not in the heartlands but in the extreme south-west, or in the area which may have been Elmet, or (in the case of Ecclefechan in Dumfriesshire) as far north-west as Galloway.[29]

The only possible evidence of British missions from outside Northumbria evangelizing that kingdom is the following account in Nennius, *History of the Britons*:

[26] Wenham, Hall, Briden et al. (1987), pp. 75–83. The map of evidence for Roman Christianity in Thomas (1981), p. 187 (fig. 29), is not only a maximal presentation of the evidence, but does not demonstrate actual continuity with the English period.

[27] Thomas (1971), pp. 17–18, notes the evidence but takes a different view.

[28] See above, pp. 31–2 and n. 32.

[29] Kenneth Cameron (1968), p. 90. On such names in what is now southern Scotland, see Geoffrey W. S. Barrow (1973b). In general, see Thomas (1981), pp. 147–9, on the difficulties of interpreting the word, and pp. 262–4 and fig. 49. For the place-name element, see A. H. Smith (1956), I, p. 145. On Eaglescliffe, see Ekwall (1960): 'perhaps'. For a contrary argument, emphasizing the evidence for the survival of British Christianity, see Thomas (1981), pp. 240–74. This is, however, by the author's own account largely inferential.

King Edwin's daughter, Eanflæd, received baptism on the twelfth day after Whitsun, and all his people, men and women, with her. Edwin was baptized at the Easter following, and twelve thousand men were baptized with him... Rhun son of Urien, that is Paulinus, archbishop of York, baptized them, and for forty days on end he went on baptizing the whole nation of the thugs (*ambrones*), and through his teaching many of them believed in Christ.

According to Bede, it was of course Paulinus who baptized Edwin at Easter, had previously baptized Eanflæd at Whitsuntide and did a stint of thirty-six days continuously baptizing. Nennius's account is thus very similar, except that it bizarrely identifies Paulinus with Rhun, son of Urien, who is presumably to be identified with the king of Rheged of this name whom Taliesin eulogized. It is possible that Nennius knew a British tradition to the effect that Rhun had baptized Edwin, a tradition which Bede had either not known or had suppressed; this finds some support in the fact that Edwin's baptism is also credited to Rhun in the *Welsh Annals*. The words 'that is Paulinus' would then have been a clumsy attempt, perhaps by Nennius, perhaps by a later scribe, to reconcile this with Bede's account. It is conceivable that both traditions were right and that Edwin was really baptized twice, once by Rhun and a second time by Paulinus, and that the discrepancies in dating Easter between the British church and Paulinus (who followed the Roman method of dating Easter) made this double baptism possible and necessary. Nennius, however, appears to know nothing else about Rhun's activities, so that it seems unlikely that a developed tradition about him existed. It seems in short much more likely that Nennius – or his informants – fabricated the information about Rhun in an attempt to give the British credit for Edwin's conversion which they did not deserve.[30]

The initial conversion of Northumbria was therefore most probably the work of the Roman missionary Paulinus. Oswald's Irish connections and Aidan's success then opened the way for very considerable Irish influence on the Northumbrian church. The influence of Rome and that of Ireland are thus the first two influences on Northumbrian Christianity which we shall need to consider. With regard to the former, what was the lasting effect, if any, of Paulinus's mission? Bede, while presenting its failure as part of the general debacle resulting from Edwin's death, concedes that one of Paulinus's colleagues, James the Deacon, remained in the church of York,

[30] Jackson (1963b), p. 33; Phythian-Adams (1996), pp. 56–7; see also Nora Kershaw Chadwick (1963b), pp. 156–9. For the texts, see Nennius, *Hist. Britons*, para. 63; Bede, *Eccles. Hist.*, bk 2 chs. 9, 14; and *Welsh Annals*, *s.a.* 626 ('Edwin is baptized, and Rhun son of Urien baptized him').

teaching and baptizing.[31] So the Christianity introduced by Paulinus had not collapsed in Northumbria after Paulinus himself had fled, the church of York had remained to some extent functional, and James the Deacon may have had some influence on the subsequent development of the Northumbrian church. Whatever that influence may have been, however, Edwin's church in York seems to have fallen into disuse, and was in such a bad way in Wilfrid's time that it was full of damp and pigeon-droppings.[32]

The immediate fate of Paulinus's mission is not in any case a guide to the level of its influence in subsequent years, since the devotion of the Northumbrian church to Rome, in common with the church in other parts of England, was largely stimulated by recollection of that mission.[33] For the anonymous author of the *Life of Gregory the Great*, Gregory was 'our apostle' who would lead the English on the Day of Judgment; and he felt able to present the definitive conversion of Northumbria as if it had been the work of Paulinus, without mentioning the apostasy following Edwin's death or the work of Aidan.[34] In the late eighth century, Alcuin presented a similar picture in his poem *The Bishops, Kings and Saints of York*, in which Aidan appears only in a miracle-story relating to King Oswald. The conversion was for him the work of Paulinus. Alcuin was of course a York man, so he no doubt had an interest in emphasizing the work of Paulinus, first metropolitan bishop of York, at the expense of that of Aidan, the first bishop of Lindisfarne, which had in Oswald's time eclipsed York, becoming the bishopric of the Northumbrians.[35]

The importance of Rome to the Northumbrian church was considerable. Rome was one of the most ancient seats of Christianity, where the culture and learning associated with that religion had been fostered by the popes for centuries. Its intellectual and cultural resources were vast: Jerome had translated the Bible into Latin there in the late fourth century, there the great Alexandrian specialist in calendrical computation, Dionysius Exiguus, had responded to the pope's demands for a better method of calculating the date of Easter. Rome's resources in terms of expertise in liturgy were equally great.[36] Contacts between England and Rome were fostered from the second half of the seventh century onwards, when the English in general

[31] Bede, *Eccles. Hist.*, bk 2 ch. 20.

[32] *Sources*, pp. 136–8 (A.4.5 = Stephen, *Life Wilf.*, ch. 16). It is conceivable, however, that this was not the result of the failure of Edwin's mission but of the transfer of the Northumbrian bishopric from York to Lindisfarne in the reign of King Oswald.

[33] Wallace-Hadrill (1960). [34] *Life Greg.*, chs. 6, 14–19. [35] Alcuin, *BKSY*, lines 291–301.

[36] Llewellyn (1971), chs. 3 and 6; Krautheimer (1980), pp. 59–87, and O Carragáin (1995).

were precocious in undertaking pilgrimages there to pray at the tombs of no less than two apostles, Peter and Paul, as well as visiting the numerous tombs of martyrs in the catacombs and basilicas around the city – and these pilgrimages were a powerful means by which cultural influences from the city entered England.[37] In the case of Northumbria, two men whose influence was to be formative on the development of Northumbrian Christianity both made pilgrimages to Rome: the founder of Monkwearmouth and Jarrow, Benedict Biscop, and the founder of Ripon and Hexham, Wilfrid. The former's career is especially instructive. He visited Rome as soon as he had decided to assume the religious life, and again on subsequent occasions, bringing back books, relics, a papal privilege for his monasteries, paintings and even the arch-chanter, John, to teach the Roman manner of chanting the Psalms and conducting the liturgy.[38]

The importance of Ireland was also considerable. Until the Synod of Whitby in 664, the bishop of Lindisfarne was the bishop for all Northumbria and was appointed, as Aidan himself had been, directly from the Irish monastery of Iona. Following the Synod of Whitby, the then bishop of Lindisfarne, Colmán, withdrew to Iona, as a result of the rejection of the teachings of the church of Iona on the calculation of the date of Easter. Nevertheless, there continued to be numerous contacts between the Irish church and Northumbria, and Bede himself comments not only on the many who came from Ireland in the time to King Oswald to preach Christianity to 'Britain and to those English kingdoms over which Oswald reigned', but also later on he emphasizes the number of English who went to study in Ireland.

> At this time there were many in England, both nobles and commons, who, in the days of Bishops Fínán and Colmán, had left their own country and retired to Ireland either for the sake of religious studies or to live a more ascetic life. In the course of time some of these devoted themselves faithfully to the monastic life, while others preferred to travel round to the cells of various teachers and apply themselves to study. The Irish welcomed them all gladly, gave them their daily food, and also provided them with books to read and with instruction, without asking for any payment.[39]

Two graphic examples of the continuing relationship between England and Ireland are provided by the history of the monastery of Mayo and by

[37] Moore (1937). [38] *Life Ceolf.*; Bede, *Hist. Abbots*, chs. 2–4, 6.

[39] Bede, *Eccles. Hist.*, bk 3 chs. 3, 25–7 (the text quoted is from ch. 27). On the Synod of Whitby, see Mayr-Harting (1991), pp. 103–13; on contacts with Ireland, see Hughes (1971).

the career of Egbert 'the Englishman'. When Bishop Colmán of Lindisfarne resigned his see and withdrew from Northumbria after the Synod of Whitby, he took with him, Bede tells us, the monks of Lindisfarne, both Irish and English, who could not accept the ruling of the synod, and established the monastery of Inishboffin off the west coast of Ireland. Tension arose between the Irish and the English, however, with the result that Colmán purchased land on the mainland at Mayo and established a monastery for the English there. 'This monastery', Bede noted, 'is still occupied by Englishmen; from small beginnings it has now become very large...and it now contains a remarkable company gathered there from England.'[40] Englishmen went to study in Ireland for much the same as one of the reasons that Englishmen were drawn to Rome: the sophistication and richness of Christian culture and learning which had developed to an extraordinary degree in Ireland. Although Ireland had never been part of the Roman Empire, and had only intermittently been exposed to Roman influences, by the seventh century it had developed ecclesiastical and scholarly centres pursuing studies of some sophistication, composing works of hagiography, computistics and biblical commentary. From Ireland moreover came not only the churchmen who established Iona and subsequently undertook evangelization in Northumbria, but also others, notably Columbanus, who went to the continent and established a formidable reputation and influence as monastic founders, reformers and scholars.[41]

It is not enough, however, just to consider the impact of the two missions which Bede describes, that of Paulinus and that of Aidan. Other areas must also have been influential on the development of Northumbrian Christian culture. In particular, Bede makes a remarkable statement concerning the influence of northern Gaul in the earlier part of the seventh century. He notes that, around 640,

> because there were not yet many monasteries founded in England, numbers of people from Britain used to enter the monasteries of the Franks or Gauls to practise the monastic life; they also sent their daughters to be taught in them and to be wedded to the heavenly bridegroom. They mostly went to the monasteries at Faremoûtier-en-Brie, Chelles, and Les Andelys-sur-Seine.[42]

[40] Bede, *Eccles. Hist.*, bk 4 ch. 4; see Nora Kershaw Chadwick (1963a).

[41] On Irish schools, see Ó Croínín (1995), pp. 169–232; on Columbanus, see Charles-Edwards (2000), pp. 344–90.

[42] Bede, *Eccles. Hist.*, bk 3 ch. 8.

This account is given in the context of the daughter and stepdaughter of Anna, king of the East Angles, who went to Faremoûtier-en-Brie, where they became abbesses; but Northumbrians were evidently enthusiastic too, for Bede tells us that Hild, who subsequently founded the monastery of Whitby (Yorkshire North Riding), wanted to go to Chelles but was prevented by Aidan.[43] No doubt there were other Northumbrians who were not so prevented.

The attendance of English women at monasteries in Gaul was only one way in which the influence of that country on Northumbria made itself felt. The routes to Rome naturally passed through Gaul, and we find Wilfrid staying for a protracted period at Lyons en route to Rome, and Benedict Biscop spending a period at Lérins in the Rhône delta. Both men's connections with Gaul went beyond this, however, for Benedict Biscop sought glaziers and masons from that country when building the church of Monkwearmouth, and Wilfrid went to Compiègne to be consecrated bishop, and was sufficiently involved in Gaulish politics to be instrumental in the return to power of Dagobert II, a member of the Frankish royal family exiled in Ireland. Connections between England and Gaul were of course of long standing and continued to be of considerable importance. Augustine's mission itself must in part have been made possible by King Æthelberht of Kent's marriage to the Frankish princess Bertha; and when, following Edwin's death, his queen Æthelburg returned to Kent, it was to Gaul that she sent her children for refuge. The bishop who played a crucial role at the Synod of Whitby was Agilbert, a Frankish bishop of Paris.[44]

From the time of Wilfrid onwards, links with Gaul were strengthened by the developing missionary activity on the continent of English churchmen, beginning with Wilfrid, but reaching its climax with Willibrord from Ripon, and Boniface from south-west England. The links between England and Gaul were most obvious in the case of Alcuin, a scholar and teacher at York, who transferred to the court of Charlemagne at Aachen and finally to the abbey of St Martin at Tours, where he died. The nature of the influence of Gaul, however, is not easy to assess. Only in the mid- to late eighth century did the scholarly activity associated with the Carolingian renaissance really develop on the continent; but it none the less seems clear that Gaul preserved some

[43] Bede, *Eccles. Hist.*, bk 4 ch. 23 (21).

[44] For a perceptive general analysis, see James Campbell (1971). See also Wood (1995). For the texts, see Bede, *Eccles. Hist.*, bk 2 ch. 20, bk 3 ch. 25; Stephen, *Life Wilf.*, chs. 12, 28; Bede, *Hist. Abbots*, ch. 5; and *Life Ceolf.*, para. 7. On Bertha, see Wood (1994), pp. 176–8.

of the scholarly resources of the Roman past and that already in the seventh century its monasteries were not inconsiderable foci of Christian culture.[45]

The appointment of a new archbishop of Canterbury in 668 introduced a new dimension into the complex of influences bearing on Northumbria. The archbishop in question was Theodore, a monk of Greek origin from Tarsus, whom the pope sent to England in company with the African churchman Hadrian. During his long and efficacious archiepiscopate (668–90), Theodore was responsible for the establishment of the pattern of English dioceses, in particular subdividing the very large sees of Mercia and Northumbria. He and Hadrian were, according to Bede, very influential also in the development of learning in England, teaching to 'a crowd of students' scripture, the art of metre, astronomy and ecclesiastical computation, and engendering competence not only in Latin but also in Greek. The identification of biblical commentaries from their school has made it possible to enlarge on this perspective, and to confirm the very considerable sophistication of the studies undertaken at Canterbury.[46]

Influences may also have reached Northumbria from farther afield, even from the Holy Land and the eastern Mediterranean, which some of Bede's contemporaries visited. These included the Frankish bishop Arculf who was shipwrecked on the west coast of Scotland and related his experiences in the east to Adomnán of Iona (d. 704); and the English missionary on the continent, Willibald, who also visited the Holy Land.[47] Evidence for the influence of such visits on Northumbria is to be found in the proportions of the crypts which Wilfrid built under his churches at Hexham and Ripon, if they are (as has been argued) based on those of the church of the Holy Sepulchre at Jerusalem as it was then. The remarkable double vine-scroll on the sculptured cross at Hexham known as Acca's Cross may also constitute such evidence, since it is paralleled in the mosaics of the Dome of the Rock in Jerusalem – although the artistic motif may in this case have been derived from Italy (ill. 6).[48]

Finally, and most obscurely, Pictish influence may have been a further factor in the genesis of Northumbrian Christian culture, although it is easier

[45] On English missionaries on the continent, see Levison (1947) and Fletcher (1997), pp. 193–227; on Alcuin, see Bullough (2002); and, on scholarship in Gaul, see McKitterick (1995), pp. 681–94, 709–57, and Riché (1976), pp. 427–46.

[46] Bede, *Eccles. Hist.*, bk 4 ch. 1–2; see *Encyclopaedia ASE*, entry 'Theodore', and for more detail Lapidge and Bischoff (1994) and Lapidge (1986).

[47] On Adomnán and Arculf, see Denis Meehan (1958); Bede, *Eccles. Hist.*, bk 5 ch. 15–17; and, for Willibald's journey, *Life Willibald* (text, pp. 92–102; trans., pp. 159–72).

[48] Bailey (1991) and Cramp (1984), p. 176.

ILLUSTRATION 6
Hexham Abbey (Northumberland), Acca's Cross, detail
Note the sophistication and quality of the double-stranded vine-scroll design on this face (face C); the design on the adjacent face (face D) is a 'complex spiral scroll' (Cramp (1984), p. 176).

to point to Northumbrian cultural influence on the Picts than vice versa. Bede reports that King Nechtan of the Picts (706–24) asked Abbot Ceolfrith of Monkwearmouth and Jarrow for 'information by letter to enable him to confute more convincingly those who presumed to celebrate Easter at the wrong time', as well as information about how to tonsure clerics, and builders so that he could build a stone church 'after the Roman fashion'. Nechtan was clearly wishing to bring the Pictish church into line with the practices which Northumbria had adopted at the Synod of Whitby, and likewise to bring that church into line with the Mediterranean-influenced building styles which dominated Northumbrian ecclesiastical architecture. Ceolfrith answered at length; builders were sent to assist the Pictish king; and we know from the

ILLUSTRATION 7
Aberlemno (Forfarshire), early (Class I) Pictish stone decorated with symbols
For a description see Romilly Allen and Anderson (1993), p. 205 (Aberlemno, no. 1). Note the serpent at the top, then the double disc and Z-shaped rod, and at the bottom the comb and mirror.

Annals of Ulster that he duly expelled the Irish clergy from his lands.[49] Perceiving any Pictish influence there may have been on Northumbria, however, is very difficult since we are dependent for our understanding of Pictish cultural development on surviving monuments, chiefly the carvings known as Pictish stones. Those presumed to be the earliest lack any Christian associations, and are decorated with 'incised linear designs, chiefly geometric and zoomorphic', probably of pagan character (ill. 7). Presumed later stones are 'shaped and dressed slabs, carved in relief and bearing a series of scenes as well as symbols', and are Christian in form. Many, like the cross in the churchyard at Aberlemno (ill. 8), are grandiose, sophisticated and evidently

[49] Bede, *Eccles. Hist.*, bk 5, ch. 21, and *Annals Ulster*, *s.a.* 717; for church architecture, see below, p. 160.

ILLUSTRATION 8
Aberlemno (Forfarshire), later (Class II) Pictish stone with cross and animal and interlace decoration
For a description see Romilly Allen and Anderson (1993), pp. 209–14 (Aberlemno, no. 2). Note the interlace and spiral design on the cross, with the interwining animals around it. The other side of the cross has symbols and a battle-scene.

the product of a cultured Christian context, and some of the motifs and designs they use can be paralleled in Northumbrian art. The dates of these stones, however, are naturally difficult to establish in the absence of written evidence to relate to them, so that it is not easy to decide whether we are seeing Pictish influence on Northumbria or simply further Northumbrian influence on Pictland.[50]

It is important to emphasize that the areas we have been discussing were not sealed one from another. On the contrary, the travels of the Irish, for example, meant that they effectively bridged Ireland, England, Gaul and

[50] Wainwright (1955), pp. 31–3; see also Isabel B. Henderson (1967); Cummins (1995, 1999); and Ritchie (1989). For Meigle and Aberlemno, see Romilly Allen and Anderson (1993), pp. 296–305, 328–40, 205, and 209–15; this work is the only corpus of the stones available. See above, p. 42.

Italy. Bede describes the career of the Irish monk Fursa, who came from Ireland to East Anglia, where he founded a monastery, and then went on to found another monastery in northern Gaul at Péronne. The career of Columbanus, who came from Ireland around the year 600 to work as a reformer in Gaul, underlines the close connections between Ireland and Gaul; he also corresponded at crucial stages in his career with the pope. The complexity of the situation is best seen, perhaps, in the fact that the Gaulish monasteries to which the English sent their daughters, Faremoûtier-en-Brie, Chelles and Les Andelys-sur-Seine, were probably all connected with the reforms undertaken by Columbanus.[51]

The Synod of Whitby is a classic example of this fusing of influences from far-flung areas of Europe. Sometimes presented as a clash between 'Roman' and 'Celtic', it was in fact something much more complex, the so-called 'Roman' party being led by Agilbert, who was as we have seen a Frank from northern Gaul, who had lived for some years as a student in Ireland, and who was from a monastic circle deeply influenced by the work of the Irishman Columbanus. Moreover, Bede tells us that before the synod itself 'one most violent defender of the true Easter was Ronan who, though Irish by race, had learned the true rules of the church in Gaul or Italy'. Presumably he was from southern Ireland, the churches of which had adopted the so-called Roman method of calculating the date of Easter at the Synod of Mag Léne in 634.[52] Another example of such fusion is the career of Egbert, which Bede outlines. Leaving England to go and pursue his religion in Ireland, Egbert settled at the unidentified Irish monastery of *Rathmelsigi* with another Englishman called Æthelhun. The latter died along with many of the monks in the plague of 664, but Egbert vowed that, if he survived, he should never return to his native land; and so it turned out. The remainder of Egbert's life was spent in Ireland, where he studied, and in Iona. He initiated the mission of Willibrord to the continent; he converted the monks of Iona in 716 to the correct calculation of the date of Easter; and he evidently had extensive dealings with the Irish and the Picts. Here was a career which spanned virtually all the areas which might have influenced Northumbrian Christianity apart from Rome and the Mediterranean lands.[53]

In view of the influences and cultural cross-currents sketched above, the form which Christian culture took in Northumbria has proved complex and

[51] On Fursa, see Bede, *Eccles. Hist.*, bk 3 ch. 19. In general, see Prinz (1988), pp. 142–3, 174–5, 296–7.

[52] Bede, *Eccles. Hist.*, bk 3 ch. 25; on Agilbert, see Charles-Edwards (2000), pp. 8–9; on Mag Léne, see O Croínín (1995), pp. 152–3. For a general discussion of the synod, see Mayr-Harting (1991), pp. 103–13.

[53] Bede, *Eccles. Hist.*, bk 3 chs. 4, 27, bk 4 ch. 3, bk 5 chs. 9–10, 22–4.

fugitive to define. Even in what we might expect to be a straightforward aspect, that of the structure of dioceses and metropolitan sees, the complexity is at once apparent. At one level, it seems clear enough that the Northumbrian church was influenced by Roman and Mediterranean examples. In one of his letters to Archbishop Augustine of Canterbury, Pope Gregory the Great had urged the establishment of a northern metropolitan see based at York with twelve bishoprics in northern Britain subject to it. Although this was only partially achieved, it is possible to discern the Northumbrian church moving towards it. The see for all Northumbria established by Aidan at Lindisfarne was transferred to York after the Synod of Whitby in 664, and then subdivided by Archbishop Theodore in the late 660s to produce the sees of York, Lindisfarne and Hexham. Whithorn had followed by 731. In 735, York was made a metropolitan see, so that Gregory the Great's plan had been partially implemented, with a metropolitan at York with three (in place of twelve) suffragans – Hexham, Lindisfarne and Whithorn. Thus far, the influence of Rome seems dominant. Gregory the Great's plans were certainly important, and so too may have been Archbishop Theodore's experience in the Mediterranean lands of small sees around which a bishop could reasonably make his way attending to the spriritual needs of all the inhabitants of the see. Bede seems to have had such a picture in his mind when in 734 he urged Bishop Egbert of York to embark on subdivision to facilitate the bishop's supplying pastoral care.[54]

It is apparent, however, that other influences were at work in the development of Northumbrian dioceses. Bishop Wilfrid had had experience in Gaul of grander, wealthier dioceses than those of Northumbria, as exemplified by the circumstances of his consecration in Gaul, when he was 'borne into the oratory aloft on a golden throne by nine bishops'; and it may be that this played a part in his persistent opposition to the subdivision of the Northumbrian see, chronicled by Stephen.[55] It may also have accounted for the rather grandiose position which the archbishops of York assumed in the eighth century. Alcuin in his poem *The Bishops, Kings and Saints of York* described the wealth and opulence of their patronage of the buildings and ornaments of the church in York. In his letters, he bore witness to their aspirations to a powerful political position, which also reflected the position of the bishops of Gaul. He accused Archbishop Eanbald II of harbouring

[54] The best discussion is Mayr-Harting (1991), pp. 135–9. See also Gibbs (1973) and, for the texts, Bede, *Eccles. Hist.*, bk 1 ch. 29; Bede, *Letter*; and Bede, *Continuations*, *s.a.* 735.

[55] Stephen, *Life Wilf.*, ch. 12; for discussion, Mayr-Harting (1991), pp. 130–5.

the king's enemies and of having too large a military force at his command.[56]

There may also have been Irish influences on Northumbrian diocesan structure. Bede tells us that the church of Lindisfarne was an unusual one, which might cause remark:

And let no one be surprised that, though we said above that in this island of Lindisfarne, small as it is, there is found the seat of a bishop, now we say also that it is the home of an abbot and monks; for it is actually so. For one and the same dwelling-place of the servants of God holds both; and indeed all are monks ... Hence all the bishops of that place up to the present time exercise their spiritual functions in such a way that the abbot, whom they themselves have elected with the advice of the brethren, rules the monastery; and all the priests, deacons, cantors, lectors, and the other ecclesiastical grades, together with the bishop himself, keep the monastic rule in all things.[57]

This arrangement, Bede tells us, had been introduced by Aidan. Clearly, it resembled to an extent Aidan's own monastery of Iona, which 'has an abbot for its ruler who is a priest'; but the resemblance was not exact since at Iona even bishops were, according to Bede, subject to the abbot, and there is no trace of this at Lindisfarne.[58] Moreover, it might be argued that there is an artificiality in trying to distinguish influences in this way, an artificiality which is underlined by considering Bede's account of Augustine's church at Canterbury. There too the cathedral church was a sort of monastery, for Augustine was a monk, Pope Gregory the Great himself had made him an abbot, and in Canterbury he and his followers 'began to imitate the way of life of the apostles and of the primitive church'. In this case, however, there is no suggestion of Irish influence: we are seeing aspirations to imitate the primitive church, coupled with the monastic tendencies of Gregory the Great who, like Augustine, was a monk.[59]

A further layer of complexity, however, is involved in assessing Irish influence on Northumbrian diocesan structure. In a classic passage in the *Ecclesiastical History*, Bede tells us that the Irish monk Columba, in addition to founding Iona, had also founded the monastery of Durrow in Ireland, and that 'from both of these sprang very many monasteries which were established by his disciples in Britain and Ireland, over all of which the island monastery in which his body lies [Iona] held pre-eminence'.[60] This passage,

[56] *Sources*, pp. 141–4 (A.4.8–10 = Alcuin, *BKSY*, lines 1222–7, 1263–70, 1488–506); for Alcuin's letter, see below, p. 205.
[57] Bede, *Life Cuth.*, ch. 16. [58] Bede, *Eccles. Hist.*, bk 3 ch. 4.
[59] Bede, *Eccles. Hist.*, bk 1, ch. 26. [60] Bede, *Eccles. Hist.*, bk 3, ch. 4.

together with the one immediately following describing the subjection of the bishop to the abbot at Iona, encouraged scholars to envisage the Irish church as being, at least in the seventh and eighth centuries, organized on the basis of confederations (*paruchiae*) of monasteries, which transcended bishoprics and were superior to them. Recent research, however, has suggested that bishoprics were in fact more important than the interpretation noted above allows; and that the organization of the early Irish church was more complex than it envisages, with a diverse and inclusive relationship between bishops, abbots and others, known as co-arbs, who had authority in it; and that the distinctions between 'monastic' and 'episcopal' have been too sharply drawn.[61] This complex Irish church organization may well have influenced the Northumbrian church. Bede shows that, until the Synod of Whitby (664), the bishops of Lindisfarne were appointed by the monastery of Iona.[62] Bishop Colmán of Lindisfarne was defeated at that synod and his withdrawal to Ireland marked the end of this authority of Iona over Lindisfarne. Nevertheless, although Lindisfarne was integrated into the sort of diocesan structure envisaged by Gregory the Great when it emerged as one of the suffragan sees of York in the late 660s, its internal organization still showed symptoms of the same complex interplay between episcopal and abbatial authority which marked the Irish church. The *History of the Kings* has under the year 854 a passage describing the accession of Bishop Eardwulf and listing a series of places which appear in our sources as early monasteries (Melrose, Tyninghame, Norham, Coldingham and Holm Cultram). These may have been monasteries which happened to be within the diocese of Lindisfarne; but in view of Lindisfarne's Irish origins it seems more likely that they were members of a confederation of monasteries of which it was the head.[63]

More striking – and at first sight more surprising – are references in Stephen's *Life of Wilfrid* which allude to the monasteries which Wilfrid either founded or patronized not only in Northumbria but also in Mercia and elsewhere. They formed, Stephen writes, a 'kingdom of churches' (*regnum ecclesiarum*), and 'almost all the abbots and abbesses of the monasteries dedicated their possessions to him [Wilfrid] by vow, either keeping them themselves

[61] Hughes (1966), chs. 4–8, emphasized the importance of monastic *paruchiae*. Fundamental criticism of this by Sharpe (1984) has now been followed by revisionist discussions by Etchingham (1999) and Charles-Edwards (2000), pp. 241–81.

[62] Bede, *Eccles. Hist.*, bk 3 ch. 25.

[63] *Hist. Kings*, *s.a.* 854 (pp. 101–2; pr. also in Johnson South (2002), p. 119); see Craster (1954), pp. 179–80. David Hall (1984), p. 67, suggests that the places listed may have been acquired in the aftermath of the Viking invasions.

in his name or intending him to be their heir after their death'.[64] Diocesan boundaries were very important to the system of dioceses envisaged by Gregory the Great, as emphasized by Archbishop Theodore at the Synod of Hertford: 'Chapter II: That no bishop intrude into the diocese of another bishop, but that he should be content with the government of the people committed to his charge.'[65] Wilfrid, however, admittedly under various circumstances of exile and adversity, had assumed episcopal responsibilities outside his diocese, in Kent, Sussex and Mercia, so that it is possible that his mind was on a 'kingdom of churches' rather than on dioceses in the conventional sense. Wilfrid may, in other words, have presided over 'a fluctuating aggregate of jurisdictional spheres' strongly reminiscent of the Irish church.[66] To complete our perplexity, such arrangements may also have been a feature of churches on the continent in the circle of the Irish monk Columbanus, and it may be that Wilfrid's approach was influenced by such examples of them that he knew from Gaul rather than from Ireland. Evidence of Wilfrid's connections with the circle of Columbanus is provided by Agilbert, the bishop of Paris whom Wilfrid spoke on behalf of at the Synod of Whitby, and with whom he was presumably closely associated. Agilbert had a cousin called Ado, who was a friend of Columbanus and under his influence, and who was further the founder of the monastery of Jouarre where Agilbert himself was buried.[67]

Our difficulties in defining the sources of influences on Northumbrian diocesan organization are not a matter for despair. On the contrary, they are themselves of the greatest importance in underlining the composite, cosmopolitan nature of the church in Northumbria. Monastic organization, for example, is equally resistant to being assigned to specific influences and equally diagnostic of an eclectic culture. Here, too, we find a strong element of Roman and Gaulish influence. Although it is unlikely that monasticism in Northumbria was formally committed to the Rule of St Benedict as later monasteries were, it is equally clear that this rule was a predominant influence on Monkwearmouth and Jarrow. Benedict Biscop's first name (that is his name in religion, his secular name was Biscop Baducing) may be an indication of his devotion to its author; several of the accounts of the activities of the abbots which Bede gives us in his *History of the Abbots of Monkwearmouth and Jarrow* seem to accord precisely with the Rule of St Benedict (the emphasis on Abbot Eostorwine's attention to manual work, for example, and on the

[64] Stephen, *Life Wilf.*, ch. 21. [65] Bede, *Eccles. Hist.*, bk 4, ch. 5.
[66] Etchingham (1999), p. 459; see also O Croínín (1995), p. 164.
[67] Prinz (1988), pp. 125–6; Wood (1990, 1994), p. 151.

free election of the abbots); and the layout of Monkwearmouth and Jarrow with their large communal buildings (presumably refectories and dormitories) is redolent of the provision of the Rule of St Benedict, as against more hermit-like, less communal ways of monastic life. Benedict Biscop claimed to have used the rules of no less than seventeen monasteries in devising that of Monkwearmouth, so it must to an extent have been what scholars call a 'mixed rule' (*regula mixta*); but the evidence suggests strongly that the Rule of St Benedict was the dominant element. Hexham and Ripon, and whatever other monasteries Wilfrid had founded in Northumbria, were presumably also following broadly the Rule of St Benedict, the introduction of which to Northumbria was one of Wilfrid's claimed achievements.[68] Even here, however, we find nothing straightforward. The Rule of St Benedict originated at the monastery of Monte Cassino in southern Italy in the sixth century, so it might be thought an example of southern Italian influence. One of its principal admirers, however, was Pope Gregory the Great, who also wrote the *Life* of its author in the second book of his work called the *Dialogues*; his support for it may have contributed to its popularity in Northumbria, thus making it an example of Roman influence. On the other hand, the monasteries in which it seems to have been most widely used in the seventh century were those monasteries in Gaul under the influence of Columbanus and the Irish monks on the continent, so that we could also see it as an example of what we can only call Irish–Gaulish influence.[69]

The Rule of St Benedict, in whatever admixture with other rules, was not, however, the only element in Northumbrian monasticism. The evidence of our written sources shows the existence of so-called double monasteries, that is monasteries for both men and women, under the government of an abbess. We hear most about Whitby, where the abbess Hild presided over a community which included nuns, such as the Deiran princess Ælfflæd, and monks, such as the future bishop of York, John. Coldingham too was a double monastery and, although Bede criticizes the loose morals of its inmates, there is no suggestion that these derived in any way from its double character. Such double monasteries were a feature neither of the Irish nor the Roman church, and were prominent only in Gaul. Indeed, the three abbeys identified by Bede as destinations for English girls seeking the religious life,

[68] The best treatment is Wormald (1976). For the layout, see Cramp (1994); for the texts, see: Stephen, *Life Wilf.*, ch. 3 (for Benedict's name); Bede, *Hist Abbots*, chs. 8, 11; and *Life Ceolf.*, paras. 6, 16. For Wilfrid's introduction of the Rule, see Stephen, *Life Wilf.*, ch. 47.

[69] On Benedict, see for example Chapman (1929); on Gregory the Great, see Markus (1997); on the diffusion of the Rule, see Fry (1981), pp. 114–18, and Prinz (1988) *passim*.

Chelles, Les Andelys-sur-Seine and Faremoûtier-en-Brie, were all so far as we can judge double monasteries.[70]

Our written sources show that another element in Northumbrian monasticism, which had little to do directly with the Rule of St Benedict, was that of the hermit-life. At Lindisfarne, Aidan himself during his time as bishop habitually withdrew to the rocky island of the Inner Farne a little to the south to live the life of a hermit; and one of his successors, Cuthbert, lived in the same place as a hermit for protracted periods before and after becoming bishop of Lindisfarne. As regards Cuthbert's life as a hermit, the *Lives* of him emphasize the austerity of his life on the Inner Farne (he subsisted in his last days on onions), his purifying the place of demonic activity, the miraculous provision of water in the form of a spring for him and the healing powers and influence over nature which his holiness permitted him to wield.[71] Aside from the fact that Aidan came from Iona, the practice of hermit-life on an offshore island is suggestive of Irish influence, for the coast of Ireland was notable in the early middle ages for its island hermitages.[72]

An emphasis on the hermit-life could lead to monasteries being organized around individual cells which were in effect hermitages within a community like those of the later Carthusians. This seems to have been the case, to judge from its archaeology, with the famous island monastery of Skellig Michael off the west coast of Ireland, where remains of individual cells, not dissimilar to the sort of cell which Cuthbert is said to have built on the Inner Farne, are grouped together without any extensive provision for communal buildings (fig. 12). The organization of the Northumbrian monastery of Coldingham seems to have resembled this to some extent, at any rate to judge from Bede's account of a vision which an Irish monk called Adomnán is supposed to have had there. According to Bede, a celestial visitor explained to this monk that he had found evidence of moral turpitude in an examination he had made of the inmates' cells and beds (*singulorum casas et lectos*), while 'the cells (*domunculae*) that were built for praying and for reading have become haunts of feasting, drinking, gossip and other delights'.[73] The fact

[70] Very useful is still Bateson (1899); see, however, the cautionary remarks of Prinz (1988), pp. 658–63, on how there were few monasteries which were strictly speaking double communities. For the texts, see (for Whitby) Bede, *Eccles. Hist.*, bk 3 chs. 24–5, bk 4 chs. 23 (21), 26 (24), and *Life Greg.*, ch. 18; and (for Coldingham) Bede, *Eccles. Hist.*, bk 4 chs. 19 (17), 25 (23).

[71] On Aidan, Bede, *Eccles. Hist.*, bk 3 ch. 16; on Cuthbert, see Anon., *Life Cuth.*, and Bede, *Life Cuth.* For discussion, see Stancliffe (1989), pp. 36–42.

[72] Herity (1995). [73] Bede, *Eccles. Hist.*, bk 4 ch. 25 (23).

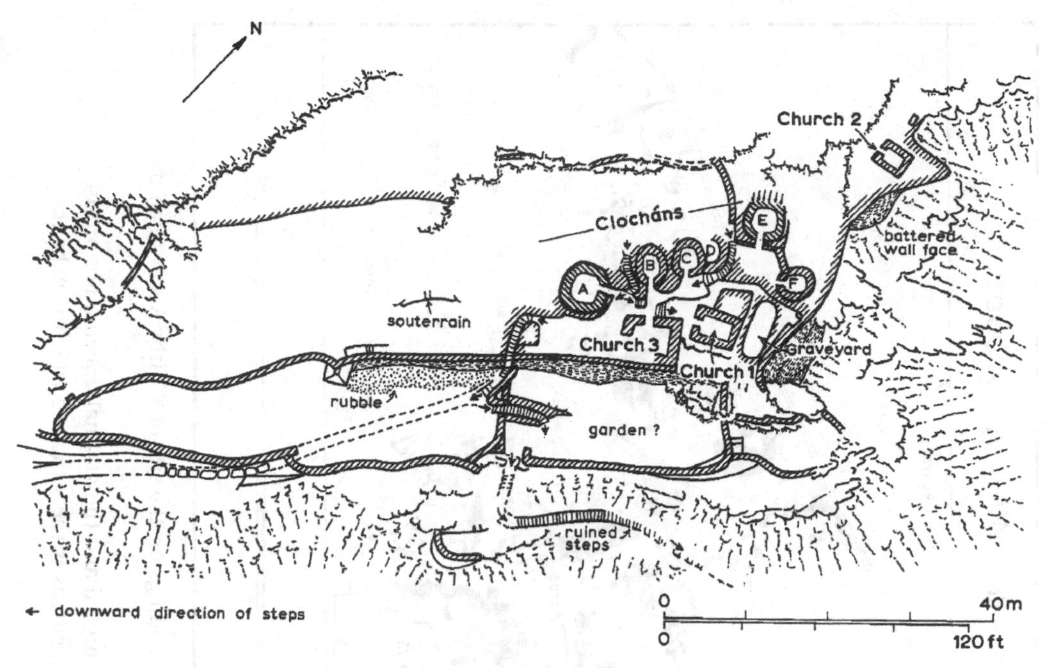

FIGURE 12 Skellig Michael (County Kerry), plan of monastery (drawing by Dr Nancy Edwards)
Note the small *clocháns* or individual cells in the shape of beehives. For a full study, see Horn, White Marshall, Rourke et al. (1990).

that the celestial visitor had searched the entire monastery and found only such small buildings as are described in this passage suggests that the layout of the abbey was different from that of Monkwearmouth and Jarrow, involving as it did great communal buildings, and that it was much more like that of Skellig Michael. No significant archaeological remains have been found at Coldingham; but on the site of the abbey of Whitby extensive remains of small early medieval buildings, or possibly rooms within larger buildings, have been found on the north side of the medieval abbey-church. These have been interpreted as workshops (because of the craft objects found in them), but also as the sort of *domunculae* or *casae* referred to in the account of Coldingham (fig. 13). It should be noted, however, that since no early medieval church has been found at Whitby it is not possible to be certain what part of the monastic complex is comprised in the excavated area. It is possible that it was a craft area rather than the religious heart of the monastery; and the same possibility exists for the excavated portion of the Northumbrian monastery of Hartlepool, where small individual buildings

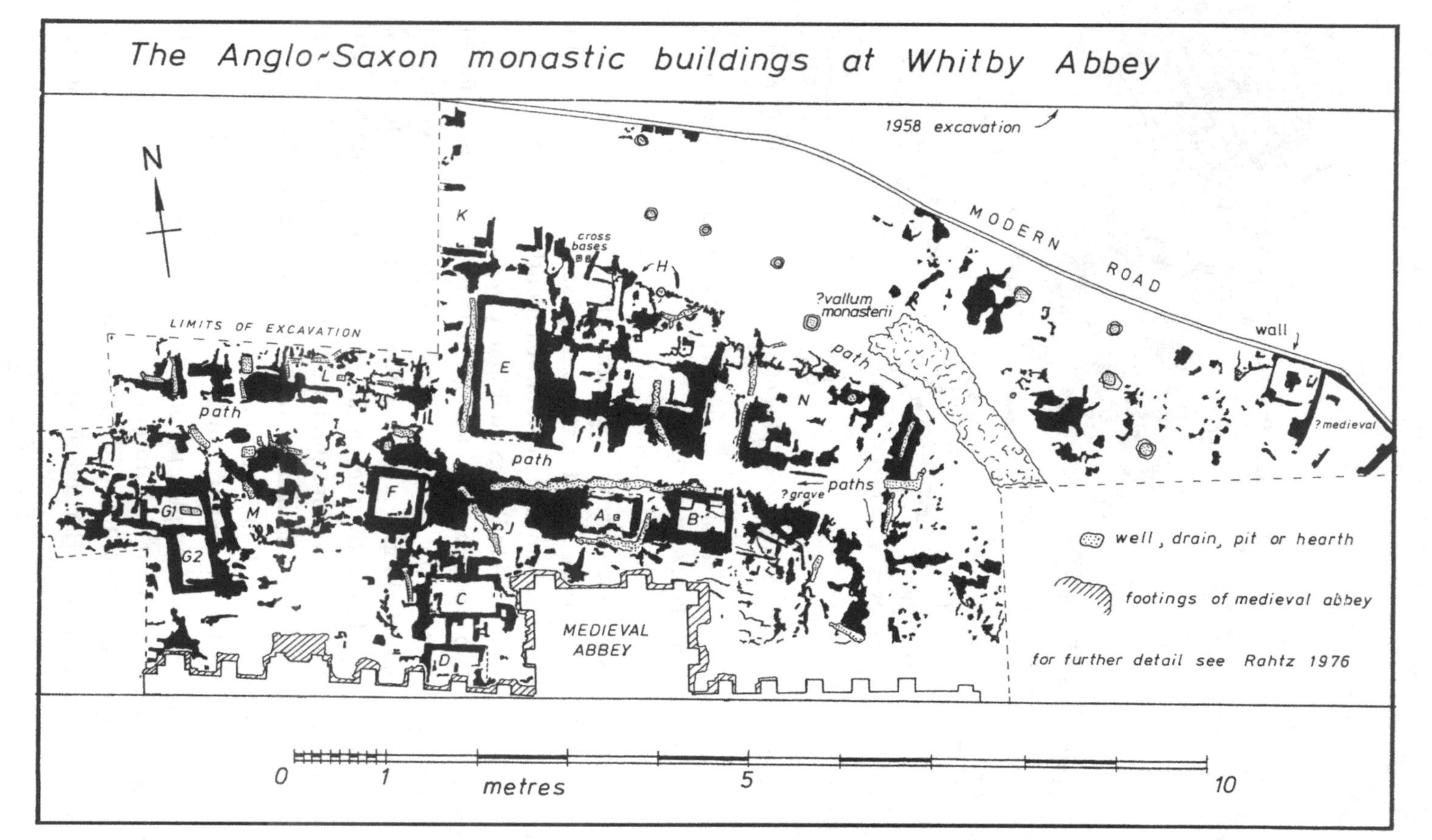

FIGURE 13 Whitby (Yorkshire North Riding), the Anglo-Saxon monastic buildings on the site of Whitby Abbey (plan by P. A. Rahtz, from Rahtz (1995), fig. 2)
Although the archaeological recording by Charles Peers in 1920–8 was confused and the original drawing not precise stratigraphically, the individual buildings (lettered alphabetically) are evident, as is the structure which may be the the *uallum monasterii* (monastic enclosure).

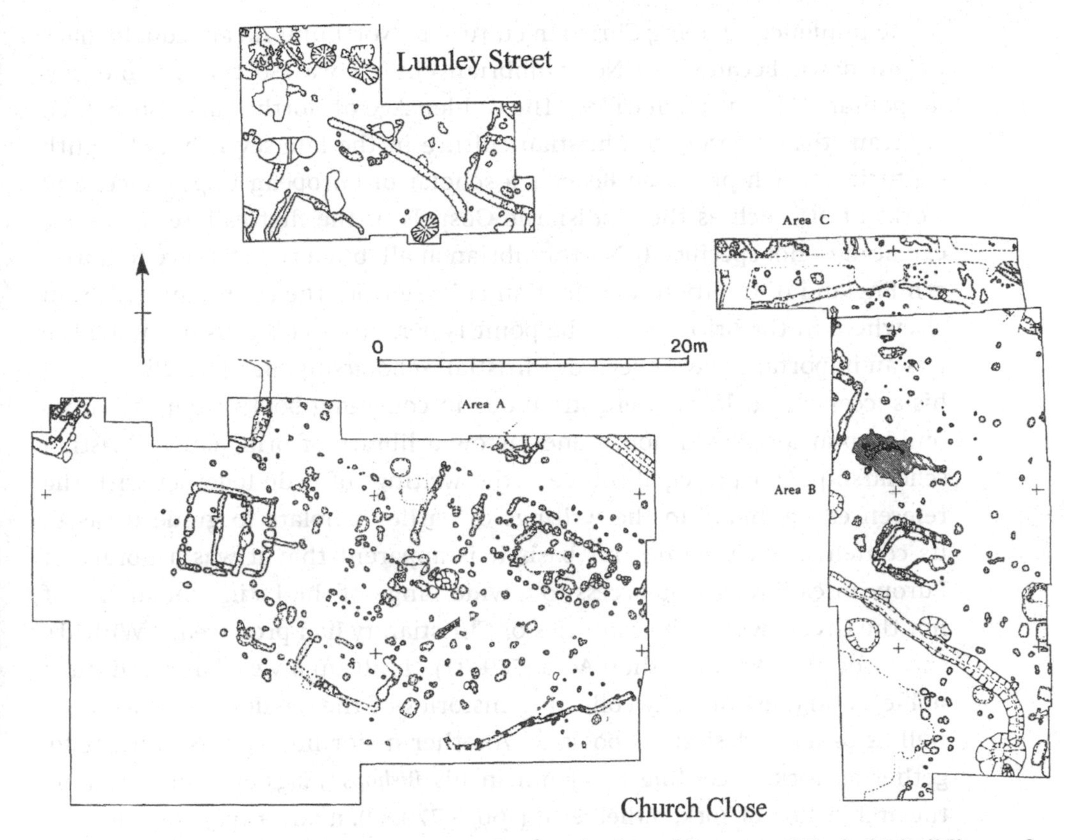

FIGURE 14 Hartlepool (County Durham), plan of features of Period I buildings of early monastery (mid-seventh–mid-eighth century), excavated at Church Close
Note the individual buildings interpreted as workshops and living quarters, and the boundary complex (A2083) interpreted as the *uallum monasterii* (monastic enclosure).

have also been recovered (fig. 14).[74] Irish influence may, however, have been part only of the picture. Hermits were prominent in Egytian monasticism, and the *Life* of the most famous, St Anthony, was influential in the west and indeed known in Ireland – it may in fact have reached Lindisfarne via Iona. In this too we find the themes of austere life, miraculous powers and influence over nature through holiness. It is entirely possible that such influences, transmitted perhaps via Gaul where Egyptian monasticism was very influential on the founder of Lérins, Cassian – the monastery where Benedict Biscop had stayed for a protracted period.[75]

[74] The original report on the 1920s Whitby excavations is Peers and Ralegh Radford (1943); for modern criticism see Rahtz (1995) and Cramp (1993). On Hartlepool, see Daniels (1988).

[75] Chitty (1966); on the *Life* of Anthony in Ireland, see Bullough (1964), pp. 126–30; on Lérins, see Nora Kershaw Chadwick (1955), pp. 142–69.

The influences shaping Christian culture in Northumbria are equally hard to pin down because the Northumbrian church was so eclectic and cosmopolitan. What is often called 'The Golden Age of Northumbria', by which is meant the flowering of Christian culture in the late seventh and eighth centuries, which produced Bede as a scholar of European importance, and works of art such as the Lindisfarne Gospels of the first calibre, was – we can argue – not specifically Northumbrian at all, but a transference to northern England of traditions of Christian culture from the continent and from elsewhere in the British Isles. The point is perhaps an obvious but none the less an important one. As regards Christian scholarship, Benedict Biscop and his successor, Ceolfrith, were strenuous in collecting books from abroad to build up at Monkwearmouth and Jarrow a library of principally Christian scholarship. Our principal source is the writings of Bede together with the references he makes to the writings of earlier scholars to guide us as to its contents; but even on this basis it is apparent that it was a library of European calibre, giving access to a wide range of the Latin scholarship, if not the Greek, which the centuries of Christianity had produced.[76] Wilfrid's successor at Hexham, Bishop Acca (709–31), 'built up a very large and most noble library, assiduously collecting histories of the passions of martyrs as well as other ecclesiastical books'.[77] Another major library was brought together at York. According to Alcuin in his *Bishops, Kings and Saints of York*, the archbishop of York, Æthelberht (766/7–779/80), made a very considerable collection of books, comprising:

> all the Roman possessed in the Latin world,
> whatever famous Greece has transmitted to the Latins,
> draughts of the Hebrew race from Heaven's showers,
> and what Africa has spread abroad in streams of light.

Alcuin gives an impressive list of writers whose works were in the collection including: church fathers such as Jerome, Ambrose, Augustine and Gregory the Great; classical writers such as Pompey, Pliny, Aristotle, Virgil and Cicero; the English scholars Aldhelm and Bede, and other Christian writers and poets such as Basil, Cassiodorus, John Chrysostom and Sedulius. The full list is immensely impressive, but, since virtually none of these books has survived, we cannot be sure that Alcuin is really giving us a library list rather than a rhetorical evocation which should not be taken as literally accurate. Nevertheless, the library of York was evidently impressive in the

[76] Laistner (1935). [77] Bede, *Eccles. Hist.*, bk 5 ch. 20.

eighth century, since its resources were sought after to supply copies on the continent. Its contents were equally clearly a reflection of the wider world of Christendom.[78]

The scholarship which these libraries underpinned was equally part of that wider world. Bede was the dominant figure, but there is every reason to suppose that he was part of a learned group in Northumbria, and the list of his works which he supplies at the end of the *Ecclesiastical History* gives us an indication of what this scholarship was about. It comprises first commentaries on the books of the Bible (the most numerous of his works) and also some homilies on the Gospels. Here Bede took forward the commentary (or exegesis as it is technically called) of the early church fathers, especially Augustine of Hippo but also the sixth-century Isidore of Seville. The originality of his work in detail has been increasingly appreciated by scholars; but in outline it was firmly in a long Christian tradition. Secondly, Bede wrote works on science, especially the science of computing the calendar of the Christian church. He refers in the list of his writings in the *Ecclesiastical History* to 'two books, one on the nature of things and one on time; also a larger book on time'. These are, first, his short work *On the Nature of Things* (*De natura rerum*), a discussion of phenomena such as comets, tides, the phases of the moon and so forth, based on the Roman writers Pliny and Isidore; and, secondly, his *On Time* (*De temporibus*) and *The Reckoning of Time* (*De temporum ratione*). In these latter two books, Bede confronted a problem which had occupied the best mathematical and scientific minds in the classical world since at least the time of the mathematician Meton in around 430 BC. This was the problem of devising a reliable calendar, and in particular of relating the movement of the sun to the movement and phases of the moon.[79]

It is worth pausing to ponder the full implications of the Easter date problem, which was one of especial difficulty. This was because, in common with the calendars of many early peoples, the Jewish calendar was a lunar and not a solar one. Thus the feast of the Passover, on which the date of Easter depended since it was of course the day on which the Last Supper had occurred, had a lunar and not a solar date. (In fact, it had a solar date too – 27 March which appears in church calendars – but Christians nevertheless persisted in celebrating Easter according to the lunar date which was of course movable relative to the solar calendar.) The rules for calculating the

[78] *Sources*, pp. 150–2 (A.4.25 = Alcuin, *BKSY*, lines 1531–62, and see also notes to pp. 120–7).

[79] Bede, *Eccles. Hist.*, bk 5 ch. 24. For general discussion see George Hardin Brown (1987) and Ward (1990). On Bede's scientific and computational work, see Stevens (1986) and C. W. Jones (1943)

date of Easter could be derived from the book of Exodus in relation to Passover, and they are expressed with great clarity in the letter which Bede's abbot Ceolfrith wrote to King Nechtan of the Picts, and which Bede included in the *Ecclesiastical History*. Put very simply, they were, as they still are, that Easter must be celebrated on the first Sunday after the first full moon after the vernal equinox. This was because the feast of Passover was in the Jewish (lunar) month Nisan, which was defined as beginning with the first moon after the vernal equinox, and Easter had to be on a Sunday. The difficulty of course lay in predicting the phases of the moon so that the date of Easter could be fixed in advance, partly because it was important that all Christendom should celebrate the principal Christian feast on the same day, partly because so much of the Christian year was tied to Easter – it was necessary, for example, to know when to begin Lent. From a mathematical and astronomical point of view, the whole problem was fraught with difficulties, and there was a series of controversies in the Christian world about how these difficulties were to be overcome. The best solution was to use an eighty-four year cycle for predicting the date of Easter, so that in effect the movements of the moon and the sun could be predicted over that period. This cycle was refined and given authority by the Alexandrian scholar, working in Rome, Dionysius Exiguus, and it was essentially his work which Bede drew on to produce *The Reckoning of Time*, which was such a clear and well-developed text-book on the whole subject of the computation of time that it remained the principal source of reference throughout Latin Christendom until the end of the middle ages. Bede not only drew on a Christendom-wide tradition of scholarship; he influenced its subsequent development.[80]

There were other developments of Christian scholarship in Northumbria which were to a greater or less degree innovative. In hagiography (writing about saints), although Northumbrian work was firmly within a Christendom-wide tradition, certain aspects are not easy to parallel at this date. The form of the *Lives of the Abbots of Monkwearmouth and Jarrow*, which comprises a series of biographies with no accounts of miracles such as we might expect from other hagiography, may be influenced by the hagiography of the Gaulish monastery of Lérins, but it may to an extent be innovative. So too may Stephen's *Life of Wilfrid*, which, although probably drawing on a tradition of hagiographical writing going back to the *Life of Saint Columbanus* by Jonas of Bobbio, also appears out of the ordinary in its biographical and

[80] The letter to Nechtan is Bede, *Eccles. Hist.*, bk 5 ch. 21; for Bede's work in translation, see Bede, *Reckoning*, which also has general discussion. See also C. W. Jones (1994).

almost legalistic approach.[81] Bede did, however, make one real innovation in this tradition. The 'martyrology of the birthdays of the holy martyrs', which he lists among his works, has survived, albeit in a much expanded and modified form; and this is so precisely because it formed the basis of subsequent martyrologies of this type in western Europe, including those of Florus and Usuard. Its characteristic was, as Bede himself expresses it, that it gave brief accounts of the various saints rather than just specifying the area from which they came and their feast-day as was the case with the earlier martyrology attributed to the fourth-century scholar Jerome (the so-called Hieronymian Martyrology). Bede was thus literally the creator of a new genre, the so-called 'historical martyrology', which was highly influential in subsequent centuries.[82]

Nor should we underestimate Bede's innovation in his historical writing. Not only was his work remarkable in helping to establish the use of *Anno Domini* dating as opposed to that derived from the Roman system of fifteen-year tax cycles or Indictions, and other methods;[83] but it is important to emphasize how original a work the *Ecclesiastical History* was in other respects. Aside from Gregory of Tours's *History of the Franks*, completed in the late sixth century, western Europe had really seen no historical composition on this scale or to this degree of sophistication since Orosius's *History* in the fifth century, a work produced in a Roman classical context under the influence of no less a figure than Augustine of Hippo. As recent scholarship has emphasized, no previous work was quite like the *Ecclesiastical History* in seeking to present the ecclesiastical history of a people, as opposed to Christendom in general.[84] None the less, with Bede – and with other Northumbria writers like Stephen and the anonymous author of the *Life of Gregory the Great* – we are not dealing with anything characteristically Northumbrian. These writers drew on, and were a product of, Latin learning, with Bede deploying, for example, the rhetorical devices of Cicero as if they were as natural to him as to their originator,[85] and drawing so deeply at the wells of Christian patristic scholarship that he clearly thought of himself as directly in line with scholars such as the church fathers Jerome and Augustine, as indeed to all intents and purposes he was.

The ability to reproduce and develop the culture and learning of the Latin and classical world in Northumbria is most apparent in the most famous

[81] David Rollason (1989a), pp. 60–82, citing *inter alia* Thacker (1977), chs. 4–5. On Stephen's work, see Foley (1992).

[82] Quentin (1908); see also Dubois (1978).

[83] Harrison (1976).

[84] Markus (1976) and Tugène (1982). See now also Tugène (2001a).

[85] Ray (1987, 2002).

manuscript produced at Monkwearmouth and Jarrow, the book now called the Codex Amiatinus.[86] It is written in a style of handwriting called uncial, a thoroughly Roman handwriting, used in the Roman world and derived ultimately from Roman inscriptions. Its execution in the Codex Amiatinus is so accomplished that it never occurred to anyone until relatively recently that the book was from anywhere but the Mediterranean area. This seemed to be confirmed by the contemporary dedication inscription in it (ill. 9), which may be translated:

> To the *monastery* of the sublime *Saviour* justly to be revered, whom ancient faith declares head of the church, I, *Peter of the Lombards*, abbot from the farthest ends of *the earth*, send pledges of my devoted affection; desiring that I and mine may ever have a place amidst the joys of so great a father, a memorial in heaven.

Scrutiny of illustration 9 shows that this inscription has been tampered with, the words, printed here in translation in italics, having been written over erasure in a different hand and a different ink. The key to restoring the inscription to its original form is provided by the *Life of Ceolfrith*, published for the first time shortly after the Codex Amiatinus began to attract scholarly attention, and which contains an account of how Abbot Ceolfrith set out to travel to Rome after he had resigned his abbacy. According to this, the abbot had caused to be transcribed three Bibles, two of which he left at Monkwearmouth and Jarrow, the third he resolved to take on his post-retirement journey to Rome to give to St Peter's. In the event, he never completed that journey, for he died at Langres in Burgundy, and it was left to some of his followers to take to the pope his presents, among which was the Bible in question. According to the author of the *Life*, this had the following verses at the beginning:

> To the *body* of the sublime *Peter* justly to be revered, whom ancient faith declares head of the church, I, *Ceolfrith*, abbot from the farthest ends of *England*, send pledges of my devoted affection; desiring that I and mine may ever have a place amidst the joys of so great a father, a memorial in heaven.[87]

This is the inscription in the Codex Amiatinus, except that the words shown here in italics identify the donor as Ceolfrith and the church to which the book was destined as St Peter's in Rome. Not only do these words better fit the space available on the page than those now in the inscription, but they also make better sense. Viewed from Rome, Ceolfrith was certainly

[86] For what follows, see the brilliant exposition of Bruce-Mitford (1968).
[87] *Life Ceolf.*, paras. 35–7.

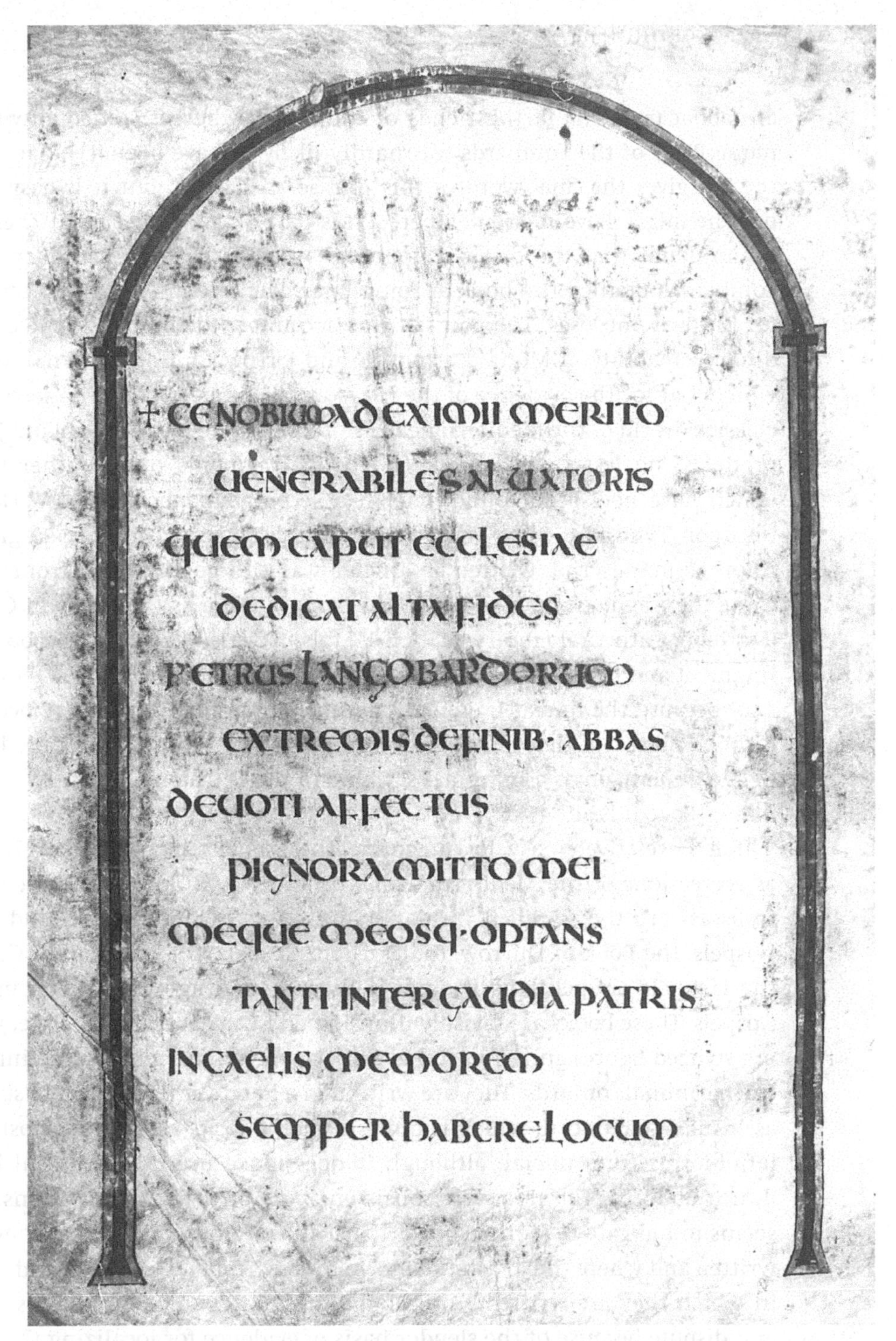

+ CENOBIUM AD EXIMII MERITO
UENERABILE SALUATORIS
QUEM CAPUT ECCLESIAE
DEDICAT ALTA FIDES
PETRUS LANGOBARDORUM
EXTREMIS DE FINIB· ABBAS
DEUOTI AFFECTUS
PIGNORA MITTO MEI
MEQUE MEOSQ· OPTANS
TANTI INTER GAUDIA PATRIS
IN CAELIS MEMOREM
SEMPER HABERE LOCUM

ILLUSTRATION 9 Codex Amiatinus: dedication by Ceolfrith, abbot of Monkwearmouth and Jarrow (Florence, Biblioteca Medicea Laurenziana, MS Laur. Amiatino 1), fol. 1v

an 'abbot from the farthest ends of the earth', whereas someone with the name Peter of the Lombards was hardly likely to have been. (The author of the *Life* gives the final word of this phrase as *England*, but it is easy to see how he might have misremembered this part of the dedication.) There can be no doubt that the original donor was Ceolfrith, and that Peter of the Lombards obtained the book at some later date and altered the inscription for his own purposes. The point of this account is to emphasize the extraordinary capability of Monkwearmouth and Jarrow to produce a manuscript which, but for the evidence of the *Life*, was taken for a book produced in the classical world of the Mediterranean. Moreover, Monkwearmouth and Jarrow produced no less than three such books, fragments of the other two of which have now been found, one in a second-hand bookshop in Newcastle upon Tyne, the other in the roof of a cottage in southern England.[88] Another manuscript written in uncial was the gospel book from which came the fragment now bound in the same volume as the Durham Gospels (Durham, Cathedral Library, MS A.II.17, fols. 103–11). We know that both this fragment and the Durham Gospels were at Chester-le-Street in the tenth century because the uncial fragment contains a marginal scribble referring to Bishop Ealdred of that church (? before 946–?968) and a priest called Boge, in the same hand as a marginal scribble in the Durham Gospels which also names Boge (ills. 10–11).[89]

If it is relatively easy to categorize these books in uncial handwriting as recreations of the Mediterranean world in Northumbria, more complex to assess are the so-called Insular manuscripts, including the Lindisfarne Gospels, the Book of Durrow, the Durham Gospels, the Echternach Gospels, the Lichfield Gospels, the Book of Kells and the fragmentary Corpus-Otho Gospels. These books are lavishly illuminated in a characteristic style, involving stylized figures and extensive use of ornamental patterns, sometimes featuring animals or birds. They are written in a handwriting known to scholars as Insular half-uncial, which is quite different in its shape and most of its letter-forms from uncial, although it does make use of the uncial letters d, n, r and s.[90] The relatively consistent style of their illuminations alone seems to indicate that these books constitute a group; but where they were written and where the artistic style which they embody – or the handwriting in which they are written – originated are matters of scholarly discussion and dispute because of the slender basis of evidence for localizing the books and defining their histories.

[88] Webster and Backhouse (1991), no. 87a–c.

[89] Brown, Verey and Coatsworth (1980); see below, pp. 245–7.

[90] Dumville (1999), pp. 1–2.

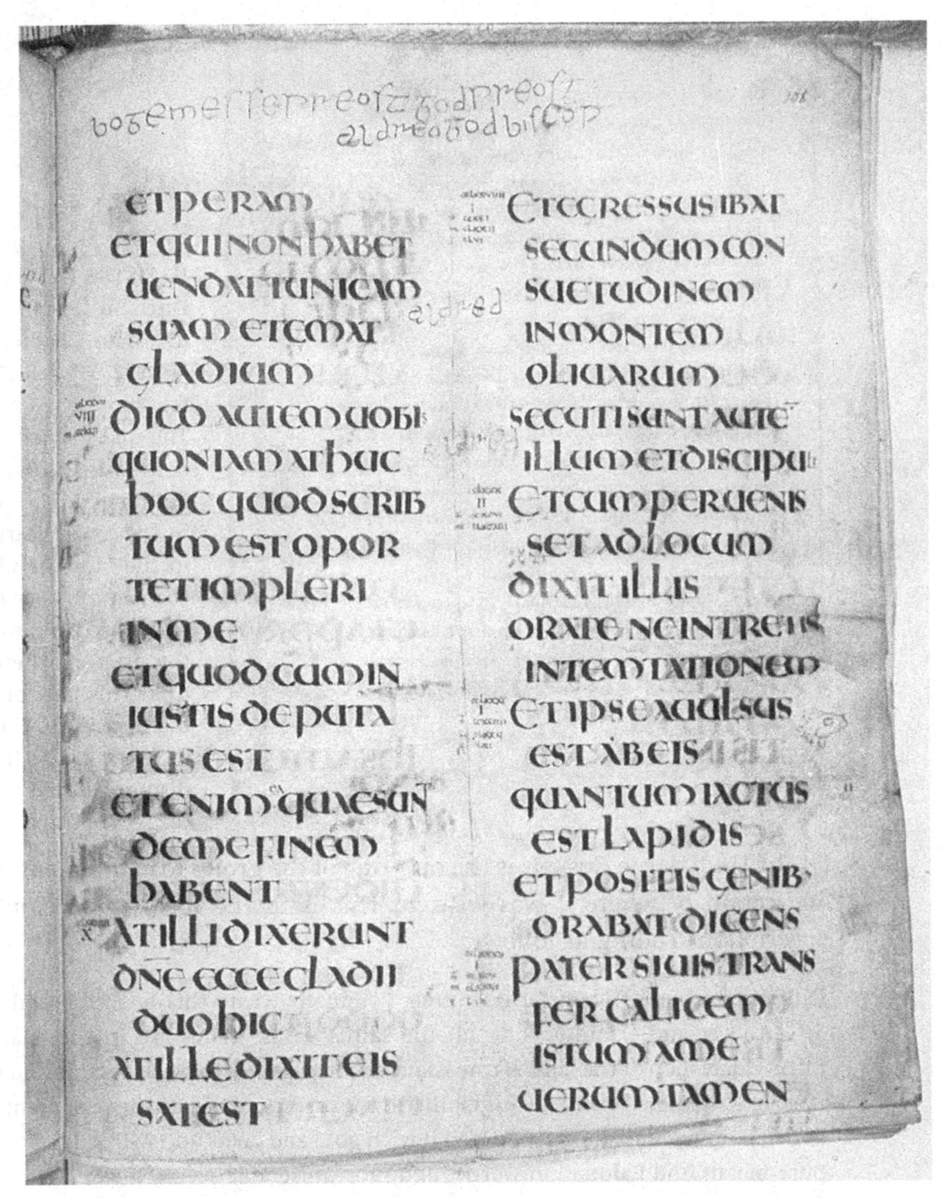

ILLUSTRATION 10 The Durham Gospels (Durham Chapter Library, MS A.II.17), fol. 106r
The main script is uncial like the Codex Amiatinus (see ill. 9). The scribble in the top margin reads 'Boge the priest, a good priest; Ealdred a good bishop'. Note the name Ealdred repeated in the central margin.

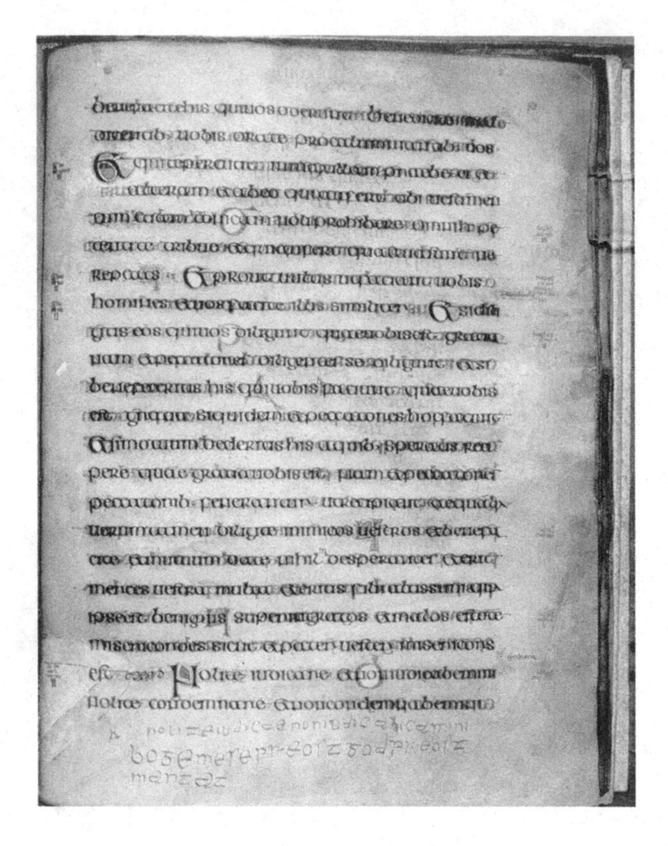

ILLUSTRATION 11 The Durham Gospels (Durham Chapter Library, MS A.II.17), fol. 80r
The main script of this section is Insular half-uncial. The scribble in the lower margin is in the same hand as that on fol. 106r (ill. 10) and also refers to Boge the priest. It therefore links this part of the manuscript to Chester-le-Street in the time of Bishop Ealdred.

The Lindisfarne Gospels is the only one of the group to contain any explicit indication of where it was written. This is in the form of a colophon or inscription reading as follows:

> Eadfrith, bishop of Lindisfarne church, originally wrote this book for God and for St Cuthbert and – jointly – for all the saints whose relics are in the island. And Ethilvald, bishop of the Lindisfarne islanders, impressed it on the outside and covered it as he well knew how to do. And Billfrith the anchorite forged the ornaments which are on it on the outside and adorned it with gold and gems and with gilded-on silver – pure metal. And Ealdred, unworthy and most miserable priest, glossed it in English between the lines with the help of God and St Cuthbert.

Both the handwriting of this colophon and its reference to Ealdred fix its date, for Ealdred was provost of the church of Chester-le-Street in 970; and his gloss (or interlineated translation) of the gospels into English is indeed found in them, in a fine tenth-century hand, identical to that of the colophon. The person he credits with writing the book was Bishop Eadfrith of Lindisfarne (d. 721); and the person said to have been responsible for the binding (which is now lost) was Æthelwald, his successor as bishop from 721 to 740. Billfrith is otherwise unknown, but his name does appear in the list of names of persons to be commemorated at Lindisfarne, the *Liber Vitae*. This was written down in the ninth century, but the position of Billfrith's name in the list makes it probable that he flourished in the eighth century and to that extent corroborates the colophon in the Lindisfarne Gospels.

There is no way of knowing how well informed Ealdred was when he wrote this colophon two and a half centuries after he believed the book to have been written. It can be urged on behalf of the accuracy of his information that the sheer obscurity of his reference to Billfrith suggests that he really did have before him some authentic information, perhaps a label in the binding now lost; and that the reference to 'all the saints whose relics are in the island' might also have derived from an original inscription or label written while the book was still on Lindisfarne rather than centuries later at Chester-le-Street when the saints were no longer 'in the island'.[91] Moreover, the fact that the Lindisfarne Gospels was at Chester-le-Street in the tenth century tends to give credence to the colophon. According to the *History of St Cuthbert*, the church of Lindisfarne was evacuated in 875 under the threat of Viking attack and transferred to Chester-le-Street for over a century. The presence of the book there must therefore create a presumption that it had come from Lindisfarne. In 995, the church and its possessions were transferred again, this time definitively to Durham; and it is clear that the Lindisfarne Gospels came to Durham, because there the early twelfth-century Durham historian Symeon saw the book and related a tradition concerning how, when the members of the church of Lindisfarne were wandering across northern England after abandoning their original home, it was accidentally dropped in the sea, being miraculously recovered unharmed apart from some stains of sea-water.[92]

Against all this, however, it can be argued that the colophon is too late in date to be admissible evidence, and represents only a legend preserved or perhaps invented at Chester-le-Street. Moreover, the surviving stone sculpture from Lindisfarne is not of a standard consistent with the artistic achievements of the Lindisfarne Gospels and therefore does nothing to support a Lindisfarne provenance for that book; even the lettering of some of the early memorial stones recovered from the site has only very general

[91] For the colophon and commentary on it, see Kendrick, Brown, Bruce-Mitford et al. (1956–60), II, p. 10; for a revised text, see Dumville (1999), pp. 76–7; on Ealdred, see Bonner (1989), pp. 392–5; for the entry for Billfrith in the *Liber Vitae*, see Gerchow (1988), p. 305.

[92] *Hist. Cuthbert*, para. 20; Symeon, *Origins*, bk 2 ch. 12; see also David Rollason (1987).

ILLUSTRATION 12
Codex Amiatinus, the scribe Ezra writing (Florence, Biblioteca Medicea Laurenziana, MS Laur. Amiatino 1), fol. Vr

resemblances to that in the Lindisfarne Gospels. Nor are there similarities between the incised wooden coffin of St Cuthbert, apparently made at Lindisfarne in 698, and the gospels.[93]

Even if we were to dismiss the evidence of the colophon, however, we should still be arguing for a Northumbrian origin for the book since the illumination of St Matthew at the beginning of that evangelist's gospel seems so closely related to an illumination of the scribe Ezra in the Codex Amiatinus that it is hard to believe that there was not a close connection between them. It is possible to argue, in fact, that the Lindisfarne Gospels is easier to interpret as having been produced at Monkwearmouth and Jarrow (ills. 12–13). This would explain the similarity of the Matthew and Ezra

[93] Dumville (1999), p. 78. On the sculpture, see Cramp (1984), pp. 194–208. See also Cramp (1989), pp. 221–6 ('If one looks for fine quality Insular elements in early sculpture one finds them not amongst the surviving pieces from Lindisfarne but at Monkwearmouth', p. 221). On the coffin, see Kitzinger (1965).

ILLUSTRATION 13
Lindisfarne Gospels, the evangelist Matthew (London, British Library, MS Nero D.IV), fol. 25b

illuminations, and it would also explain why the text of the Lindisfarne Gospels is so close to that of the Codex Amiatinus. It would also fit with the development of handwriting at Monkwearmouth and Jarrow, although it would have to be dated in or after the mid-eighth century, rather than the early eighth-century dating suggested by the colophon, if its origin were to be assigned to that church and it were to fit in with the known chronology of handwriting there.[94] On the other hand, there were close links between Lindisfarne and Monkwearmouth and Jarrow, shown for example by Bede's authorship of his *Life of St Cuthbert* and the contacts with Lindisfarne which he acknowledges in the preface to that work.[95] Such uncertainties and complexities, however, should not obscure the fact that, whether at Lindisfarne in the early eighth or at Monkwearmouth and Jarrow in the mid-eighth century or later, it does seem as clear as anything can do that we are dealing with a product of the Northumbrian church.

[94] Dumville (1999), pp. 64–80. [95] Bede, *Life Cuth.*, preface.

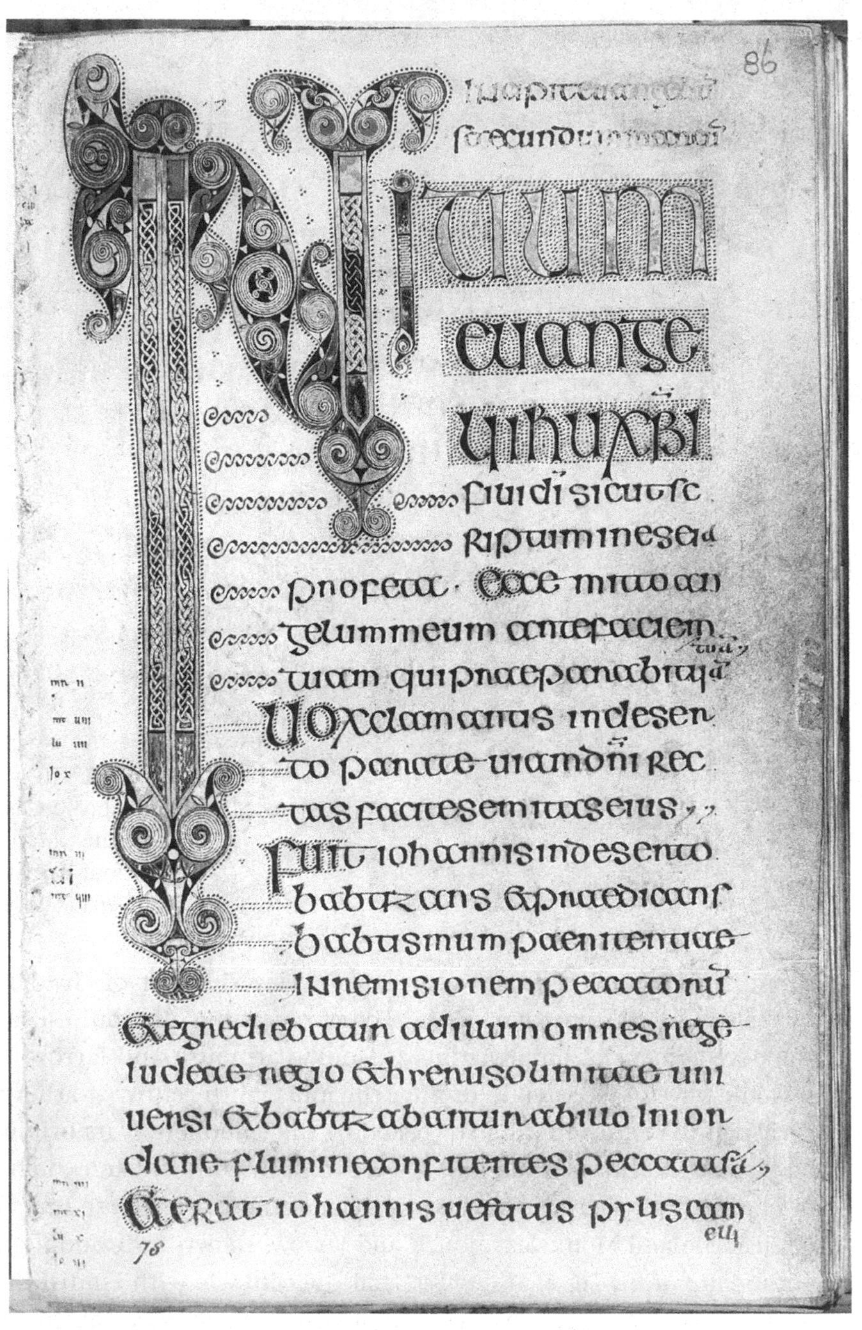

ILLUSTRATION 14 Book of Durrow, fol. 86, beginning of Gospel of St Mark
The page is the beginning of the Gospel of St Mark ('Initium euangelii Iesu Christi, filii Dei'). Note the 'diminuendo' of the lettering by which it becomes progressively smaller from the very large IN to the third line of text; the spiral and interlace decoration of the IN, and the use of red dots as a decorative feature behind the first three lines of text. Compare ill. 15, where all these features are found in the parallel page in the Lindisfarne Gospels.

ILLUSTRATION 15 Lindisfarne Gospels, beginning of Gospel of St Mark (London, British Library, MS Nero D.IV), fol. 95r
Compare with ill. 14.

Another gospel book which was definitely with the Lindisfarne Gospels at Chester-le-Street in the tenth century, as we have seen, was the Durham Gospels (ill. 11). These presumably also moved to Durham when the community itself moved, and have remained there ever since. That they were produced in Northumbria is further suggested by the fact that the same scribe who corrected parts of the Lindisfarne Gospels is thought also to have corrected them.[96] Even so, doubts are admissible. The scribe of the Durham Gospels was either identical to or very closely related to the scribe who wrote a single page in Insular half-uncial in the Echternach Gospels which were in the middle ages at the monastery of Echternach (Luxembourg), founded by the Northumbrian missionary Willibrord. The parchment on which the Echternach Gospels and some of the Durham Gospels are written was prepared in the manner used on the continent rather than in the British Isles. Were both manuscripts produced on the continent in the context of the English mission there? Were both produced in Northumbria but under some sort of continental influence? Did the scribe produce one in Northumbria and one on the continent? No certainty is possible.[97]

In the case of the other manuscripts of the group, the clues available to us are equally problematic. A colophon shows that in the eleventh and twelfth centuries the Book of Durrow was at the monastery of that name in Ireland; but the resemblances of its decoration on the one hand to various objects from the treasure in the early seventh-century ship-burial at Sutton Hoo, and on the other to the Lindisfarne Gospels and related books associate it with England, and perhaps more particularly with Northumbria (ills. 14–15). Durrow was a foundation of St Columba, closely linked with Iona which had close Northumbrian links, and it is by no means impossible that the Book of Durrow was produced in Northumbria and taken to Durrow, perhaps via Iona, in the face of Viking attacks in just the same way as the Lindisfarne Gospels ended up in Chester-le-Street and then Durham. But equally it might have been produced in Iona under Northumbrian influence.[98] There is, however, no certainty about this or about its date, often given as *c.* 675 on the grounds that it is simpler and therefore should be earlier than the Lindisfarne Gospels.

The Book of Kells, now in Trinity College Dublin, is much more complex in execution, but it shares features of its text with the Book of Durrow and

[96] Fols. 80r, 106; facsimile: Brown, Verey and Coatsworth (1980). For the corrector, see Christopher D. Verey in T. J. Brown (1972), pp. 243–5.

[97] Bruce-Mitford (1989) and Dumville (1999), pp. 93–4.

[98] George Henderson (1987), pp. 19–55; see also Neuman de Vegvar (1987), pp. 83–105, where a provenance at Iona is suggested.

is clearly related to it in some way. There has been no certainty as to where and when was it produced. The arches over its canon tables (that is tables representing the relationships of the four gospel texts) have been compared with those in manuscripts produced at the court of Charlemagne, which if correct would place its production *c.* 800 at the earliest. Other authorities have argued for a date around 750. As for its place of origin, the book was at Kells in Ireland from the late eleventh century, when property transactions were recorded in it. Scholars have been reluctant to accept that it was made there because of its close relationship with the other books in this 'Northumbrian' group: maybe, they have suggested, it was made at an Irish church in close contact with Northumbria, Iona for instance, or even at a Pictish church under strong Northumbrian influence.[99] None of the arguments advanced is more than plausible. We know that many Northumbrians went to Ireland; why should not Kells have been equally under Northumbrian influence and therefore be a plausible place of origin? As for the other books in the group, we are even more in the dark. The Lichfield Gospels was definitely in Wales in the tenth century when documents were copied into its margins, so it is possible that it originated in the English midlands rather than in Northumbria.[100]

All this uncertainty, however, should not blind us to the immense significance of the developments which these books represent: developments in the text of the Bible, in the art of illumination and in handwriting. Indeed, the uncertainty should paradoxically be a source of enlightenment and not of frustration, since it stems precisely from the fact that these books are so eclectic and cosmopolitan and that they also represent considerable advances in the areas just mentioned. In the art, first of all, we are seeing a highly sophisticated fusion of different elements. These include: the patterns developed on British and Irish metalwork which are so evident in the so-called 'carpet' pages of the manuscripts; the stylized representation of forms which derive, like the symbol of Matthew in the Book of Durrow, from metalwork or like the eagle in the same book or in the Corpus-Otho Gospels from Pictish stone-carving; and the inspiration of imported oriental figure-silks which may lie behind some of the animal and bird patterns which are characteristic of these manuscripts.[101] We cannot be certain about where the style comprising these elements originated. Maybe it was adumbrated

[99] See, for example, George Henderson (1987), pp. 131–98, and O'Mahony (1994); for a useful summary see Bernard Meehan (1994a), pp. 90–2.
[100] George Henderson (1987), pp. 122–9.
[101] George Henderson (1987), and Neuman de Vegvar (1987), pp. 168–203.

ILLUSTRATION 16 Cathach of St Columba (Dublin, Royal Irish Academy, s.n.), fol. 6r This is the beginning of Psalm 36 ('Noli emulari in malignantibus, neque zelaueris facientes iniquitatem'). Note the diminishing size ('diminuendo') of the lettering and compare ills. 14–15.

in the much simpler decoration around a cross in a manuscript now in Dublin, Trinity College, A.4.15 (55), the so-called Codex Usserianus Primus, which has nevertheless some of the same features; or in the Cathach of St Columba (Dublin, Royal Irish Academy, s.n.), where the 'diminuendo' effect of the handwriting is reminiscent of that in the Lindisfarne Gospels and related manuscripts (ill. 16, and compare ills. 14–15). The origins of the Codex Usserianus Primus and the Cathach of St Columba are themselves obscure, however: possibly Ireland, possibly a church on the continent.[102] What is important to us is that Northumbria fully shared in the style and, to judge

[102] Neuman de Vegvar (1987), pp. 73–5; Nordenfalk (1977), pp. 11–13; for discussion of origins, see Dumville (1999), pp. 35–9.

only from the Lindisfarne Gospels and the Durham Gospels, carried it forward in various ways. These included combining it with influences from Mediterranean art in the evangelist portraits in the Lindisfarne Gospels which are notably classical in style. We have already noted the resemblance between the picture of Matthew and the miniature of Ezra in the Codex Amiatinus, which would seem to be closely modelled on a Mediterranean manuscript, almost certainly the Codex Grandior which the late Roman scholar Cassiodorus had at his monastery at Vivarium in southern Italy.[103]

Much the same admixture of Mediterranean influence is observable in the Bible texts in the books. These show variation between the 'Old Latin' Bible, used in the Irish church until shortly before the production of these manuscripts, and the purer, more up-to-date Vulgate Bible, of which the Codex Amiatinus is a copy, derived almost certainly from a book brought by Abbot Ceolfrith from Italy. The Book of Durrow has 'Old Latin' summaries but a Vulgate text, for example, whereas the Durham Gospels and the Lindisfarne Gospels have full Vulgate texts.[104] Other aspects of the books also point to Mediterranean or continental influence, for example, the saints' cults from Naples (especially that of St Januarius) which are noted in the preliminary materials in the Lindisfarne Gospels.[105]

Finally, we see a considerable advance in Insular half-uncial handwriting in these books. This probably originated in Ireland if the prayer-book called the Cathach of St Columba is correctly associated with that saint and if the wooden tablets recovered from Springmount Bog in Ireland, on which the handwriting occurs, were indeed written (as seems likely) in Ireland itself. To judge from the Durham Gospels and the Lindisfarne Gospels, however, what happened in Northumbria was that this handwriting was modified to produce a more formal disciplined handwriting under the influence of the Roman uncial handwriting in which the Codex Amiatinus and other books were written. Minuscule (lower case) forms of handwriting were also developed as less grand versions of Insular half-uncial.[106]

A similar fusion and development of forms in Northumbria within an eclectic and cosmopolitan culture is observable in the relatively numerous carved stone crosses of the pre-Viking Northumbrian church. The three-dimensional, humanistic figure-sculpture found on them derives from the

[103] Bruce-Mitford (1968), pp. 11–17 (repr. pp. 199–205).
[104] Verey (1989); George Henderson (1987), pp. 21–2.
[105] Kendrick, Brown, Bruce-Mitford et al. (1956–60), II, pp. 34–5.
[106] The most useful critical survey of this area of study is Dumville (1999), which deals in particular with the seminal work of Julian Brown (see, for example, Bately, Brown and Roberts (1993)),

ILLUSTRATION 17
Ruthwell (Dumfriesshire), cross, Christ and the blind man
Christ on the left with cruciform halo is healing the blind man by touching his eyes with a rod. Note the three-dimensionality and the naturalistic, classical Mediterranean style of the figure carving.

Mediterranean world, and the vine-scrolls likewise emanated from Mediterranean carving and mosaic work; but there were also advances, not least in the fact that these sources of inspiration had not been applied in their areas of origin to free-standing stone crosses. The great crosses of Ruthwell and Bewcastle are especially important in this respect. Not only is the style of carving distinctively Mediterranean, but the iconography has been shown to be particularly sophisticated and closely related to developments in the liturgy of the church of Rome in the late seventh century (ill. 17).[107] Even the Old English poem, an early version of the poem called *The Dream of the Rood*, which is inscribed in runic letters on the edges of the Ruthwell

[107] O Carragáin (1978). See also Cassidy (1992).

ILLUSTRATION 18
Ruthwell (Dumfriesshire), cross, 'inhabited' vine-scroll
The 'inhabited' vine-scroll with birds and animals is typical of Northumbrian sculpture. Note also the runic inscription of the version of *The Dream of the Rood*.

Cross, belongs to a Christendom-wide culture (ill. 18): despite its Old English language and its rather 'barbarian' heroic imagery, it can be argued that it addresses issues of Christendom-wide significance, notably the heresy of Monotheletism which preoccupied the papacy in the second part of the seventh century, and which was refuted at the Synod of Hatfield in 679. For it emphasizes the will of Christ in ascending the cross, which was exactly what was at issue in the discussion of Monotheletism.[108] Stone sculpture then was very much within the broad tradition of Christian art and, at least in the case of the Ruthwell and Bewcastle Crosses, addressing themes of Christendom-wide significance.

[108] I am grateful to Prof. E. O Carragáin for letting me see the text of his forthcoming book on the Ruthwell Cross and the *Dream of the Rood*; for the synod, Bede, *Eccles. Hist.*, bk 4 ch. 17 (15).

Much the same can be said of church architecture. Surviving church buildings from early Northumbria, which include Jarrow, Monkwearmouth and Escomb, are distinguished from southern English churches, such as Reculver and Brixworth, by, amongst other things, their high narrow proportions (ill. 19). Although these churches could be envisaged as built 'in the Roman manner', as King Nechtan of the Picts wrote to Ceolfrith, the principal influences on them seem to have been from Gaul, and we know that Monkwearmouth at least was built by Gaulish masons. They may, however, have been influenced by Germanic building styles in wood, as evidenced by buildings at Yeavering.[109] As for the churches built by Wilfrid at Hexham and Ripon, these have not survived apart from their crypts, but they are enough to show that we are dealing with highly sophisticated architectural design. In the case of Ripon and Hexham, as we have seen, it is possible that the crypts were influenced by the proportions of the Holy Sepulchre in Jerusalem (fig. 15, ill. 20).[110] To judge from Stephen's description, the building above the crypt at Hexham was an impressive structure complete with columns, side-aisles (*porticus*), walls of notable height and length, 'surrounded by various winding passages with spiral stairs leading up and down'. Designed by Wilfrid himself, it was – Stephen emphasizes – the only church he had heard of on such a scale north of the Alps. The comparability of Hexham with the Mediterranean world was thus explicit.[111]

York was evidently also an architectural centre. Although nothing of the building has survived from this period, we hear of work on the Minster in the seventh century,[112] and in the eighth of the construction of the church of the Holy Wisdom which seems to have been to a round or polygonal design of some splendour. According to Alcuin, whose poem on the *Bishops, Kings and Saints of York* is our only source of information about it,

> This lofty building, supported by strong columns,
> themselves bolstering curving arches, gleams
> inside with fine inlaid ceilings and windows.
> It shines in its beauty, surrounded by many a chapel
> with its many galleries in its various quarters,
> and thirty altars decorated with different finery.

Nothing more is known, but it may be that something like an early Christian baptistery was being described, something perhaps like St Constanza

[109] Bede, *Eccles. Hist.*, bk 5 ch. 21; Bede, *Hist. Abbots*, ch. 5; and *Life Ceolf.*, para. 7. On the Gaulish parallels of these churches, see Fernie (1983), pp. 56–9.

[110] Bailey (1991); see also Fernie (1983), pp. 59–63.

[111] Stephen, *Life Wilf.*, ch. 22. [112] *Sources*, pp. 136–8 (A.4.5 = Stephen, *Life Wilf.*, ch. 16).

ILLUSTRATION 19 Escomb (County Durham), St John's Church, exterior of nave looking east
Note the high narrow proportions and the traces of a western *porticus* (chapel), remains of which have also been found by excavation.

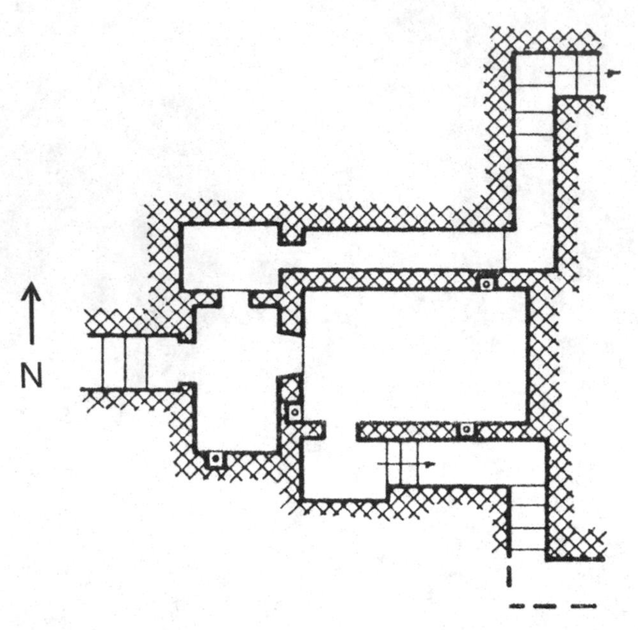

FIGURE 15 Hexham Abbey (Northumberland), plan of the crypt (drawing from Taylor and Taylor (1965–78), III, fig. 740)
It has been suggested that the crypt permitted circulation of lay pilgrims from the western passage into the antechamber, then out by the northern passage, whereas churchmen had access to the central chamber itself via the south passage (Taylor (1969)). Note the tortuous and confusing shape of the passages, possibly intended to replicate the catacombs in Rome and to give the pilgrim a sense of disorientation. For the parallels between the proportions of the central chamber and the Holy Sepulchre, see Bailey (1991).

in Rome, the original oval church of St Gereo in Cologne or the rather later baptistery at Fréjus in Provence.[113]

Conversion to Christianity thus drew the developing kingdom of Northumbria into the culture of western Christendom as a whole, a culture which it took to itself and, in some respects, came for a time to dominate. How far did this culture penetrate into Northumbrian society, or how far was it really the culture of a tiny, learned and clerical elite, cut off from the rest of society? It is, needless to say, almost impossible to answer that question when our sources are so exclusively ecclesiastical in origin and naturally emphasize the pervasiveness of Christian culture. Even within this constraint, however, some lines of approach may be possible.

[113] *Sources*, pp. 155–7 (A.4.27 = Alcuin, *BKSY*, lines 1507–20). For the possibility that the church most resembled, and may have influenced the design of, the slightly later Palatine Chapel in Aachen, see Norton (1998a), p. 23.

ILLUSTRATION 20
Hexham Abbey (Northumberland), central chamber of crypt looking west
The opening on the left leads to the southern passage, that directly ahead to the western (note the original steps beyond the antechamber leading to the nave). The lamp niches with funnel-shaped cavities in their roofs to create a draught are a characteristic feature which, together with the plan, proportions and other features, parallel Wilfrid's other crypt at Ripon.

First, and very striking to our modern eyes, we cannot help but be impressed by the extent to which the church had very rapidly obtained a dominating position in Northumbria, as elsewhere in early medieval Europe. It seems clear, in the first place, that there were many more ecclesiastical foundations than we can name or locate. Casual references to what were evidently numerous monasteries occur, for example, in Stephen's *Life of Wilfrid*, and often when Bede names a church it is in an incidental rather than a systematic way. He would never have told us, for example, that an early monastery existed at Dacre in Cumberland if he had not had occasion to describe a miracle there involving St Cuthbert's hair.[114] These churches,

[114] Stephen, *Life Wilf.*, ch. 25, for example: 'the many thousands of monks he was forced to leave behind'; Bede, *Eccles. Hist.*, bk 4 ch. 32 (30) (on Dacre, see above, pp. 28–9).

or at least the monasteries amongst them, could be very large: when Abbot Ceolfrith set off on his last journey to Rome, he left behind him no less than 600 monks of Monkwearmouth and Jarrow, some of whom gathered to bid him farewell. Some of these may have been labourers, the equivalent of the later Cistercian lay brothers; but none the less the figure is an impressive one.[115]

Aside from the scale of the Northumbrian church, and even allowing for the likely bias of our sources, we seem also to see a very influential group of churchmen who were closely in touch with kings and their courts. The kings often visited Lindisfarne. Bishop Aidan died at a royal vill where he often spent time preaching. Benedict Biscop felt he had to appoint Eostorwine as his deputy in governing the abbey of Monkwearmouth because he was so often called away to advise kings. Wilfrid was so deeply implicated with the royal house that this was part of the reason for his repeated expulsions from his see, and we find him in effect sponsoring Osred, one of the claimants to the throne, after the death of King Aldfrith. Bede, as we have seen, sent the *Ecclesiastical History* to King Ceolwulf.[116] Indeed, we cannot fail to be struck by the evidence we have for the influence of the church on kings. Unless we are to dismiss Bede's account as the fantasy of an ecclesiastical writer, there were evidently traditions, which must have seemed plausible at the time, that King Oswald's conversion to Christianity while in exile on Iona had been sufficiently deep for him not only to invite that church to send a missionary to his kingdom, but also for him to assist by simultaneously translating Aidan's preaching.[117] King Aldfrith, who as an illegitimate son of King Oswiu had probably not been expected to succeed to the throne, had therefore been sent to Iona to study, and was as a result sufficiently literate to want to exchange a book (said to be a book 'of the Cosmographers') with Benedict Biscop in return for a landed estate.[118] In the eighth century, King Eadberht was the brother of the bishop, then archbishop, of York, Egbert, and King Alhred patronized and supported the Northumbrian missionaries on the continent.[119]

More telling still is the willingness of kings, especially in the seventh century, to found monasteries and to endow them on what can only be

[115] Bede, *Hist. Abbots*, ch. 17.

[116] Bede, *Eccles. Hist.*, bk 3 chs. 17, 26; *Life Ceolf.*, para. 12; Stephen, *Life Wilf.*, ch. 59; and Bede, *Eccles. Hist.*, preface.

[117] Bede, *Eccles. Hist.*, bk 3 ch. 3.

[118] Bede, *Hist. Abbots*, ch. 15; on Aldfrith's position, see Moisl (1983), pp. 120–4.

[119] On Egbert and Eadberht, see Alcuin, *BKSY*, lines 1248–88; on Alhred, see Boniface, *Letters*, no. 121 (trans. *EHD I*, no. 187); and Anskar, *Vita sancti Willehadi*, in Pertz (1829), p. 380.

described as an enormous scale. The lands of Lindisfarne, as described in the *History of St Cuthbert*, were huge, and they were augmented in the eighth century by further extensive grants, including that of Warkworth in Northumberland 'with all its appurtenances', by King Ceolwulf.[120] Monkwearmouth and Jarrow had similarly vast estates, although these were not augmented on a comparable scale, so far as we know, after the time of King Ecgfrith (670–85).[121] The reasons for the massive transfer of lands involved in monastic endowments in this period may have been complex. It is possible that the monasteries were serving the royal interests by functioning as centres of governance, which would explain why we find Coldingham and Hexham at the head of the administrative districts called shires. But we cannot prove that, and it is in any case hard to avoid the conclusion that the foundations and endowments point to Christian belief having had a real impact on the kings, who acquired at least some level of the Christian culture we have been discussing.

Perhaps the most direct evidence for the impact of Christianity on kings as individuals is provided by the cases of the kings who abdicated to enter the church, of which we know of a number in the eighth century. It is not, of course, always easy to distinguish cases where kings abdicated and entered the church of their own volition, and cases where they were compelled to do so by opponents wishing to neutralize them politically. This latter was certainly the motive which led the Frankish ruler Charlemagne to impose the tonsure on his enemy, Duke Tassilo of Bavaria, and confine him to the monastery of St Wandrille in the Seine Valley. It was presumably the motive of those who in 731, according to the Continuation of Bede's *Ecclesiastical History*, 'captured' and tonsured King Ceolwulf of Northumbria. Clearly there was an attempt by his enemies to remove him permanently from rule by confining him to the church, but he successfully resumed his throne; presumably his supporters had rallied to his cause, although we are in possession of none of the details. In 737, however, the Continuation notes that he was 'tonsured at his own request and resigned the kingdom to Eadberht'. We cannot of course be sure how voluntary this second renunciation of power really was, although that Ceolwulf's aspirations to the monastic life were genuine may be proved by the fact that he was remembered at Lindisfarne, where he became a monk, as a generous benefactor, who had amongst other things allowed the monks to drink beer. It should be noted, however, that – to judge from a discrepancy in the dates given

[120] *Hist. Cuth.*, paras. 3–6, 8; see Craster (1954). [121] Bede, *Hist. Abbots*, chs. 4, 7, 15.

by our sources – there may have been an interregnum between Ceolwulf's resignation and the coming to power of his successor, possibly suggestive of political upheaval. Nevertheless, there seems little reason to doubt that religious commitment was important in this case.[122]

As we shall see in the next chapter, all our evidence suggests that Northumbria was dominated, at any rate from the late seventh century, by a well-developed aristocratic elite.[123] Our sources provide a considerable amount of incidental evidence for the impact of Christianity on the members of that elite who appear in miracle-stories as beneficiaries of miracles and as asking for saintly intervention, as, for example, in the case of the prefect whose wife was exorcized by St Cuthbert. They appear also as founders of churches, for example in the miracles of John of Beverley which quite incidentally inform us of two counts (*comites*) whom John of Beverley visited to consecrate churches they had built.[124] Very notable is Bede's condemnation of the foundation by nobles of what he regarded as spurious monasteries, which he considered to be numerous. Whatever their motives, the nobles who founded these spurious monasteries were evidently sufficiently imbued with Christian culture for such foundations to have seemed an appropriate course of action, even if they did not meet with Bede's approval.[125]

The involvement of the aristocracy with Christianity and the consequent interaction of aristocratic culture and Christian culture may have left their mark on the famous Franks Casket. Nothing is known of the medieval provenance of this whale-bone box, which is called the Franks Casket after the collector who first acquired it (one panel was missing and is now in Florence). The view that the casket was made in eighth-century Northumbria is based on: the form of the Northumbrian dialect of Old English used in the inscriptions on it; the character of the runic letters in certain of the inscriptions; and the artistic style of the scenes represented on it. The scenes carved on the casket are very instructive in connection with the influences making up Northumbrian culture. In part, the casket is a witness to Christian culture in that kingdom. The back panel shows the capture of Jerusalem by the emperor Titus as described by the historian Josephus, a subject which, although part of Jewish history, was regarded as crucial to the Christian

[122] In general, see Stancliffe (1983). For the texts, see: Scholz and Rogers (1972), pp. 66–7 (on Tassilo); Bede, *Continuations*, *s.a.* 731, 737, and *Vita sancti Oswaldi*, ch. 21 (Arnold (1882–5), I, pp. 360–1) (on Ceolwulf). For the chronological discrepancy, see *Sources*, p. 52.

[123] See below, pp. 180–2.

[124] David Rollason (1989a), pp. 83–104; for the texts, Anon., *Life Cuth.*, bk 2 ch. 8; Bede, *Life Cuth.*, ch. 15; and Bede, *Eccles. Hist.*, bk 5 chs. 3–5.

[125] Bede, *Letter*; see below, pp. 188–90.

ILLUSTRATION 21 The Franks Casket (London, British Museum), front
Note the runic inscription in the Northumbrian dialect of Old English and the damage where the precious metal fittings have been torn away. For discussion of the scenes, see pp. 167–8. The inscription, which is a riddle on the whale-bone of which the casket is made, may be translated: 'The fish beat up the seas on to the mountainous cliff; the king of ? terror became sad when he swam on to the shingle. Whale's bone' (R. I. Page (1999), pp. 173–5).

tradition. The front shows the adoration of the Magi, with the Virgin and Child enthroned and the Magi bowing to present their gifts, the star of Bethlehem clearly visible above them – and the word *Magi* is added in runes above (ill. 21). But there are other elements which indicate quite different cultural influences. Alongside the adoration of the Magi, on the same panel, there is represented a quite different scene. A man sits awkwardly holding a cup and a pair of tongs, other blacksmith's tools around him, a headless corpse at this feet. Two women approach him, one reaching out towards the cup. On the right another man is engaged in strangling birds. There are no runes to help with the interpretation of this scene, for those around this panel concern the origin of the whalebone from which the casket is made ('The fish beat up the seas on to the mountainous cliff; the king of ? terror became sad when he swam on to the shingle. Whale's bone'). Rather the key to the scene lies in Scandinavian mythology first recorded later in the middle ages. This includes the story of Weyland, a marvellous smith, who was kidnapped by a king who wanted his services and had the tendons of his knee cut to prevent his escape – the figure with the tongs is presumably Weyland and his awkward attitude is due to this injury done to him. According to the story, Weyland took a dreadful revenge by killing the king's son and making his skull into a drinking cup – this may be what Weyland is holding, or it may be the cup of drugged drink which he gave the king's daughter so that he could rape her. It is presumably she who is represented approaching Weyland; the other woman may be Weyland's wife, a swan-maiden. According to the story, Weyland then made good his escape with the aid of his brother, who made him a marvellous flying cloak out of the feathers of birds – the brother may be the figure represented on the casket strangling birds in preparation for making the cloak, or that may be the king trying in vain to seize Weyland before he could fly away. At all events, there is no doubt that this scene derives from Germanic mythology, the story of Weyland being found in the Old English poem *Deor* and in a more developed form in Old Norse literature. Its presence on the casket, however, does not bear witness to the residual paganism or the imperfect conversion to Christianity of its patron. It can be shown that 'the casket's themes are organized contrapuntally, providing three pairs of scenes, in each of which a Christian topos offers a commentary on a pagan Germanic one'. Thus the Weyland scene is a counterpoint to the Magi scene; the 'humility, magnanimity, and deference of the king of heaven' are compared with the wicked ruler's treatment of Weyland.[126] If, as seems likely, the casket was made for a church, it provides vivid evidence of the fusion of Christian and English aristocratic culture in ecclesiastical contexts. So too may the Old English epic poem *Beowulf*, which – it has been argued – can be best understood as a work using aristocratic secular stories in an essentially Christian context, probably a monastery.[127]

To what extent Christian culture reached from the elites of the church and the secular world farther down the social scale is naturally very difficult to establish. It was certainly Bede's aspiration that it should do so. He lays emphasis in the *Ecclesiastical History* and his *Life of Cuthbert* on the role of churchmen in preaching even to the most humble people; and his commentaries on scripture emphasize the role of churchmen as teachers (*doctores*) in

[126] For the interpretation advanced here and for the quotations, see Webster (1999) and R. I. Page (1999), pp. 173–5. For *Deor*, see Malone (1977). See also Lang (1999) and Neuman de Vegvar (1987), pp. 259–73. The casket has been connected with Ripon, but such precision is really not possible, given that nothing specific is known about its origin (Wood (1990)).

[127] Wormald (1978).

spreading the word of Christ to create a fully developed Christian society. Whether this remained more than an aspiration until the tenth century at the earliest is more difficult to establish. Since we have no evidence relating to the lower echelons of society themselves, we can only approach this question by considering what provision the church was able to make for actually extending Christianity to them, as distinct from having aspirations to do so. Our ignorance on this point is considerable. In his *Letter to Bishop Egbert*, Bede seems to envisage that the principal agents for the spread of Christianity and for pastoral care will be the bishops themselves; he observes that Bishop Egbert's diocese is too big for him to reach all parts of it and he therefore urges him to appoint assistants: 'priests should be ordained and teachers established who may preach the word of God and consecrate the holy mysteries in every small village, and above all perform the holy rites of baptism wherever opportunity arises'. There are, Bede avers, some dioceses in which in the more inaccessible parts 'a bishop has never been seen over the course of many years performing his ministry and revealing the divine grace'. If pastoral care was dependent in this way on bishops responsible for travelling throughout dioceses based on York, Hexham, Lindisfarne and Whithorn,[128] it would appear on the face of it that the machinery for Christianizing the lower echelons of society must have been inadequate indeed. It has been argued, however, that the burden of pastoral care in fact fell on monasteries (which the supporters of this argument prefer to call by the Old English word *mynstre* to avoid more modern connotations). According to this argument, monasteries had territories attached to them for which they had pastoral responsibility, so that a framework of pastoral care existed even if it was not the one which is most familiar in more modern times. This argument has been widely accepted, but it is difficult to substantiate, especially in the case of Northumbria. Bede never mentions such functions of monasteries, except in his eulogies of the preaching activities of the Irish monks who came in the context of the conversion of Northumbria, and his silence about them in the *Letter to Bishop Egbert* is especially striking.[129]

Although the mechanism by which Christianity penetrated down the social scale remains obscure, that it eventually did is not in doubt. Aside from sporadic references in the penitentials and some passages in a text called

[128] See above, p. 131.

[129] For the use of 'minster', see Foot (1992); for the arguments generally in favour of the importance of 'minsters', see John Blair (1988) and John Blair and Sharpe (1992); against, see David Rollason and Cambridge (1995) and David Rollason (1999b). For Bede on *doctores*, see Thacker (1983). For judicious commentary, see Hadley (2000), pp. 292–7. For the text, see Bede, *Letter.*

the *Northumbrian Priests' Law* probaby relating to the tenth century and the aftermath of Viking influence, we hear little of paganism after the mid-seventh century.[130] Moreover, we see the beginning of the acceptance of Christian culture at relatively low social levels in changes in burial customs. The cessation of grave-goods altogether may not, as was once supposed, be a symptom of the penetration of Christian culture, but the process by which burial came to be removed exclusively to the cemeteries of Christian churches certainly was.[131]

The conclusions of this chapter are then that the predominant culture of Northumbria in the pre-Viking period was Christian. A secular culture no doubt existed; but apart from some decorated objects we have no evidence to illuminate it in any detail.[132] Christian culture in Northumbria was, except possibly in certain limited respects such as the emphasis placed on free-standing stone crosses and the decorative art of the Lindisfarne Gospels and related manuscripts, a cosmopolitan, eclectic culture, which differed in no essential from that in neighbouring kingdoms or on the continent. The cultural identity of Northumbria was not one therefore to create the identity of a people. Rather it was a much more wide-ranging one, an identity which embraced western Christendom at large.

[130] For the penitentials, see John T. McNeill and Gamer (1990), esp. pp. 198, 228; for the *Northumbrian Priests' Law*, see Liebermann (1903–16), I, p. 383; *EHD I*, no. 53, paras. 48, 54. For a new text, see Tenhaken (1979).

[131] Neuman de Vegvar (1987), pp. 245–6; in general see Richard Morris (1983), pp. 49–62.

[132] The principal piece of evidence discussed in this connection by Neuman de Vegvar (1987), pp. 238–73, is in fact the Franks Casket (on which, see above).

CHAPTER 5

The framework of power: government, aristocracy and the church

Having looked in the previous chapters at the ethnicity and culture of Northumbria, let us now turn to the third aspect of the identity of a kingdom which we identified at the outset: the framework of governance which held it together. We need to evaluate the relative roles of what we may (crudely and for the convenience of clarity) perceive as three elements. First, governmental machinery, in our case that of the kings of Northumbria. By this we mean the existence of some form of central government organized so as to have procedures capable of functioning independently of the personality or personal intervention of the king; the existence of some form of local government capable of implementing the decisions of central government; and the existence of some form of communication between the two. Secondly, the role of political elites, in our case the Northumbrian aristocracy. By this we mean the extent to which power resided in the personal relations between king and aristocracy, and how far political stability and effective government depended on the attitudes and cooperation of that aristocracy. Thirdly, the role of ideologies and their associated organization and rituals, in our case those emanating from the Christian church. By this we mean, the extent to which the church provided the underpinning for government, both ideologically in terms of buttressing and strengthening the moral authority of kings (for example, through rituals of king-making), and practically through putting its own framework of organization and government at the disposal of the kings.

We may of course find these elements in combination with each other, but the degree of importance which our evidence leads us to attach to one or other of them will profoundly affect our view of the sort of state we are dealing with. If our evidence points to Northumbria as a kingdom with a developed governmental machinery, evenly covering the entire area of the kingdom, then we are dealing with a state similar to that of the Roman Empire, or indeed of European states in the modern period. To evaluate this, we shall need to ask, how sophisticated was the machinery of government and how far it was capable of constraining and controlling all of the state's population evenly and impersonally (or at least those elements of the population not too low down the social scale to be effectively beyond its purview). If our evidence points to Northumbria as a kingdom based on the attitudes and good-will of the aristocracy, we are clearly dealing with a political unit which looks much less Roman, much less 'modern', and for which a different set of questions must be pursued, namely: what was the relationship of the king to that aristocracy and to what extent did king and aristocracy owe their respective positions to each other? What was the attitude of that aristocracy towards the royal line and how far was it conditioned to identify itself with the kingdom of Northumbria and its political well-being? And did the aristocracy possess the power and the procedures for influencing or disrupting the rule of the kingdom of Northumbria? To what extent was the king's position simply that of leader and collaborator with the aristocracy, which put its power at his disposal because it was bound to him by links of fealty (*fidelitas*)? If, finally, our evidence points to Northumbria as a kingdom heavily dependent on the church, then we are dealing with a political unit which has elements of theocratic power, and rather resembles states in which the church assumed a prominent – if not a predominant – role the Carolingian and Ottonian empires, for example. In this context, we shall need to enquire into the relationship between royal power on the one hand, and on the other the church's theories about royal power, its rituals as they affected the position of kings and more prosaically the relationship between the power of the church and that of royal government.

In assessing the sophistication of royal governmenal machinery in Northumbria, we can perhaps apply five tests: first, the existence of a hierarchy of officials with defined roles; secondly, the nature of the royal itinerary; thirdly, the extent of the kings' use of written documents; fourthly, the kings' ability to levy tribute; fifthly, coinage; and, sixthly, the extent to which they were able to control trade in their kingdom.

First, a hierarchy of royal officials. At the highest level, we find in the *Ecclesiastical History*, as well as in Stephen's *Life of Wilfrid*, figures called 'sub-kings' (*subreguli*) or 'princes' (*principes*). Thus when Ecgfrith was attacked by the Picts, 'he quickly mustered a troop of cavalry and putting his trust in God, like Judas Maccabeus, set off with Beornhæth, his trusty sub-king' (*audaci subregulo*). It is not clear, however, how far all or any of these men were royal officials rather than members of the royal families of previously independent kingdoms, or perhaps members of the royal family of Northumbria, possibly ruling whole provinces of the kingdom as kings in their own right as was certainly the case with the rulers of Deira in the mid-seventh century.[1] At a similar level of power but more clearly royal officials were the 'patricians' (*patricii*), four of whom are referred to in the Northern Annals and in a letter of 757–8 addressed to King Eadberht of Northumbria and Archbishop Egbert of York: Moll, who was presumably the later king of Northumbria, Æthelwold Moll (759–65), Sicga who was forced into exile, Bearn, and Osbald, who subsequently seized the kingship and ruled for a month in 796 before being expelled by his own household. Since, on the admittedly slender basis of this evidence, only one 'patrician' was in post at any one time, he may have been the king's principal official. We also find references to 'dukes' (*duces*), who were clearly very important figures but whose role is unclear. The next step down in the hierarchy appears to have been occupied by prefects (*prefecti*). In Stephen's *Life of Wilfrid*, a prefect appears as responsible for the *urbs* of Dunbar and another for that of *Inbroninis*, both of which were suitable places for the king to imprison an important man such as Wilfrid. The prefect of Dunbar clearly possessed resources sufficient to do the job properly, for the king ordered him to keep Wilfrid 'bound hands and feet with fetters', which the prefect duly ordered smiths to make. At *Inbroninis*, Wilfrid was kept 'under guard in hidden dungeons', under the supervision of the prefect, who is also described as a count (*comes*). Below the prefects may have been the 'counts' (*comites*, rendered *gesiths* in the Old English translation of Bede's *Ecclesiastical History*), although the fact that the prefect of *Inbroninis* was also described as a count raises a question as to whether the term referred to social status rather than royal office, to which we shall return. A similar question applies to the men who were apparently below 'counts' in the hierarchy, the 'ministers' (*ministri*) or 'soldiers' (*milites*) of the king, both terms being rendered as 'thegn' in the Old English translation of Bede's *Ecclesiastical History*. We have no information

[1] See James Campbell (1980), and below, p. 180; for the texts, Stephen, *Life Wilf.*, ch. 19; Bede, *Eccles. Hist.*, for example, bk 3 ch. 21, bk 4 ch. 12, bk 5 ch. 24. For Deira, see below, p. 191.

as to their role apart from as attendants on the king and, as we shall see, warriors.[2]

Secondly, the itinerary of the Northumbrian kings, that is their practice of moving from royal centre to royal centre. Stephen's *Life of Wilfrid* must be referring to the royal itinerary, when it describes the 'king and queen... making their progress through the cities, fortresses, and villages (*ciuitates et castella uicosque*) with worldly pomp and daily feasts and rejoicings'. Bede gives a similar picture of Edwin, with more detail as regards the pomp:

> So great was his majesty in his realm that not only were banners carried before him in battle, but even in time of peace, as he rode about among his cities (*ciuitates*), estates (*uillas*) and *prouincias* with his thegns (*ministris*), he always used to be preceded by a standard bearer. Further, when he walked anywhere along the roads, there used to be carried before him the type of standard which the Romans call a *tufa* and the English call a *thuf*.

We could argue that the itinerary of the Northumbrian kings is evidence for the primitive nature of their government. They simply had to be always on the move, partly to impose their will on people whom they lacked the mechanisms to control from a distance, and partly to consume the produce of their estates which they were unable to have transported to a fixed residence. On the other hand, the pomp emphasized in the passages quoted above suggests that the itinerary, which was an activity common to almost all medieval kings, was for the Northumbrian rulers an effective mechanism of power. It appears to have been underpinned by many 'royal vills' (*uillae regales*), such as Yeavering, the unnamed vill where Edwin was staying when he held the council to discuss the merits of adopting Christianity, and the unidentified royal vills at which Aidan is said to have preached. That such royal vills were surrounded by lesser settlements is suggested by Bede's account of the people flocking to hear Paulinus's preaching at Yeavering 'from every village and district' (*de cunctis uiculis ac locis*). The use of the diminutive (*uiculi*) suggests that these villages were dependent on the royal vill of Yeavering. It is therefore not unreasonable to argue that royal vills were centres for lesser and dependent settlements spread over quite wide areas and owing food rents to a royal vill for royal consumption in the course

[2] Hector Munro Chadwick (1905), esp. pp. 308–54, with some modifications in Thacker (1981). See also Loyn (1955). For the texts: Bede, *Eccles. Hist.* (for refs. Putnam Fennell Jones (1929)); Stephen, *Life Wilf.*, chs. 36, 38; *Hist. Kings*, *s.a.* 788 (p. 52), 796 (p. 57); for the letter to King Eadberht and Archbishop Egbert, Haddan and Stubbs (1869–71), III, pp. 394–6 (trans. *EHD I*, no. 184). On *urbes* and other names for places, see James Campbell (1979) and references therein.

of the royal itinerary. There is admittedly no evidence from Northumbria proving the existence of such food-rents, but it is reasonable to suppose that something like those known somewhat later in the south as the royal 'farm of one night' (*feorm, firma unius nocti*) existed there. Their imposition on the *uiculi* would presumably have been assessed in hides, and organized around the structure of a *uilla* with its dependent *uiculi*, which may perhaps have constituted a *regio* (region). For example, Stephen's *Life of Wilfrid* refers to the *regio* which Wilfrid obtained from the queen to found Hexham; and the anonymous *Life of St Cuthbert* mentions the *regio* of *Ahse* between Hexham and Carlisle. Bede elsewhere uses *prouincias* to mean 'kingdoms' so in the text quoted here it may refer to Deira, Bernicia, Elmet and so forth; but it may more naturally mean 'administrative districts' in this context, although presumably quite wide administrative districts are intended. Moreover, the references in the texts quoted above to 'cities', 'fortresses' and 'provinces' suggest a hierarchy of organization above the level of royal vills, and with which the itinerant kings were concerned.[3]

Thirdly, the use of written documents. Although there is no evidence that the Northumbrian kings promulgated laws, it is fairly certain that they issued charters granting lands and privileges. Stephen's *Life of Wilfrid* seems to refer to such documents in describing Wilfrid's dedication of the church of Ripon:

Then the holy bishop Wilfrid stood in front of the altar, and, turning to the people, in the presence of the kings, read out in a clear voice the names of the lands which the kings had previously given him for the good of their souls, with the consent and signature of the bishops and all the princes (*principes*).

[3] For itinerant kingship, see Charles-Edwards (1989), pp. 28–33, and above, p. 23. For the texts: Stephen, *Life Wilf.*, ch. 39; and Bede, *Eccles. Hist.*, bk 2 ch. 16. On royal vills, see Sawyer (1983); on food-rent, see, for a later period, Stafford (1980); on *regiones*, see Stenton (1971), p. 295. For the function of Yeavering, see Hope-Taylor (1977) and for a revised view Alcock (1989), p. 26. For the texts, see Bede, *Eccles. Hist.*, bk 2 ch. 14 (Yeavering), bk 3 ch. 17 (Aidan and royal vills); Stephen, *Life Wilf.*, ch. 22 (*Inaegustaldesae, adepta regione a regina*: Colgrave misleadingly translates, 'having obtained an *estate* from the queen', Colgrave (1927), p. 44); and Anon., *Life Cuth.*, bk 4 ch. 5. It is possible that such royal organization left its traces in the 'shires' or 'multiple estates' which appear in *Domesday Book* (1086), *Boldon Book* (1183) and other documents as largely in the hands of the church and the aristocracy. We have discussed already the strength of the case that the arrangements described in these sources dated from the period before the English, and would therefore have been in force in Northumbria before the Viking period. It is not impossible that they originated under the kings of Northumbria, and that the prevalence of units made up of twelve or six vills points to a system of royal organization imposed from above. The evidence, however, is far from consistent, and the prevalence of the number twelve may be simply another instance of the popularity of duodecimal counting. In short, the later sources are best treated as informing us about landed organization in their own period rather than preserving relicts of royal administration from an earlier age. See above, pp. 91–3; for the pattern of twelve vills, see Johnson South (2002), pp. 124–9. On *prouincias*, see for example Stenton (1971), pp. 293–4.

The reference to 'consent and signature' must be to the witness-lists of charters, which in this period would often have been given in the form, 'I [name] consent and subscribe'. There is a similar allusion in the letter which Bede wrote at the end of his life to Bishop Egbert of York, which relates how laymen had fraudulently obtained land to found spurious monasteries, doing this by means of 'royal edicts' bearing the subscription of bishops, abbots and secular persons. Aside from those possibly preserved in the *History of St Cuthbert* and a twelfth-century copy of a dubious charter of King Ecgfrith in favour of Lindisfarne, nothing has survived.[4]

Fourthly, the Northumbrian kings were capable of levying tribute on extraneous peoples, certainly on the Mercians according to the evidence of Stephen's *Life of Wilfrid*, possibly on the peoples south of the Humber generally if the Tribal Hidage is indeed a Northumbrian document. The second of these texts suggests that tribute was levied on the basis of assessment in hides, that is units of land, each of which was notionally large enough to support one family (*terra unius familiae* in Bede's Latin). The fact that Bede gives precise figures for the assessment in hides of Anglesey and Man, which Edwin is supposed to have conquered, suggests that tribute may have been imposed on these islands according to such a scheme. We know from Bede's *Lives of the Abbots of Monkwearmouth and Jarrow* as well as Stephen's *Life of Wilfrid* that lands within Northumbria were also assessed in hides, and it is reasonable to suppose that this was a system of assessment for levying dues to the king as appears in the charter-record for elsewhere in England and indeed Europe.[5]

Fifthly, coinage. Here, the evidence is much slighter. No Northumbrian king is known to have minted coins before Aldfrith (686–705), who issued silver pennies of quite high face value, but which may have been more for prestige than for practical use. After his death, no further coins are known to have been minted until the silver pennies of Eadberht (737/8–58), from whose reign onwards there was a more or less continuous Northumbrian coinage (ill. 22). This declined steeply in precious metal content in the

[4] On the early development of charters in England, see Wormald (1985). For the texts, see Stephen, *Life Wilf.*, ch. 17, Bede, *Letter* (quoted below, p. 188), *Hist. Cuth* (on which see Hart (1975), pp. 117–42) and, for charters, above, pp. 11–12. Ecgfrith's charter is Sawyer (1968), no. 66.

[5] On hides, see Maitland (1897), pp. 416–62, and *Encyclopaedia ASE*, entry 'hide'. On tribute, see Charles-Edwards (1989). For the texts, see Stephen, *Life Wilf.*, chs. 8 (the Latin term used is *mansiones*), 20; Bede, *Eccles. Hist.*, bk 2 ch. 9 (Anglesey and Man, and see above, p. 43), and Bede, *Hist. Abbots*, chs. 4, 7, 15 (*terra unius familiae*). For the Tribal Hidage, see above, pp. 39–40.

ILLUSTRATION 22 Penny (*sceat*) of King Alhred of Northumbria (765–74), obverse and reverse (Cambridge, Fitzwilliam Museum)
The obverse of this silver coin has the name ALCHRED with a cross; the reverse has a crudely executed and fantastic quadruped with horns, raised foreleg and long tail. It is similar to the beast found on the coins of Kings Aldfrith and Eadberht. See Grierson and Blackburn (1986), no. 1181; and also North (1994), no. 179.

ninth century, however, resulting in the so-called *stycas*, which were effectively bronze coins, produced in considerable quantities (ill. 23).[6]

Sixthly, control of trade. It is possible to argue that the Northumbrian kings had some measure of direct control over commerce in their kingdom on the basis of excavations at Fishergate just downstream from the walls of York. This site has been interpreted as a royally controlled entrepôt (known as an *emporium* or in Old English *wic*), in the light of a general model of the development of trading centres such as Dorestad in the Netherlands, *Hamwich* (modern Southampton), Ipswich in Suffolk and Birka in Sweden. According to this model, such places were of royal foundation and under royal control. They were, it is argued, invariably connected with a royal centre, as Dorestad is presumed to have been with the Carolingian palace of Nijmegen, *Hamwich* with the West Saxon city of Winchester, Ipswich with the East Anglian royal vill of *Rendlesham* and Birka with the Swedish royal centre of Old Uppsala. They were nevertheless quite separate from such centres, and formed specialized settlements often at some distance. Thus the Fishergate

[6] The most accessible reference works are North (1994), pp. 70–2, and Grierson and Blackburn (1986), pp. 166, 295–303; for the ninth century see further Metcalf (1987a). Some gold coins from York, possibly of King Ecgfrith's reign, were probably issued only for prestige; see Pirie (1992) and Tweddle and Moulden (1992).

ILLUSTRATION 23 *Styca* of Archbishop Wigmund of York (837–54), obverse and reverse (C. E. Blunt collection, Cambridge, Fitzwilliam Museum)
The obverse of this essentially copper coin has the name VIGMUND AREP (the second word an abbreviation for *archiepiscopus*, 'archbishop'). The reverse has the name of the moneyer. This coin is very similar to the *stycas* of the Northumbrian kings from Eanred onwards. See also North (1994), no. 196.

site (identified with the *Eoforwic* mentioned in some of our sources) would have been associated with the presumed royal centre within the walls of York, identified with the *Eoforburh* of the sources, but separated from it in being outside the fortifications and beyond the River Foss (map 8). In this it is thought closely to parallel London where a trading centre on the Strand outside the old Roman walls (identified with *Lundenwic*) is presumed to have been attached to the royal centre within them (*Lundenburh*). In the case of Fishergate, however, there is reason for doubting whether the evidence is enough to justify the idea that it was royally planned, rather than being a spontaneous settlement of merchants. Instructive in this context is Altfrid's *Life of St Liudger*, a near-contemporary text which describes how Liudger (d. 809) was sent from Frisia to study at York but his studies were not to be undisturbed. 'For, when the citizens went out to fight against their enemies, it happened that in the strife the son of a certain noble of that province was killed by a Frisian merchant, and therefore the Frisians hastened to leave the land of the English, fearing the wrath of the kindred of the slain young man.' The colony of Frisian merchants referred to in this passage may very well have been at Fishergate, in which case the unhappy events described may have been the reason for the first abandonment of the site as revealed in the archaeological record. At all events, there is little sign of royal control or even protection in this passage. The merchants do not

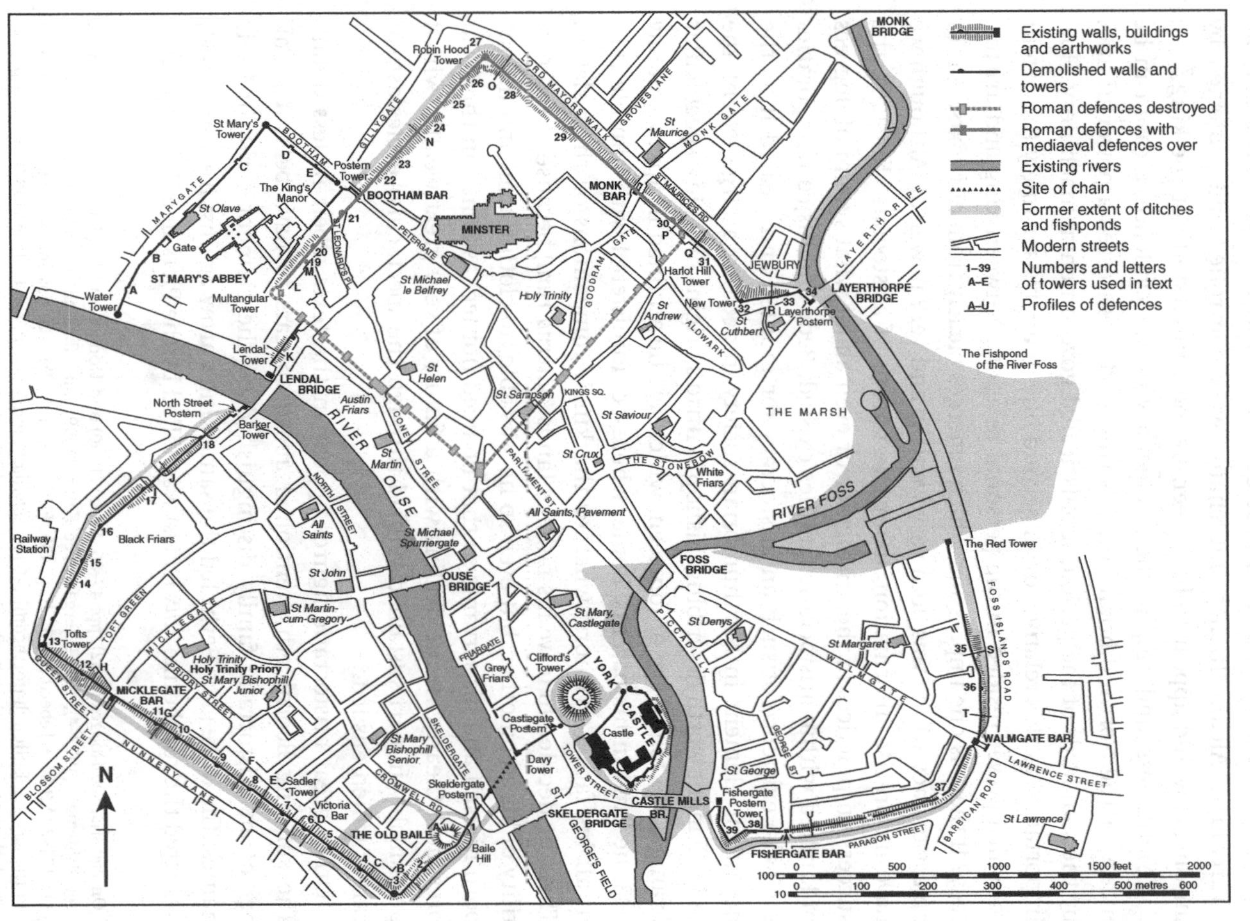

MAP 8 Street-plan of York in the nineteenth century with walls of the Roman fortress superimposed (from Royal Commission on Historical Monuments (England) (1972))
Note that Lendal Bridge is modern, as also is Parliament Street. Duncombe Place represents a drastic widening of the former Lop Lane.

appear to have been under royal protection in a royally controlled *emporium*. Rather, they acted independently or in association with the 'citizens', and the person who intervened to protect Liudger was no royal agent but the churchman, Alcuin. If these merchants were controlled and protected by anyone, it would appear to have been by the church and not the king.[7]

Nevertheless, and despite the inadequacy of our evidence, a case can be made for the Northumbrian kings as rulers who possessed a powerful and, by the standards of the period, sophisticated governmental machinery. How important was this relative to our second line of analysis, namely the extent to which power resided in the personal relations between king and aristocracy? If there is one impression which emerges clearly from our texts, it is of an aristocracy which was both very wealthy and very powerful. Although we have no precise information about the nature of aristocratic lineages, the succession practices of the aristocracy or even about what system of degrees of kindred within aristocratic families existed, Bede and his contemporaries took the existence of such an aristocracy for granted and treated it as the dominant element in Northumbrian society, indeed to a large extent the only element worth describing. Bishop Wilfrid's origins provide a spectacular example. Up to the age of fourteen, which age he attained in approximately 648, he lived in his parents' home, where he 'ministered with humble skill to all his father's visitors, whether the king's companions or their slaves'. Then, he decided to leave his father's lands (*paterna rura*) because of the hostility of his stepmother, but even so he was able 'to clothe, arm, and mount both himself and his servants so that he need not feel ashamed in the royal presence'. Clearly, we are dealing with a family of very great wealth, which presumably in part sustained Wilfrid through the spectacular ecclesiastical career which ensued.[8]

Other great aristocratic figures appear in the texts albeit in less detail. The sub-king Beornhæth mentioned above appears to have been a member of an aristocratic family distinguished by the element *Berht-* in their names. The duke (*dux*) Berht led the same king's army against Ireland in 684; Berhtred the royal duke (*dux regius*) was killed by the Picts in 698; the prefect

[7] On *emporia*, see Hodges (1982), pp. 47–86, Vince (1990) and Hodges and Hobley (1988); for *Hamwich*, see Addyman (1973–4); for Fishergate, see Kemp (1999), and for the *Life of St Liudger*, see *Sources*, pp. 131–2 (A.3.2 = Lebecq (1983), II, p. 109, *EHD I*, no. 160). For the possible relevance of Fishergate to the *Life*, see Bullough (forthcoming). For *Eoforburh* and *Eoforwic*, see Palliser (1984), written before the Fishergate excavations took place, and, guided only by documentary evidence, concentrating on the other bank of the River Ouse.

[8] In general, James Campbell (1989); for Wilfrid, see Stephen, *Life Wilf.*, chs. 2–3.

Berhtfrith fought against the Picts in 711 and was, according to Stephen's *Life of Wilfrid*, 'second only to the king'. Nothing except the name-forms proves that this was a single aristocratic group, but that evidence is convincing in the context of early medieval naming practices, where surnames were unknown and shared name-forms were used to indicate family connections.[9]

Aristocratic lineages are easier to perceive later in the eighth century, when we find them competing for the throne itself. We possess a letter from the Northumbrian churchman Alcuin to the former 'patrician' Osbald, apparently written after the latter's expulsion from the throne after his short reign of one month. Alcuin, who had it seems been advising Osbald, was angry with him because he had not taken the churchman's advice to enter a monastery, and in upbraiding him, he wrote: 'Do not add sin to sin by ravaging your country and shedding blood. Think how much blood of kings, princes and people has been shed through you and your family. It is an unhappy line which has brought so much evil upon our country!' We are clearly dealing here with a politically active aristocratic lineage of great power, which had for some time had ambitions to assume the throne. It was unsuccessful, but the family of the 'patrician' Moll, who is mentioned in the letter from Pope Paul I to King Eadberht and Archbishop Egbert, succeeded. Moll himself was presumably the Æthelwold Moll who became king in 759, ruling until his deposition in 765; his son Æthelred reigned from 774 until his expulsion in 779, and was then recalled to the throne in 790, reigning until his murder in 796.[10]

The appearance of the 'patricians', Moll, Osbald, Bearn and Sicga, mentioned above as royal officers, is especially interesting. *Patricius* was a perfectly good Latin word with a background in the Roman period; but it may be significant that in the course of the eighth century in the Frankish kingdom its use was revived by Charles Martel and it came to be applied exclusively to the mayors of the palace. Charles Martel, after whom the Frankish family of the Carolingians is so-called by modern historians, held the office of 'mayor of the palace' (*maior domus*) at the court of the Merovingian kings of Frankia. On the strength of vast estates, his family rose to a position of political dominance where the Merovingian kings were politically subject to them from at least the last quarter of the seventh century,

[9] Thacker (1981), pp. 215–16, who includes Beornhæth sub-king in the group (Stephen, *Life Wilf.*, ch. 19); for the texts, Bede, *Eccles. Hist.*, bk 4 ch. 26 (24), bk 5 ch. 24; and Stephen, *Life Wilf.*, ch. 60. For early medieval name-forms, see Werner (1979), pp. 151–2.

[10] In general, see Kirby (1991), pp. 142–62; for the dates, *Sources*, pp. 52–4; for the texts, Alcuin, *Letters*, no. 109, and Haddan and Stubbs (1869–71), III, pp. 394–6 (*EHD I*, no. 184).

until in 751 Charles Martel's son Pippin succeeded in making himself the first Carolingian king. This was more or less contemporary with the struggle of the Moll family and that of Osbald and Sicga to obtain possession of the throne of Northumbria. If the parallel between the Frankish kingdom and Northumbria holds good, a Northumbrian 'patrician' may, like the Frankish mayor of the palace, have been a royal officer from a great aristocratic family seeking either to control the king or to become king himself.[11]

Our sources naturally tell us most about the church, but the picture they present of that is of an institution dominated by aristocrats. Like Wilfrid, Benedict Biscop, founder of Monkwearmouth and Jarrow, was of noble extraction, and so too were his successors, including Bede's abbot Ceolfrith. Indeed, it appears that so aristocratic was the composition of Jarrow, that Ceolfrith was forced to withdraw from the abbey because of 'the jealousies and most violent attacks of certain nobles who could not endure his regular discipline'. Almost every churchman who appears in the sources, where social class is mentioned, turns out to be an aristocrat. It has sometimes been argued that Cuthbert, bishop of Lindisfarne, was of humble extraction; but the fact that he arrived at Melrose to become a monk, mounted, carrying a spear and attended by a servant suggests that this was not the case, as also do references to his foster-mother and his time in the royal army. Of Bede's own ancestry we know nothing; but there is no reason to suppose that it was not aristocratic. Virtually, the only non-aristocratic churchman to appear in our sources is the Whitby cow-herd Cædmon who was miraculously inspired to compose religious songs, which sufficiently accounts for his prominence in the church despite his lowly social class. That aristocratic churchmen had a major role in the government of the kingdom is suggested by the account in our sources of how Benedict Biscop felt it necessary to appoint Eostorwine as his deputy as abbot because he was so frequently called on to advise the king.[12]

In the case of the monastery of Coldingham, the aristocratic life-style was considered by Bede to have surpassed the bounds of decency. Under its royal abbess Æbbe, sister of King Ecgfrith, the female inmates were 'weaving elaborate garments with which to adorn themselves as if they were

[11] Thacker (1981), p. 215, who notes that Bede's only use of the word *patricius* (admittedly derived from his source) is as the title of Eorconwald, a mayor of the palace in the western area of Neustria in the seventh century. On the rise of the Carolingians, see Riché (1993).

[12] On Cuthbert, see David Rollason (1989a), p. 93. For the texts, see Bede, *Hist. Abbots*, chs. 1, 8; *Life Ceolf.*, paras. 2, 8, 12; Bede, *Life Cuth.*, chs. 6, 14; Anon., *Life Cuth.*, bk 2 ch. 7 and bk 1 ch. 7; and Bede, *Eccles. Hist.*, bk 5 ch. 24, bk 4 ch. 24 (22), bk 3 ch. 26.

brides'. Coldingham was not alone in this; in the early eighth century the English churchman on the continent, Boniface, was especially critical of English churchmen who wore garments decorated with 'dragons', presumably a reference to oriental silks. Alcuin was critical not only of rich dress in churchmen, but also of religious communities which were accustomed during mealtimes to hear readings not from the Bible or religious works, but rather from sagas and heroic adventures, a markedly aristocratic taste. 'What', Alcuin famously demanded, 'has Ingeld to do with Christ?' The view that this letter was for the church of Lindisfarne has been disputed; but it was certainly intended for a church and, given its generalized tone, Alcuin probably considered it applicable to Northumbrian churches as well as to the church (possibly that of Leicester) to which it was actually addressed. Since Ingeld was a hero of pagan mythology, the letter suggests that the aristocratic inmates of churches wished to preserve some of the culture and interests of the aristocratic world from which they had come. Ingeld appears in the Old English epic poem *Beowulf*; and it is not impossible that the poem itself was composed for use in religious communities, and for just the sort of aristocratic audience we have been invoking. 'A late seventh- or an eighth-century monastery', it has been observed, 'often had many of the aspects of a special kind of nobleman's club.'[13]

Very pertinent in this connection are the results of archaeological excavations at Hartlepool and Whitby, together with a site at Flixborough (Lincolnshire) just to the south of the Humber. In all three cases, the sites have been interpreted as those of monasteries: in the case of the first two because early monasteries are known to have been located there, in the case of the third because the discovery of a written plaque of possibly religious character on the site is held to indicate that it was ecclesiastical. The striking thing, however, is that all of them are characterized by the presence of rich objects, some clearly intended for personal adornment, and by a difficulty in determining with any certainty that they are in fact religious sites (ill. 24). The conclusion must be that in some cases at least it can be very difficult to distinguish a monastic site from an aristocratic, secular centre; and this conclusion in itself underlines the degree of dominance of the Northumbrian church by the aristocracy.[14]

[13] For Coldingham, see Bede, *Eccles. Hist.*, bk 4 ch. 25 (23); Alcuin, *Letters*, no. 124 (for the addressee, see Bullough (1993)); on *Beowulf*, see Wormald (1978); for other references and trenchant discussion, see James Campbell (1989), pp. 12, 14.

[14] For Hartlepool, Daniels (1988); for Whitby, Peers and Ralegh Radford (1943); and for Flixborough, Webster and Backhouse (1991), no. 69 (a–w), esp. no. 69(a).

ILLUSTRATION 24 Whitby (Yorkshire North Riding), strap-end from the site of the pre-Viking monastery (Whitby Literary and Philosophical Society, Strickland loan, W.52)
Dated to the mid-ninth century on grounds of the style of the animal, the strap-end is made of silver with niello decoration (that is using alloys to fill in an engraved design to produce a decorative effect). See Webster and Backhouse (1991), no. 107b.

We need now to enquire more closely into how this aristocracy related to the kings, beginning with what little we know of the raising of Northumbrian armies. Bede describes how at the Battle of the Trent, a young man called Imma, a member of the king's *militia*, was wounded and rendered unconscious. On coming to his senses, he was captured by men of the enemy (i.e. the Mercians) and taken to 'their lord' (*ipsorum dominum*), who was a count (*comes*) of the Mercian king. This point is itself of interest, since it shows something of the organization of the Mercian army under 'counts', and it is reasonable to suppose that the Northumbrian army was similar. But the really interesting part of the story follows: 'On being asked who he was, he [the wounded man] was afraid to admit that he was a soldier (*miles*); but he answered that he was a poor peasant (*rusticum... et pauperem*) and married; and he declared that he had come to the army in company with other peasants to bring food to the soldiers.' The Mercian count accordingly kept him in custody but, 'those who watched him closely realized by his appearance, his bearing and his speech that he was not of common stock' (*de paupere uulgo*), as he had said, but of noble family (*de nobilibus*). His captor then questioned him, promising to spare his life in any case, and Imma duly revealed that he was a minister (*minister*) of the king. At this the Mercian count observed: 'I realized by every one of your answers that you were not a peasant, and now you ought to die because all my brothers and kinsmen were killed in the battle; but I will not kill you for I do not intend to break my promise.' 'Minister' seems to be used here more as a designation of social status than as an office, and the implication of this passage is that warfare in the seventh century was a matter first and foremost for the aristocracy; the peasants were cast in the role of porters and provisioners.[15]

Moreover, the rules of warfare were those of the aristocratic elite: warfare entailed vendetta, just as did private feuds. This is apparent not only in the passage quoted above but also in Bede's account of Archbishop Theodore's intervention in the aftermath to the Battle of the Trent in 679. The killing of King Ecgfrith's brother Ælfwini at that battle, Bede notes, provided good reasons for 'fiercer fighting and continued hostilities between the kings and between these warlike peoples', presumably that is the operation of the blood-feud or vendetta. As a result of the archbishop's mediation, however, 'no further lives were demanded for the death of the king's brother, but only the usual money compensation which was paid to the king to whom the duty of vengeance belonged'. The 'money compensation' was of course the *wergild*, the payment referred to in the law-codes from southern England as necessary to compensate the aggrieved party and so to avert the feud.[16] If the aristocracy thus dominated warfare, it was equally firmly ensconced in the business of ruling. The prefects who were in charge of *urbes* were clearly aristocrats, and so too were the sub-kings.[17]

We need now to consider the source of power of the aristocracy, and in particular whether its power derived from the kings and was in their gift. For, if the aristocracy was dependent on the king for its position, the king's power to act would have been very much greater than if he had been confronted by an entrenched aristocracy which owed nothing to him and could please itself whether it cooperated with him or not. One clue is provided by the distinction in our sources between certain words for aristocratic status: count (*comes* in Latin, *gesith* in Old English) and minister or knight (*minister* and *miles* in Latin, 'thegn' in Old English). It looks as if 'counts' differed from 'ministers' in being established, propertied, married men. Thus, for example, we read of the bishop of York, John of Beverley, visiting the *uilla* of a count called Puch, who had invited him to dedicate a church which he had built, and whose wife was ill. Puch was clearly a man of consequence, able to build a church of his own; and we meet in the subsequent chapter another count called Addi, who also had a church for the bishop to dedicate, and one of whose retainers (*pueri*) was ill. Wherever *ministri* or *milites* appear, on the other hand, they seem always to have been attached to the king as was Imma in the story cited above; no doubt these were what constituted the

[15] Bede, *Eccles. Hist.*, bk 4 ch. 22 (20); for discussion, John (1966a), pp. 136–7.
[16] Bede, *Eccles. Hist.*, bk 4 ch. 21 (19); on the church's influence on feud, see Wallace-Hadrill (1962).
[17] See above, p. 173.

'young nobles' (*nobilium iuuentus*) who accompanied the future king Oswald into exile after Edwin's victory at the Battle of the River Idle. They were in effect the household knights of the kingdom of Northumbria, or perhaps rather the *uassi*, that is the vassals or military retainers, who appear in the Carolingian documents on the continent. Moreover, those we know anything of display a fairly consistent career pattern of royal service followed by establishment on land. The implication is that they began as 'ministers', and after a period of service would have been endowed with land and established as 'counts'. The clearest account of this process is that of Bede concerning Benedict Biscop who was 'about twenty-five and a "minister" of King Oswiu when the king gave him possession of the amount of land due to his rank'. In the event, he preferred to 'put behind him the things that perish', and to devote his life to the church; but the development of the career that was expected of him is clear enough. We are less well informed about his relative and successor Eostorwine; but we do know from Bede that he was the minister of King Ecgfrith until at the age of twenty-four he laid down his arms and, 'girding himself for spiritual warfare', entered the monastery of Monkwearmouth and Jarrow. It looks very much as if he too would have followed the minister-to-count progress but for his vocation.[18]

There is a striking similarity between this career progression, whereby the king established his 'ministers' as 'counts' with lands and households in their mid-twenties, and the pattern revealed in the Old English poem *Beowulf*, which may be eighth century in date, although not necessarily Northumbrian. At the end of the poem, the hero, Beowulf, then king of the Geats, is called upon to fight a dragon, and goes to confront the beast accompanied by his young retainers (the word thegns is not used, but the vocabulary is generally obscure and archaic). When the hero confronts the dragon, however, all take refuge from it and will not stand by their lord, except for one, Wiglaf, who was moved to go to the assistance of his lord, because: 'He remembered then the favours he had formerly bestowed on him, the wealthy dwelling-place of the Waymundings, confirming him in the landrights his father held.'

[18] For the texts, see Bede, *Eccles. Hist.*, bk 1 ch. 3 (Oswald's exile), bk 5 chs. 4–5 (Puch and Addi), and Bede, *Hist. Abbots*, chs. 1, 8 (Benedict's and Eostorwine's careers). Cf., in Mercia, the career of Guthlac, who became a monk (after a life of quasi-military violence) presumably in his mid-twenties, or at least nine years after 'his youthful strength had increased' enough for him to take command in warfare; see Colgrave (1956), pp. 80–1. For prefects and princes, see above, p. 173. The most interesting discussion of the Carolingian evidence for social status is Odegaard (1945), who suggests a relationship between *uassi* and *fideles* similar to that between *ministri/miles* and *comites* in the material discussed here.

Now, it emerges from other parts of the poem that Wiglaf was of the family of the Waymundings. The passage therefore means that Wiglaf was not in a position to hold his ancestral (indeed paternal) landrights until the king confirmed them to him. A somewhat later passage is of particular note in this connection. After the dragon has been killed and Beowulf himself has died of injuries sustained in the conflict, Wiglaf upbraids the king's other retainers who were cowardly. First, he recalls the benefits they have received from:

> a lord of men who allowed you those treasures, who bestowed on you the trappings that you stand there in – as, at the ale-bench, he would often give to those who sat in hall both helmet and mail-shirt as a lord to his thegns, and many things of the most worth that he was able to find anywhere in the world.

It appears therefore that Beowulf has been responsible for equipping his retainers with armour and trappings and that they have in a very real sense assumed the duty of assisting him in battle by way of return.

This is strongly reminiscent of household knights, and particularly of the custom of *heriot*, known from eleventh-century documents in England, by which a lord would give his retainer or vassal armour and weapons. At a later date *heriot* found some reflection in feudal service in the shape of the relief, that is the payment, which a son made to inherit the land or fief which his father had held: he was, notionally, repaying the cost of the weapons and armour provided for his father by the lord. Just like a 'feudal' lord of later days, Beowulf had, it seems, been responsible also for providing his retainers with land, for Wiglaf goes on with his reproof:

> Your kinsmen every one,
shall become wanderers without land-rights
as soon as athelings over the world
shall hear the report of how you fled,
a deed of ill fame.

Just as with Wiglaf himself, Beowulf would have been responsible for granting or confirming lands to them. The career pattern implied is much like that of Benedict Biscop and Eostorwine: from unlanded household 'minister' to landed count; it is a career pattern in which the influence of the king over the progress of his retainers up the social scale is paramount.[19]

[19] See *Encyclopaedia ASE*, article '*Beowulf*' and references therein; see, in particular, Chase (1981). For *heriots*, see Brooks (1978), pp. 85–91; for the later middle ages, Poole (1946), pp. 94–5. For the texts, see *Beowulf*, lines 2606–8, 2865–70, 2887–90.

The texts discussed here must be related to another, the letter which Bede wrote at the end of his life to Bishop (later Archbishop) Egbert of York. In this, he analysed the evils of the condition of the English church in his own day, and one particular abuse he highlighted was that of secular men founding monasteries which were not really monasteries. The passage is worth quoting in full:

> There are others, laymen who have no love for the monastic life nor for military service, who commit a graver crime by giving money to the kings and obtaining lands under the pretext of building monasteries, in which they can give freer reign to their libidinous tastes; these lands they have assigned to them in hereditary right (*ius hereditarium*) through written royal edicts (*regalibus edictis*).

Why should laymen, including particularly 'almost everyone of the prefects' and 'the king's ministers and servants' have founded monasteries at all when by Bede's account they had no interest in monastic life and no intention of pursuing it? Two answers, not entirely incompatible with each other, are possible. The first is that the monasteries were not as bad as Bede maintained and he was simply criticizing establishments which did not conform to his own standards but which were devout enough according to their lights.[20]

A second possibility, and a more interesting one in the present context, is that the laymen were trying to get something practical out of their monastic foundations. Spiritual benefits cannot have been paramount if we accept Bede's criticism of the state of these monasteries. It seems more likely that what the laymen were in fact seeking to obtain was hereditary right to land as mentioned by Bede in the passage cited above. This fits well with the texts relating to 'ministers', and their progress to establishment as 'counts' through being confirmed by the king in their ancestral rights; by founding monasteries on the land, this process was circumvented, and full hereditary right obtained, with consequences for the king's power to which we must return. The evidence of Bede's *Letter* needs to be seen in the context of the wider issue of land-tenure in early England. Charters from southern England, particularly a much discussed text relating to a place called Wassingwell in Kent, seem to imply that there were two types of tenure: 'bookland' and 'folkland'. 'Bookland' was literally land granted by charter (*boc* in Old English). There has been much discussion as to what the rights attaching to 'bookland', known as 'bookright', were and how these differed from 'folkright', the rights attaching to 'folkland'. But it appears most likely

[20] Sims-Williams (1990), pp. 126–30; see also Peter Hunter Blair (1970), pp. 135–6, and Wormald (1978), pp. 51–4. For the text, Bede, *Letter*.

in view of the wording of the charters that 'bookright' was distinguished from 'folkright' precisely because it conferred hereditary, perpetual tenure. Now, the earliest charters were all in favour of churches, and the church was in principle an undying corporation. Only perpetual tenure would meet its needs. It can be argued therefore that with the establishment of the church in England and the introduction of charters as a means of granting land, perpetual or hereditary tenure was introduced for the first time. Laymen wishing to benefit from it for the aggrandizement of themselves and their immediate families could only do so by establishing churches or monasteries of which their families held the abbacies; or which were effectively spurious monasteries, over which they retained control and in which (as Bede asserted) they practised their former life-styles.[21]

This interpretation conforms well with another observation made by Bede in his letter to Egbert, to the effect that one result of the establishment of spurious monasteries is that

> there is nowhere that sons of the nobles or retired soldiers (*emeriti milites*) can take possession of. In consequence, wandering and without a spouse, ... they live without a commitment to continence, and in consequence they either leave their homeland for which they ought to be fighting in order to go overseas or, with even greater wickedness and lack of shame, because they lack the intention of being chaste, they give themselves up to indulgence and fornication.[22]

Here, then, we may suppose, are the 'ministers', who could reasonably expect to be established on land by the king; but were denied this because the introduction of bookright with its hereditary tenure meant that the land that should have been confirmed to them remained in the hands of the founders of the spurious monasteries.

The interpretation presented above is by no means certain, given the slender body of evidence which underpins it and the intractable character of some of the texts involved, but it is worth standing back and examining its implications for the development of Northumbria. The first is that before the introduction of charters there was no hereditary tenure such as was embodied in charters, but only 'folkright', so that there was no such thing as an hereditary aristocracy either, but simply a group of men wholly dependent on the king for their status as 'ministers' and their subsequent endowment as 'counts'. Thus, the introduction of 'bookright' and the

[21] John (1964, 1966b). For the Wassingwell charter, see Sawyer (1968), no. 328; for a classic commentary see Stenton (1971), p. 311.
[22] Bede, *Letter*.

emergence of spurious monasteries actually created an aristocracy. If this is so, the process presumably took place in the second half of the seventh century when monasteries proliferated, or perhaps more in the early eighth, since Bede particularly identifies the period after the death of King Aldfrith (705) as that in which the foundation of spurious monasteries really took hold. This is a radical view indeed, and in denying the existence of an hereditary aristocracy in earlier periods it seems an implausible one. For, if there was no such aristocracy it is hard see how Wilfrid around the year 648 was part of a family which was evidently aristocratic. Moreover, Bede especially notes that King Oswine of Deira was 'bountiful to nobles and commons alike', so that 'noblemen from almost every kingdom flocked to serve him as retainers'. Such a comment, assuming that Bede was well informed, implies that an aristocratic class was already present in the first half of the seventh century, and that its members sought service with kings, in much the same way as Bede envisaged the disappointed warriors of his letter to Egbert doing, and more strikingly in just the same way as warriors (including Beowulf himself) are presented as coming to serve King Hrothgar of Denmark, as described in *Beowulf*: 'Then to Hrothgar was granted glory in battle, mastery of the field; so friends and kinsmen gladly obeyed him, and his band increased.'

It may be that the situation was more complex; and what bookright did was not so much to create hereditary right but to modify the way in which land was inherited. In some early laws, it appears that land was regarded as the property of the kindred as a whole, and could only be inherited by that kindred rather than by individuals or families within it. Thus the effect of 'bookright' would have been to enable particular lines within a kindred to build up wealth from generation to generation. Even so, the effect on the structure of society of the emergence of really powerful and wealthy aristocrats may nevertheless have been very considerable, the proof for which may be the political instability of the kingdom of Northumbria in parts at least of the eighth century, to which we must now turn.[23]

Once the Bernician line of Æthelfrith had been established as the ruling family of Northumbria from the time of King Oswald (634–42), the seventh century was notable for the stability and continuity of this family's power,

[23] For the chronology of monastic foundation, see James Campbell (1971), p. 51; for tenure and the kindred, Wormald (1985), pp. 22–3; for the political impact of the aristocracy, Yorke (1990), pp. 91–4. For the texts, see Bede, *Eccles. Hist.*, bk 3 ch. 14, and *Beowulf*, lines 64–7. On Wilfrid, see above, p. 180 n. 8.

even though it had been responsible for subsuming under its rule the rival kingdom of Deira (fig. 1). There was an almost unbroken succession from Oswald, to his brother Oswiu (642–70), to his son Ecgfrith (670–85), and so to his brother Aldfrith (686–705). In reality, the political situation may have been less tranquil than this outline suggests. Although Oswald himself appears to have ruled without any challenge from the Deiran royal family, that family was not eliminated at once. Bede tells us that Oswiu 'had as a partner in the royal dignity a man called Oswine, of the family of King Edwin, a son of Osric', who ruled the kingdom of Deira for seven years, while Oswiu ruled Bernicia. Dissension increased between the two rulers; armies were raised and, although battle was not joined, Oswiu caused Oswine to be murdered in 651. As far as is known, Oswine was the last king of Edwin's line; but separatist tendencies in Deira, despite Oswald's mother being Edwin's sister Acha and Oswiu's wife being Edwin's daughter Eanflæd, may still have been a cause of political instability. Oswiu's son Alhfrith and Oswald's son Œthelwald probably ruled Deira as sub-kings under Oswiu, and it is noteworthy that Œthelwald certainly revolted against his uncle, preparing to side with King Penda of Mercia at the Battle of the *Winwaed* in 655; and Alhfrith's disappearance from the historical record after the Synod of Whitby has suggested the possibility that he too may have revolted and been killed or exiled. Stability may therefore have been dearly bought in these ways, and in other ways which Bede does not dwell on, for example by eliminating more or less nearly related family members of the Bernician royal house, who had a potential claim to the throne because there was in this period no expectation of the son of a king necessarily being his successor. To make matters worse, King Oswiu had two wives – Rieinmellt, daughter of Rhun of Rheged (according to Nennius, at any rate) and Eanflæd, daughter of Edwin – as well as a concubine, an Irish princess with whom he fathered Aldfrith. It is unlikely that we are being informed about all the children of these marriages, and very likely that rivalries between them existed. In the case of Aldfrith's accession after the death of Ecgfrith at the hands of the Picts at the Battle of *Nechtanesmere* in 685, Bede may be concealing from his readers a less stable political situtation than he wanted them to assume existed. Aldfrith was illegitimate, and the fact that he was spending his time studying on Iona strongly suggests that he was preparing for a career in the church. His accession was therefore unexpected; discrepancies in Bede's dating of that accession suggest that there may have been an interregnum after Ecgfrith's death; so that it is possible that the accession was a politically divisive one, perhaps sponsored by the Picts who had been victorious over his predecessor, and that the interregnum was a period of conflict.[24]

Oswald's line faltered after the death of Aldfrith (705), and it is striking that the first real indication of political instability appeared after that time, which was also exactly the period to which Bede assigned the greatest evil arising from the foundation of spurious monasteries. Bede does not describe this instability but Stephen's *Life of Wilfrid* does. According to Stephen, an otherwise unheard-of king called Eadwulf succeeded to Aldfrith but, failing to reconcile himself with Wilfrid and indeed threatening to expel him (a move in Stephen's view fatal to his political position), there was a conspiracy against him 'and he was driven out after a mere two months' reign', to be replaced by Aldfrith's son Osred. Although Bede referred in favourable terms to Osred as a Josiah (the king of Judah responsible for purification of worship), there were clearly considerable stresses in his reign, and strongly worded criticism of his actions. Boniface, the English churchman on the continent, represented him as 'driven by the spirit of wantonness, fornicating, and in his frenzy debauching throughout the nunneries virgins consecrated to God', and the author of the ninth-century Latin poem *De abbatibus* accused him of killing his nobles or forcing them into monasteries.[25]

Osred's death in 716 marked the end of the dominance of the house of Æthelfrith, for he was succeeded by Cenred (716–18), who came from a family claiming collateral descent from King Ida of Bernicia (547–59). He in turn was succeeded by Osric (718–29), who was probably a son of Aldfrith, and whose reign thus marked a temporary return to the house of Æthelfrith (fig. 16). At his death, however, Osric appointed as his successor Cenred's brother, Ceolwulf, to whom Bede dedicated the *Ecclesiastical History*. Ceolwulf's reign (729–37) was not entirely a happy one. Bede refers darkly to it as 'filled with so many and such serious commotions and setbacks that it is as yet impossible to know what to say about them or to guess what the outcome will be'. As we have seen, Ceolwulf was temporarily deposed in 731, and his abdication of his own accord in 737 may have been followed by an interregnum resulting from political turbulence.[26]

[24] In general, see Kirby (1991), chs. 4–5; see also Moisl (1983) and the perceptive commentary by Kirby (1974b); for the text, Bede, *Eccles. Hist.*, bk 3 ch. 14. On Rieinmellt, see above, p. 88.

[25] Bede, *Letter*; Stephen, *Life Wilf.*, ch. 59; Bede, *Met. Life Cuth.*, preface; Boniface, *Letters*, no. 73; and Æthelwulf, *De abbatibus*, chs. 2–4 (Campbell (1967)). For the dates of the reigns of these and succeeding kings and justifications for them, see *Sources*, pp. 51–7; for further discussion, see Kirby (1991), pp. 142–62.

[26] Bede, *Eccles. Hist.*, preface and bk 5 ch. 23; on Ceolwulf's expulsion and subsequent abdication, see above, pp. 165–6.

DESCENDANTS OF IDA THROUGH ÆTHELFRITH, KING OF NORTHUMBRIA (d. 616)	DESCENDANTS OF IDA THROUGH CUTHWINE	DESCENDANTS OF IDA THROUGH EATA	DESCENDANTS OF IDA THROUGH BYRNHAM	KINGS NOT CLAIMING DESCENT FROM IDA
Aldfrith (686–705)				
				Eadwulf (705–6)
Osred I (706–16)				
	Cenred (716–18)			
Osric (718–29)				
	Ceolwulf (729–37)			
		Eadberht (737/8–58)		
		Oswulf (758–9)		
				Æthelwold Moll the patrician (759–65)
			Alhred (765–74)	
				Æthelred I (son of Æthelwold Moll; first reign 774–9)
		Ælfwald I (779–88)		
			Osred II (788–90)	
				Æthelred I (second reign 790–6)
				Osbald the patrician (796)
				Eardwulf (796–806)

FIGURE 16 Chart of the succession of the kings of Northumbria in the eighth century
The descent of the branches of the family of Ida is based on the ninth-century genealogies and may reflect claims rather than reality (Dumville (1976). See the very helpful genealogical charts in *EHD I*, tables 4–5.

Eadberht's long reign (737/8–58) was looked back on as something of a golden age by Alcuin in his poem *On the Bishops, Kings, and Saints of York*, and it clearly was a return to the greatness of the reigns of his seventh-century predecessors – although not to the rule of King Æthelfrith's line, for Eadberht was a cousin of Ceolwulf, and therefore a descendant of Ida by another branch of the family. His reign was not free of dynastic strife. In 750, we find him pursuing Offa, the son of his predecessor Aldfrith, and forcibly removing him from Lindisfarne, where he had taken sanctuary. Like Ceolwulf, Eadberht is said to have resigned voluntarily in 758 to become a monk, but such a calm state of affairs seems implausible since his son Oswulf was killed 'by his household' (*a sua familia*) only a year after his accession. He was in turn succeeded by an aristocrat apparently with no pretence to descent from Ida, Æthelwold Moll (759–65), presumably the Moll referred to as 'patrician' on whose position we have already commented. There must be a strong suspicion that this man and his supporters had engineered the downfall of Oswulf and perhaps also the resignation of Eadberht. Æthelwold Moll had to defeat and kill a certain Oswine, perhaps a claimant to the throne, and he himself 'lost the kingdom' in 765. His successor was

Alhred, who claimed descent from Ida by yet another collateral branch. His reign (765–74) was clearly not tranquil, for a letter from him and his queen Osgifu to the continental churchman Bishop Lul refers to disturbances in Northumbria, and in 774 he was deposed and fled into exile with the Picts. The Northumbrians then accepted as king Æthelred I, son of the former king Æthelwold Moll. In 778, he ordered the killing of three dukes (*duces*) and in 779 he was driven into exile, to be replaced by a member of the line of Eadberht, namely Ælfwald I, son of the former king Oswulf. Ælfwald himself was murdered in 788 as a result of a conspiracy formed by his 'patrician' Sicga. During the reign, the Northern Annals report the killing at Christmas 780 of an earlier 'patrician' of the king, Bearn, at the hands of two dukes (*duces*). Alcuin indeed regarded the reign as inaugurating a period of moral decline: 'From the days of King Ælfwald fornications, adulteries, and incest have flooded the land, so that these sins have been committed without any shame and even with the handmaids of God.'

Osred II (788–90) then ruled; he was the son of King Alhred and his queen Osgifu, and thus descended from Ida by two branches of the family, that is via Ida's son Ocga on his mother's side, and via Ida's son Eadric on his father's side. In 790, he was deposed, tonsured at York and expelled to be replaced by Æthelred I returning from exile. According to the Northern Annals, Osred was on the Isle of Man in 792 when he relied on the proffered support of certain nobles to attempt to return to Northumbria. He was deserted by his 'soldiers', however, and captured by King Æthelred who had him killed. In 796, Æthelred was himself murdered at Corbridge; certain 'princes' (presumably those responsible for the murder) made Osbald, the 'patrician', king, but he was deposed after only twenty-seven days. His successor was Eardwulf, whose father was called Eardwulf or Earulf, but whose family background is otherwise obscure. He had evidently been involved in royal affairs, for the Northern Annals record how in 790 he had been arrested by King Æthelred who had ordered him to be put to death. He survived to become king, however, and in 798 he defeated a revolt led by Ealdorman Wada, one of the murderers of King Æthelred, who may have had the support of the exiled Osbald (if a letter of Alcuin to the latter belongs to 798 and was intended to dissuade him from joining the revolt). In 799, an ealdorman called Moll, possibly a descendant of the former king Æthelwold Moll, was put to death at the king's urgent order; and in 800 Ealhmund, said by some to have been son of the former king Alhred, was seized and put to death by Eardwulf's guards. Expelled in 806, according to the *Anglo-Saxon Chronicle*, he was (according to the *Royal Frankish Annals*, corroborated by papal correspondence) restored to the throne in 808 through the intervention of the Frankish emperor Charlemagne, apparently after the two-year reign of a certain Ælfwald II, otherwise entirely unknown.[27]

It is clear from the foregoing that the eighth century was characterized by much greater political instability than was the seventh. In view of the quite different character of our sources for the two centuries, the contrast may not be as great as this sketch suggests. Bede's *Ecclesiastical History*, our principal source for the seventh century, aimed to paint as rosy a picture as possible of it in order to throw into relief by comparison the deficiencies which Bede perceived in his own time. The Northern Annals tended to give prominence to disruptive events such as lend themselves to representation in annalistic form. Nevertheless, the contrast is too stark for it not to have some basis in reality. It is indeed a very real possibility that the instability of the eighth and early ninth centuries arose because of the growth of aristocratic power, for the foregoing account shows how often members of the aristocracy were involved in the politics of the kingdom of Northumbria in that period, as claimants to the throne, as violent actors in the political drama and as members of the councils which deposed kings.

That the Northumbrian church was also involved in the dynastic disputes is beyond doubt. It appears from a letter of Alcuin that Eanbald II, archbishop of York, had received and protected the king's enemies, while seizing the lands of others, so that he found it advisable to go about with a large armed force. Some churches clearly had an axe to grind in the political struggles. It can be no accident that when, as we have seen, Offa, son of King Aldfrith, was dragged from sanctuary at the tomb of St Cuthbert at Lindisfarne, the then king Eadberht incarcerated the bishop of Lindisfarne, presumably because he was implicated in Offa's activities. Mercia too was hostile; in 801 King Eardwulf of Northumbria led an army against King Cenwulf of Mercia allegedly because the latter had received his enemies. This campaign was admittedly followed by a peace accord between the two kingdoms.[28]

The deficiency of our sources hinders us in assessing how far this instability continued into the ninth century. After 806, the continuation of the Northern Annals was either never attempted, or the results of such an

[27] Alcuin, *BKSY*, lines 1273–87; *Hist. Kings*, entries for 750, 758, 759, 765, 774, 778, 779, 780, 790, 792, 796, 798, 800 (pp. 39–63); for the letter of Alhred, Tangl (1916), no. 121, trans. *EHD I*, no. 187; Alcuin, *Letters*, nos. 16, 109. On Moll, see above, p. 173, and, on Eardwulf, *ASC* (D), *s.a.* 806, *RFA*, *s.a.* 808, below p. 196, and Story (2003), pp. 145–56.

[28] For Offa, see *Hist. Kings*, *s.a.* 750 (pp. 39–40); for discussion, see David Rollason (1996), pp. 97–9; for the 801 campaign, see *Hist. Kings*, *s.a.* 801 (p. 65).

attempt have been lost. At any rate, the historical sources available from then until the Viking capture of York in 867 are very much inferior to those for the eighth century. They consist chiefly of Symeon of Durham's *On the Origins and Progress of this the Church of Durham* and Roger of Wendover's *Flowers of History*. Although the writers of these may have had access to authentic information relating to ninth-century Northumbria, they are relatively late in date (Symeon writing in the early twelfth century, Roger in the thirteenth) and they appear not to have been independent of each other – at best, Roger's sources were the same as Symeon's – so they cannot be regarded as corroborating each other. Moreover, neither writer was aware of the second reign of Eardwulf, which means that their dating of the reigns of subsequent kings was almost certainly wrong. In view of this, a case has been made for discarding their dating after the reign of Ælfwald II in favour of one based on numismatic evidence, that is on estimating the probable lengths of reigns on the basis of the volume of surviving coinage, using the coinage of the archbishops to confirm the dating of the kings' coinage; and deploying the dating evidence of the hoards as a whole in which coins have been found. Although this approach is still subject to scholarly confirmation, and will perhaps always retain an element of uncertainty, the corrections it proposes to the dates of the reigns of ninth-century kings (which is all we really know about them) make those reigns look somewhat more settled than does the dating derived from the historical sources. Eardwulf's second reign for example, looks longer and more settled (808–*c.* 830, rather than *c.* 810). The brief reign of Rædwulf (*c.* 858 on the basis of the numismatic evidence) is notable; but according to Roger of Wendover (who admittedly may not have been well informed) he was killed fighting Vikings rather than being the victim of political turbulence. Even the *Anglo-Saxon Chronicle*'s claim that dispute between two rival kings, Osberht and Ælle, preceded the capture of York by the Viking Great Army in 866–7 may not, as we shall see, have been well founded.[29]

Whatever the truth about the ninth century, however, there can be no doubt that aristocratic ambition and political turbulence were features of

[29] See Pagan (1969); see further Dumville (1987) and Lyon (1987); for a useful tabulation, see Grierson and Blackburn (1986), p. 302. For Rædwulf, see below, p. 212 and n. 4, and for the capture of York, below, p. 212. The dates of the ninth-century kings with the dates derived from both historical and numismatic sources are as follows: Ælfwald II (806–8); Eardwulf (second reign: historical: 808–*c.* 810; numismatic: reign may extend as late as *c.* 830); Eanred (historical: *c.* 810–40; numismatic: succeeded as late as *c.* 830; died in 850s, perhaps 854); Æthelred II (first reign: historical: 840–4 or 841–5; numismatic: succeeded in the 850s, perhaps *c.* 854; died perhaps 858); Rædwulf (historical: 844; numismatic: *c.* 858); Æthelred II (second reign historical: 844–8; numismatic: *c.* 858–62); Osberht (historical: 849–62 or 848–67; numismatic: *c.* 852–67); Ælle (historical: 862–7; numismatic: 867).

at least substantial parts of the eighth century. It does not follow, however, that the aristocracy was anarchic in its aims and actions. In this connection it is worth considering the precise wording, which may well be contemporaneous, of the Northern Annals when describing King Alhred's deposition in 774. The Annals state that 'by the counsel and consent of all his people he was deprived of the society of the royal household and princes' (*consilio et consensu suorum omnium, regie familie ac principum destitutus societate*). The account of the expulsion of Osbald in 796 is expressed in virtually the same words: 'deprived of the society of the royal household and princes' (*omni regie familie ac principum est societate destitutus*). This is strongly suggestive of a council or assembly, an interpretation reinforced by the 774 annal in the *Anglo-Saxon Chronicle*, D version, which reads: 'In this year the Northumbrians drove their king Alhred from York at Easter, and took as their lord Æthelred, Moll's son.' The reference to York and Easter suggests an assembly held in the city, perhaps one which met regularly at Easter, and the second half of the annal suggests that that assembly, which had deposed Alhred, went on to elect Æthelred Moll. The words should be compared with another annal in the Continuation of the *Ecclesiastical History*, which describes how, after the killing of Oswulf in 759, Æthelwold 'was elected (*electus*) by his people (*plebs*) and began to reign'. In this case, we are told, Oswulf had been 'treacherously killed by his soldiers' (*ministri*) – so the implication would appear to be that a council had been called following the killing. The account of the deposition of Æthelwold Moll in 765 may also be a reference to a council: 'Æthelwold lost the the kingdom of the Northumbrians at *Pincanhalh* on 30 October.' This place, which is unidentified, occurs in the Northern Annals and the *Anglo-Saxon Chronicle* as the place of an ecclesiastical council in 787 and 798.[30]

There are grounds here for seeing Northumbrian politics in the eighth century as more ordered than they appear at first glance, and for seeing the kings as dependent for their position on their nobles in council, and the influence of that council as considerable. Such 'constitutional' arrangements were naturally not always effective, but that they existed and were accepted is corroborated by distinctions made by the annalist for occasions when they were not respected. Thus in 796 Osbald was 'raised to the throne by certain princes of this people' (*quibusdam ipsius gentis principibus*), presumably

30 For Alhred, *Hist. Kings*, *s.a.* 774 (p. 45), and *ASC* (D) *s.a.* 774; for Oswulf, Bede, *Continuations*, *s.a.* 759; for Osbald, *Hist. Kings*, *s.a.* 796 (p. 57); for Æthelwold, *Hist. Kings*, *s.a.* 765 (p. 43). On *Pincanhalh*, see Cubitt (1995), pp. 291, 292. For the case that the wording of the annals in *Hist. Kings* can be treated as contemporary with the eighth century, see Story (2003), pp. 93–133.

the work of an inadequately constituted council; Ælfwald I was killed 'as a result of a conspiracy'; and Osred II was 'betrayed by the treachery of his princes, captured and deprived of his kingdom'. The record of the *Anglo-Saxon Chronicle* for tenth-century Northumbria also suggests a 'constitutional' role for a council in making and breaking the Viking kings, and agreeing or refusing to cooperate with the kings of southern England. Our evidence therefore suggests that the aristocracy, as represented by the council, whose function appears to have been recognized, had a major role in sustaining or disrupting political stability in Northumbria.[31]

It appears, however, that the action of the church in shaping Northumbrian kingship, that is the third element in our discussion, may have been even more important. Discussion of this is rendered difficult by our ignorance of what English kings – and Northumbrian kings in particular – were like before the coming of Christianity. It may be that paganism had elevated their position by associating them with religious rites and practices. Something of this sort is discernible in Tacitus's first-century account of the peoples east of the Roman Empire, *The Germania*, in which he describes kings who had a role in the rituals of their peoples' pagan religion. It may even be that pre-Christian English kings had something like the sacral religious functions analysed in Frazer's classic study, *The Golden Bough*, and that this affected many aspects of their behaviour and their rule. In particular, it can be held to explain why recently Christianized kings became saints, as certainly happened in early Northumbria: the pagan priest-king, in other words, evolved into the Christian saint. Oswald had a widespread cult following his death in battle against the pagan Penda, Oswine was venerated as a martyr-king following his murder by Oswiu and Edwin was venerated as a martyr in battle at least to the extent of his body being translated to the abbey of Whitby and his head being enshrined at York Minster. Moreover, death by violent means was a particular means to sanctification, because kings who died violently were (according to clues found in later Scandinavian saga material) equated with sacrificial kings, whose lives were offered up for the good of their people. Related to this may have been the way in which the pagan king of Mercia Penda, having killed King Oswald in battle, cut off his arms and head and impaled them on stakes. It is possible to interpret this as a sacrifice to a pagan god, so that Oswald's status as a saint-king was the church trying to christianize his real status in the eyes

[31] In general, see Bund (1979), esp. pp. 625–61. For Osbald's elevation, *Hist. Kings*, *s.a.* 796 (p. 57); for Ælfwald and Osred II, *Hist. Kings*, *s.a.* 790 (p. 52). On the council in the tenth century, see below, p. 230.

of his people as a pagan sacrificial king. Corroboration of the pagan connotations of this can be found, it can be argued, in the twelfth-century *Life of St Oswald*, which describes how Oswald's right arm was left on the battlefield, where his brother Oswiu, after his accession to the throne, sent a messenger to retrieve it. As this messenger went to take the arm, a great bird of the crow family took it up into an ash tree; when the bird was persuaded to drop the arm, a curative spring emerged from the place where it fell to the ground. The great bird may well have represented a raven, which was sacred to the pagan god Woden, as also was the ash tree; and it is therefore possible that we have here an indication of the original pagan, sacral character of kingship. The implication of all this would be that kingship was shaped ideologically by paganism, to which the church had to adapt its ideas of the basis of royal power, treating sacrificial kings as saint-kings, for example. As for the kings, their tracing of their ancestry, as set out in the early ninth-century collection of English genealogies, back to the German god Woden, could be seen as an indication that they acknowledged the pagan origins of the ideological basis of their power even in a Christian world.[32]

Such arguments cannot be dismissed lightly, even though the sources are so exiguous or late, and the arguments often based on inference. Our sources derive exclusively from church writers after the conversion to Christianity, and may in consequence have deliberately masked the real (pagan) ideological basis of kingship from us. Against these arguments for the pagan origins of the ideology of kingship, we possess an impressive array of evidence for the role of the church in shaping that ideology. First, the evidence of royal saints. Although as we have seen this can be interpreted in pagan terms, there are strong grounds for seeing it as an entirely Christian phenomenon, stemming in part from the church's desire to bolster the political position of kings, to differentiate them from the aristocracy whose ambitions (at least from the late eighth century) could threaten their position perhaps to the church's detriment, and to restrain attacks on them. A Christian saint-king was not the same as a pagan sacral king; the latter's powers were supposed to derive from his blood and were exercised in his life-time, whereas the former's derived from his deeds or the manner of his death and his status was accorded to him posthumously by the church in the form of a cult. By elevating a king to sainthood, the church could therefore make a political point – in the case of Oswald, that Christian devotion coupled with

[32] Chaney (1970). See also Hoffmann (1975), pp. 16–58, Folz (1984) and, specifically on Oswald, Jansen (1995). For *The Golden Bough*, see Frazer (1922), chs. 2, 6, 24. For the texts, see Bede, *Eccles. Hist.*, bk 3 ch. 12, and, for *Life of St Oswald*, Arnold (1882–5), I, pp. 352–8.

resistance to pagans was the highest form of merit; in the case of Oswine, that the king's virtues (which Bede extolled) were worthy of sainthood and that the crime of his murder should never be forgotten. Significant in this respect is the concentration of murdered royal saints in the late eighth and early ninth centuries, shortly after papal legates had visited England in 786 and promulgated canons, including one which described the king as the 'Lord's anointed' and forcefully condemned those who killed kings. King Ælfwald, who presided over the council, was shortly afterwards murdered and venerated as a saint, and so was Ealhmund, murdered in 800, and recorded in a ninth-century list of the resting-places of the saints. King Eardwulf too may have been regarded as a saint, since he was supposed to have miraculously survived the attempt to execute him outside Ripon Minster. There was not at this period any formal process of canonization and the cults of these saints may often have been quite localized; nevertheless they seem to represent an example of the church seeking to shape the ideology and status of kingship.[33]

Another area in which the church may have been shaping the ideology of kingship was that of the rituals by which kings were inaugurated. The important text here is the account in the Northern Annals of the inauguration as king of Eardwulf in 791: 'Eardwulf...was raised to the insignia of the kingdom (*regni infulis est sublimatus*), and was consecrated (*consecratus*)... in York in the church of St Peter at the altar of the blessed Apostle Paul.' The setting for the ceremony and the use of the word 'consecrated' suggest that this was an example of the ritual of anointing with holy oil which had been used for the inauguration of the first Carolingian king Pippin in 751, and became a regular and necessary part of Carolingian inauguration rituals in the ninth century. Although it does not seem to have become a regular part of English royal inauguration rituals until the tenth century, it does seem to have occurred in Northumbria at least on this occasion. It emphasizes the importance of the church in influencing kingship because anointing was an ecclesiastical monopoly, although the aristocracy is likely to have been involved also in the role of electors.[34]

We have seen already the impact that Christianity had on kings, so we should perhaps not be surprised that the church was in a position to influence the ideological basis of their position in such ways as this. Already in

[33] See Nelson (1973) and David Rollason (1983), with references. Thacker (1985) emphasizes the local character of such cults in Mercia.

[34] On Carolingian anointing, see Enright (1985) and Nelson (1987). For the text, see *Hist. Kings*, *s.a.* 791 (pp. 57–8).

the seventh century, Bede was developing an image of Christian kingship in the treatment he gave in his *Ecclesiastical History* of Oswald, 'the most Christian king' (*rex Christianissimus*): his piety; his deference to the church; his defence of Christianity; above all the tangible demonstration of his sanctity in the miracles which followed his death. Bede was in effect developing a 'mirror of princes' through Oswald, and to some extent Edwin and Oswine. The purpose of this may not have been deliberately to elevate the image of kingship and to emphasize to kings how far their position was linked to their piety and the support of the church, but this was very much the effect of it, for not the least striking aspect of the *Ecclesiastical History* is its treatment of Christianity as a means to worldly success. As the pagan priest Coifi said to King Edwin when about to renounce paganism: 'None of your followers has devoted himself more earnestly than I have to the worship of our gods, but nevertheless there are many who receive greater benefits and greater honour from you than I do and are more successful in all their undertakings.' When Oswald was leading his army against the British king Cædwalla who had invaded Northumbria, the prayers of his army around a makeshift cross he had set up were, in Bede's presentation of the events, the key to success. King Oswiu was, according to Bede, able to defeat the pagan king of Mercia Penda who had thirty legions of soldiers because he trusted in Christ and had vowed to 'dedicate his daughter to the Lord as a holy virgin and give twelve small estates to build monasteries' in the event of victory. We find the same attitude in Stephen's *Life of Wilfrid*, in which King Ecgfrith is represented as enjoying worldly success while he was favourable to Wilfrid: 'The pious King Ecgfrith and Queen Æthelthryth were both obedient to Bishop Wilfrid in everything, and their reign was marked by fruitful years of peace and joy at home and victory over their enemies.' When the same king ejected Wilfrid from his see, and his second queen, Iurminburg, stole holy relics from him, the results were very different: the queen was stricken with illness and the king threatened with punishment. The refusal of his successor King Aldfrith to favour Wilfrid brought on him an irreversible doom: he did not recover from the illness which afflicted him. The attitudes reflected in these examples are perhaps not surprising and were certainly not restricted to Northumbria; but the prominence and explicitness of the presentation of Christianity as a success-religion are striking and should be taken seriously. Allowing for the ecclesiastical character of our sources, it is nevertheless hard to avoid the conclusion that if only a proportion of what they present affected the kings, Christianity was a major force in shaping royal policy and attitudes. Away in Wessex, Bishop Daniel of Winchester, one of Bede's correspondents, wrote to the English missionary on the continent, Boniface, to suggest searching questions that the latter might address to the pagans and which went even further in this direction. Amongst them was a demand to know why 'the Christians possess fertile lands, and provinces fruitful in wine and oil and abounding in other riches' while the pagans have 'lands always frozen with cold'.[35]

This attitude to Christianity as a guarantee of earthly success has to be seen in the context of the close relationship between king and church emphasized by our ecclesiastical writers. The church perceived kings as its most important source of assistance (as Bede in his letter to Bishop Egbert urged him to lean on the support of King Ceolwulf) and sought to influence their behaviour. Nothing is more striking in this respect than the preface to the *Ecclesiastical History* addressed to the same King Ceolwulf, in which Bede sets out with absolute explicitness the didactic purpose of his work:

Should history tell of good men and their good estate, the thoughtful listener is spurred on to imitate the good; should it record the evil ends of wicked men, no less effectually the devout and earnest listener or reader is kindled to eschew what is harmful and perverse, and himself with greater care pursue those things which he has learned to be good and pleasing in the sight of God.

Nowhere is this perceived inter-dependence of kings and church more clearly articulated than in Alcuin's treatment of the period when Eadberht was king (737/8–58) and his brother Egbert was archbishop of York (735–66):

These were fortunate times for the people of Northumbria, ruled over in harmony by king and bishop: the one ruling the church, the other the business of the realm. On his shoulders the one wore the *pallium* sent by the pope, on his head the other bore his ancestors' ancient crown. The one was powerful and energetic, the other devout and kindly, both lived in peace together as kinsmen should.

We do not know how Eadberht came to power, but we cannot fail to be struck by the fact that that his brother became archbishop before his accession to the throne.[36]

The balance between royal government, aristocratic power and the position of the church may have been epitomized by York, the city at the heart of the kingdom as it developed through the eighth century and beyond.

[35] See Wallace-Hadrill (1971), pp. 72–97. For the texts, see Bede, *Eccles. Hist.*, bk 2 ch. 13, bk 3 chs. 2, 24; Stephen, *Life Wilf.*, chs. 19, 39, 58; and (for Daniel's letter), Tangl (1916), no. 23; trans. *EHD I*, no. 167, and Emerton (2000), no. 15.

[36] Bede, *Letter*; Bede, *Eccles. Hist.*, preface; Alcuin, *BKSY*, lines 1277–83 (on which, James Campbell, pers. comm.).

Capital cities are an important ingredient in the making of a quasi-modern state, and scholars have often assumed that York in the pre-Viking period was in some sense the capital of Northumbria. In this they have been encouraged above all by two references in the writings of Alcuin. In a letter from Frankia to King Æthelred I of Northumbria concerning the 793 sack of Lindisfarne by the Vikings, he used the words 'head of the whole kingdom' (*caput totius regni*) in connection with York. Then again in his *Poem on the Bishops, Kings, and Saints of York* he described Edwin's baptism in 627 as being in York within the walls of the city, 'whose heights he then raised to greater eminence, by choosing to make it the *metropolis* of his realm'. Natural as it is to link these passages with York's putative status as a royal centre, it is potentially misleading. The expression *caput totius regni* needs to be seen in its context, which is Alcuin's account of a rain of blood which fell on York and which he regarded as a presage of the catastrophe which befell Lindisfarne: 'What is the meaning of the bloody rain which we saw in Lent in the city of York in the church of St Peter, the chief apostle, which is the head of the whole kingdom, falling in a clear sky menacingly from the top of the roof at the north end of the building?' It is not absolutely clear whether the clause, 'which is the head of the whole kingdom', applies to the city of York or the church of St Peter. In her translation, Dorothy Whitelock opted for the former, but the natural reading of the sentence, especially in view of the reference to the roof of the building, is to see it as qualifying the church of St Peter. The context of Alcuin's expression is therefore entirely ecclesiastical. He meant that St Peter's was the metropolitan church of the north, not that York was the capital of Northumbria. The word *metropolis* in the lines quoted above may have had a similarly ecclesiastical significance. It may be that the fortress walls themselves were given over to ecclesiastical use, in part at least as the precinct of an ecclesiastical quarter, as they have been in later centuries.[37]

Our sources generally have few references to reigning kings being present in York at all. As we have seen, Edwin was there in 627 for his baptism, and possibly later if the account in the *Life of Gregory the Great* relates to York. The oft-repeated statement that King Osric (633–4) was killed defending York is based on the assumption that Bede's reference to his last stand in an unnamed 'fortified city' refers to York – an assumption which is almost certainly erroneous. King Oswald is said to have completed Edwin's

[37] Alcuin, *BKSY*, lines 202–4, and Alcuin, *Letters*, no. 16 (cf. trans. in *EHD I*, no. 193). See the suggestive reconstruction proposed by Norton (1998a), and cf. the somewhat different interpretation in Tweddle, Moulden and Logan (1999), pp. 151–67.

stone church, but, even assuming that this task required his presence, there is no evidence that he visited York on any other business but that of the church. The next reference to a king being in York is not until 774 when King Alhred is said to have been expelled 'from York at Easter', presumably – as we have seen – by a council which was meeting there. Then in 796 Eardwulf was consecrated king there but the ceremony – as we have also seen – was a markedly ecclesiastical one and the emphasis is on his presence in the church of St Peter rather than on York as a royal centre. The next reference does not come until the Viking capture of York in 866/7. In this York does indeed appear as crucial to the control of Northumbria when the two kings Ælle and Osberht united to recapture and attempt unsuccessfully to hold it, but this is the first time we get any such impression, and it is notable that the sources in question, the *Anglo-Saxon Chronicle* and Asser's *Life of King Alfred*, may well not have been well informed because they were written in the south.[38]

This modest record of reigning kings being present in York should be compared with the record of dead kings there. Two of Edwin's children were buried in the church of St Peter, and his head was enshrined in the *porticus* of St Gregory in the same church, although significantly his body was at Whitby. The sub-king Ælfwini was likewise buried there, although we should notice that the context for this was in a sense ecclesiastical – the king had met his end at the Battle of the Trent in fulfilment, according to Wilfrid's biographer Stephen, of a prophecy uttered by Wilfrid in response to a perceived injustice done to him by King Ecgfrith. The burial of Osbald in St Peter's does not really count, since he had long ceased to be king at the time of his death, nor for the same reason does that of Eadberht – also, the sources seem to go out of their way to emphasize that he was buried there together in the same *porticus* with his brother Egbert, archbishop of York, a family rather than a royal burial. By contrast, we hear extensively of other royal burial-places: Oswine at Tynemouth, Œthelwald at Lastingham, Oswald at Bardney, Bamburgh and Lindisfarne, Oswiu and Edwin at Whitby, Ecgfrith on Iona, Ceolwulf at Lindisfarne. For many of the eighth- and ninth-century kings, we simply do not have any information about their place of burial, but that is in itself significant. If they had been buried at York, the annalists would surely have said so. In short, York was by no means a royal mausoleum like St Augustine's Canterbury was to the Kentish kings,

[38] *Sources*, pp. 126–7, 132–6 (A.1.2, A.4.1, A.4.4 = Bede, *Eccles. Hist.*, bk 2 chs. 14, 20, bk 3 ch. 1; see above, p. 13, and on the Viking capture below, p. 212.

St Germain-des-Prés to Frankish rulers or St Denis to the Capetian kings of France. When kings were buried at York, the reasons behind it seem to have been exceptional, and to have had a markedly ecclesiastical flavour. Repeated accounts in our sources show that kings, and also important aristocrats or royal officers, were living in York in retirement or as political refugees. King Eadberht's death and burial there in 768 came a full ten years after his abdication, in which period he had been a tonsured member of the church of York. King Osbald's short reign in 796 was followed by a period in which he was an 'abbot', perhaps in York, before his burial in the city. Two royal servants, presumably great nobles, had similar careers. In 794 and 796, we hear respectively of the deaths at York of Æthelheard and Ælric, both described as 'formerly duke, then a cleric' (*quondam dux, tunc clericus*). King Osred II, however, was less fortunate; although tonsured in York after his deposition, his enemies would not allow him to remain and he was driven into exile. Even less lucky were King Ælfwald's sons who took refuge in York Minster, but were taken thence by force and killed by King Æthelred.[39]

These records of the presence of ex-kings and aristocrats in the church of York suggest not that York was a royal centre, but that the political and military independence of the church of York and its dominance of the city made it a suitable place for such people to take refuge, as the monastery of Lindisfarne sheltered the prince Offa from King Eadberht's wrath (ineffectually as things turned out). This aspect of the activities of the church of York is suggested in a letter of Alcuin to Archbishop Eanbald II, in which he accuses him of maintaining too many military retainers and sheltering the king's enemies. It is difficult to reconcile this with York as a predominantly royal centre. How could the archbishop have behaved as Alcuin accuses him of doing had it been so? The implication must be that York was essentially an ecclesiastical city ruled by its archbishop, comparable to Trier, Mainz or even Cologne. The Coppergate helmet, found in the excavations at Coppergate in York, may fit into this context. It is unquestionably a high-status object, according to Dominic Tweddle 'made for a member of the Northumbrian royal house, or one of the greater nobles of Northumbria' (ill. 25). But it may be that Tweddle's further comment that it shows the 'intertwining of secular and ecclesiastical power' is understating the importance of the

[39] For burials, see Krüger (1971), esp. pp. 35, 251–3. For the texts, see *Sources*, p. 146 (Edwin's children, A.4.14 = Bede, *Eccles. Hist.*, bk 2 ch. 14), pp. 135–6 (Edwin's head, A.4.4 = Bede, *Eccles. Hist.*, bk 2 ch. 20), pp. 146–7 (Ælfwini, A.4.15 = Stephen, *Life Wilf.*, ch. 24), p. 148 (Osbald, A.4.18 = *Hist. Kings*, *s.a.* 799 (p. 62)), p. 147 (Eadberht, A.4.16 = *ASC*, *s.a.* 768) and pp. 144–5 (A.4.12 = *Hist. Kings*, *s.a.* 791 (p. 53)). For Osred II, see *Hist. Kings*, *s.a.* 790 (p. 52).

ILLUSTRATION 25 York, Coppergate, eighth-century (*c.* 750–75) helmet, by courtesy of York Archaeological Trust
The inscription referred to below is on the bands of brass running from nose to nape; another (incomplete) inscription is on the band running from crown of the head to the right and left ears.

remarkable inscription across the crown of the helmet, now translated to read: 'In the name of our Lord Jesus Christ and of the Spirit of God, let us offer up Oshere to All Saints. Amen.' Christian symbolism on helmets is found elsewhere, for example in the representation of the cross on the nose-piece of the Benty Grange helmet from Derbyshire, or the crosses on the helmet from Plänig near Mainz, but an inscription of this length is unparalleled. Its strong Christian character and the mention of dedicating the wearer to All Saints fit extremely well into a model where York was dominated by the Christian church, wielding massive political influence and substantial military power.[40]

By contrast with the sparsity of evidence for York as a royal centre, we have a clear picture of it as an ecclesiastical centre. Established for a short period as a metropolitan see in 627, it lost its status after the death of King Edwin and the apostasy of his successors. The next Christian king Oswald established not York but Lindisfarne as the Northumbrian see. Following the Synod of Whitby in 664, however, York once again became a bishopric and in 735 an archbishopric, a status which it has retained down to the present day. Our sources show that it was important enough to have been a pre-eminent centre for the building and endowment of churches. As we have seen, King Oswald completed Edwin's stone church (St Peter's), and this then fell into disrepair to be restored by Wilfrid between 669 and 671. Bishop Bosa seems to have organized a monastic community around this church. Then in the later eighth century the archbishops enriched the decoration of the churches in York, and specifically St Peter's and their generosity is described in some detail by Alcuin. Archbishop Æthelberht founded an entirely new church, the *Alma Sophia*, in the city and, if Alcuin's description is accurate, this was a church of some magnificence, possibly based on a round or polygonal plan reminiscent of the Byzantine or south Italian models which also seem to be invoked by its dedication. It was clearly a major church, and in addition to it and St Peter's Alcuin mentions a 'cell' of St Stephen, which despite the term used was clearly a church of some importance, and a church or chapel of St Mary. Some of the churches mentioned in *Domesday Book* may well have had their origin in the pre-Viking period.[41]

At the heart of Northumbria, then, was a city of the church, comparable to Metz with its complex of churches, or perhaps to Trier with its cathedral dominating the life of the city. Kings no doubt came there, but the evidence suggests that they came on the church's terms. It is noticeable that after the time of Oswald they were not even involved in the building and endowment of churches. In York, the bishops and then the archbishops were the patrons. The city's character may have reflected that of Northumbria at large in so

[40] Alcuin, *Letters*, nos. 209 and 233 (wrongly cited as no. 232 in David Rollason (1999a), p. 134); on the helmet, see Tweddle (1992), esp. pp. 1170–1; and cf. on the inscription *Sources*, pp. 160–1 (A.4.32), citing Binns, Norton and Palliser (1990). It should be recognized, however, that the helmet's connection with All Saints Pavement is inferential, that we might expect a connection rather with St Peter's, and that in any case we do not know whether or not the helmet was made in York.

[41] On York churches, see *Sources*, pp. 132–63 (A.4) and, on this and York's development generally, David Rollason (1999a), pp. 128–38. It is notable, however, that York was not the exclusive centre for the consecration of bishops, which also took place at Sockburn (County Durham), Corbridge and Bywell (Northumberland); see *Hist. Kings*, *s.a.* 786 (p. 51), *s.a.* 796 (p. 58), and Symeon, *Origins*, bk 2 ch. 5.

far as we have seen evidence for a preponderant role of the church in that kingdom. As with Northumbria's cultural identity, the kingdom's political identity may have owed more to the church than to the kings or even to the powerful and apparently ubiquitous aristocracy. Whether because of royal government, aristocratic influence or ecclesiastical dominance, however, Northumbria in the pre-Viking period appears to have been a solid and unified kingdom. Even the political turmoil of the eighth century, when rival dynasties vied for power and reigns of kings often ended violently, seemed in no way to threaten its unity as a kingdom. From the late ninth century, however, events and processes were put in train which led not just to Northumbria's fragmentation but ultimately to its disappearance. It is to those which we must now turn.

PART III

The destruction of Northumbria

CHAPTER 6

The Northumbrian 'successor states'

The fragmentation of Northumbria, 866/7–c. 1100

The history of how the kingdom of Northumbria came to be fragmented as a result of the Viking invasions and then divided, definitively as things turned out, between the kingdoms of Scotland and England is clear enough in outline, and can be briefly summarized. Viking raids, which are first recorded in the south of England with the *Anglo-Saxon Chronicle*'s account of three ships' companies of 'Northmen' landing at Portland Bill in Dorset sometime in the period 786–802, began to affect Northumbria with a Viking raid on Lindisfarne in 793, and another in the following year on a monastery called *Donamuthe* in the *Anglo-Saxon Chronicle*'s account. In the twelfth century, the Durham historian Symeon identified this second place with Jarrow, which does indeed lie beside a small river called the Don which debouches into the Tyne a couple of hundred yards downstream of Jarrow monastery; but, as no other source refers to Jarrow by this name, it is possible that some otherwise unknown monastery by the mouth of the River Don in Yorkshire (perhaps at Doncaster itself) was the church in question.[1] These two raids are the only ones which we know to have occurred in Northumbria in this first phase of Viking attacks. How far the kingdom suffered from Viking attack in the early ninth century is impossible to establish because the historical evidence is exceptionally poor. There is a hint of Viking activity in the annal for 844 in the *Flowers of History*, a compilation made by the thirteenth-century monk Roger of Wendover. This says that the Northumbrian king

[1] *ASC*, *s.a.* 789, 793, 794; on *Donamuthe*, David Rollason (2000), p. 90 n. 27, citing Richardson (1985).

Rædwulf was killed in battle against the 'pagans', presumably Vikings; but we have no way of assessing how reliable this annal is.[2] If there were few if any Viking raids on Northumbria in the earlier ninth century, this would fit with the pattern elsewhere – for the rest of Britain, even for southern areas well covered by the *Anglo-Saxon Chronicle*, there seems to have been a pause in Viking raids. Although the evidence of Kentish charters indicates that there were more raids than the *Anglo-Saxon Chronicle* reveals, it nevertheless seems likely that the Vikings turned their main attention away from England to Gaul, Ireland and Pictland, where they made an apparently isolated raid in 839.[3] In autumn 865, however, the Vikings resumed their attacks. A Viking force known to the *Anglo-Saxon Chronicle* as the Great Army (in Old English *micel here*) landed in East Anglia, and commenced a protracted campaign in England. After making peace with the East Angles, it moved north and twice attacked York, first on 1 November 866, then on 21 March 867. After the first attack, it departed to plunder the area around the River Tyne, leaving the Northumbrian king Ælle, who was ruling jointly with another king called Osberht, to recapture the city. In the second, it again captured York and the two Northumbrian kings were taken and killed.[4]

Shortly after this, the Great Army left Northumbria and returned south, where it defeated and killed King Edmund (subsequently known as Edmund the Martyr) in East Anglia. It then invaded Mercia, which kingdom it effectively destroyed. Its campaigns in the southern kingdom of Wessex, however, were successfully resisted by King Alfred (871–99), who won a major victory over what remained of the Great Army at Edington (Wiltshire) in 878. This re-established his control over much of southern England, with eastern England, the midlands and the north generally under Viking dominance. Thus the conditions were established for Alfred's successors as kings of Wessex to launch expansionary campaigns in the early tenth century into the Viking-controlled areas, campaigns which led ultimately to the unification of England under their control. In 876, the Great Army had split into three, with one section led by one of the original leaders, Halfdan, returning to Northumbria. This led to the establishment broadly between the River Tees and the Humber of the Viking kingdom of York, which held some sort of

[2] The existence of this king, who is mentioned only by Roger, is corroborated by the survival of coins (North (1994), p. 72). For the text, see Wendover, *Flowers*, *s.a.* 844.

[3] For Kent, see Brooks (1971), pp. 79–80; for a general representation of what is known of Viking activity, see David Hill (1981), map nos. 47–55, and Haywood (1995), pp. 55, 57.

[4] For a general narrative, see Stenton (1971), pp. 246–60; for the sources and commentary, see *Sources*, p. 71. The date of 866 proposed there for the arrival of the Great Army in East Anglia is not sustainable (for correction, see David Rollason (forthcoming)).

power in the area, albeit intermittently, until the last of the Viking kings, Eric, usually identified with Eric Bloodaxe, was expelled and killed in 954.[5]

Viking activity north of the River Tees was much more sporadic and less intensive. Although, as we have seen, there were early raids into that area, our sources do not mention a lasting Viking occupation; and we see in this region two closely related political entities quite separate from the Viking kingdom of York to the south. Between the Rivers Tees and Tyne were the principal lands of the 'liberty' of the Community of St Cuthbert, which was the heir to the monastic community of the church of Lindisfarne. Within the 'liberty' that community had a degree of juridical independence, as well as being the dominant land-holder. The Community of St Cuthbert was still at the time of the capture of York in 866–7 based at Lindisfarne, but it moved, in the years 875–83, according to the *History of St Cuthbert*, to Chester-le-Street on the River Wear, and then in 995 to Durham. As for the lands north of the River Tyne and into what is now southern Scotland, we find, soon after the capture of York, kings whose identity beyond their names is utterly obscure ruling in the area;[6] and in the early tenth century we find an earl called Eadwulf (fl. *c.* 890–912) apparently ruling the same area from his seat at the old Northumbrian royal centre of Bamburgh, and being succeeded by his son Ealdred (fl. 913–27), and then by other members of the same family.

The situation was different again to the west and north-west of the Pennines, that is in Cumbria (northern Lancashire, Westmorland, Cumberland and the lands north of the Solway Firth which formed part of Northumbria at its greatest extent). This area was also spared the attentions of the Great Army, but appears to have suffered a series of incursions by Norwegian Vikings and Vikings operating from bases in Ireland such as Dublin which they had established in the first half of the ninth century. There are no chronicle references to actual Norwegian incursions north of the River Mersey, but surviving place-names in the area leave no doubt that people of Norwegian language were an important element in the population of lands to the north of the Solway Firth and in north-west England, although in the latter area Danes, presumably originating east of the Pennines, were also important. To complicate the situation, however, north-west England was ruled from the early tenth century by rulers, some of whom described themselves as 'king of the Cumbrians'. The first of these to appear in our

[5] Stenton (1971), pp. 248–53, 360–2; Haywood (1995), pp. 67, 70–1. That this king was not in fact Eric Bloodaxe, scion of the Norwegian royal house, but another Eric, possibly connected with the rulers of Dublin, is argued by Downham (forthcoming).

[6] *Sources*, p. 63.

sources is Owain (*c.* 915–*c.* 937). The meaning of the title, and the extent of the power of these kings, is disputed.

These changes to political frontiers were thus far-reaching and various, and the dismemberments of the kingdom of Northumbria drastic. They constitute in one sense the destruction of that kingdom. We need to assess, however, the extent to which the changes were as fundamental as they appear at first glance. How far did they transform the political, ethnic and cultural make-up of the former kingdom of Northumbria? Alternatively, how far did the political, ethnic and cultural elements of that kingdom survive to shape and influence what we might loosely call its 'successor states', that is the Viking kingdom of York, the 'liberty' of the Community of St Cuthbert, the earldom of Bamburgh, and Cumbria? We need to engage in the same analyses as in our consideration of the making of the kingdom of Northumbria and, although it will be more convenient in this case to consider the aspects in a different order, the analytical framework will be the same: political transformation, ethnic transformation, cultural transformation. Given the different histories of the 'successor states' which have emerged even from our brief survey, we shall give separate consideration to each of them.

The Viking kingdom of York: political transformation?

First, the influence of the former kingdom of Northumbria on the political structure of the Viking kingdom of York. At the simplest level, we should note that the capture of York by the Viking Great Army in 866–7 had resulted in the death of the reigning kings of the Northumbrians and, aside from the wholly speculative possibility that the house of Bamburgh was related to them, the royal line of Northumbria was never heard of again. It might in other words be urged that we should finish this book with that event. But how great really was the break between the rule of the last Northumbrian king and that of the Viking kings? How far in other words was the kingdom of the latter really a continuation of Northumbria between Tees and Humber? The Viking kingdom of York is often seen as a powerful and sophisticated kingdom with proper frontiers and governmental machinery, and one moreover which was deeply influenced by Vikings and Viking culture and thus quite different from what preceded it. Richard Hall, for example, referring to York as the supposed 'capital' of this kingdom, writes of York as 'one of the great cities of the Viking world'. Soon after the Viking capture of York, he continues, 'the English puppet kings were discarded ... and

Viking rulers took their place. Although it is not explicitly stated, York was their capital.'[7] In this view, the Viking kingdom of York was far from being an extension of the history of Northumbria; it represented a discontinuity, albeit a beneficial one, as a result of which this part of what had been Northumbria flourished as part of the 'Viking world'. Thus the Viking capture of York marked the end of Northumbria and the beginning of the Viking kingdom of York as a quite new political unit with a new orientation. This view owes much to the rehabilitation of the Vikings initiated by Peter Sawyer in his highly influential book *The Age of the Vikings*, in which he argued that the effect of the Vikings on western Europe was more constructive than destructive. In particular the Vikings played an important role in the promotion of commerce and urban life, York in the period of the Viking kings being an important instance of this.[8] Whatever the general merits of Sawyer's views, it is not at all clear that they can be applied to the Viking kingdom of York. We need to ask, in the first place, how stable – or indeed how real – was this kingdom?

We should note first that our sources never use the expression 'kingdom of York' in connection with the Viking kings, or indeed at any time. They simply refer to one or other of these kings 'seizing York'; and they refer also to Northumbria, although it is by no means clear what is meant by this. Of course, these sources are exiguous and fragmentary, but the omission is nevertheless a striking one. Moreover, there is no evidence that the Viking capture of the city in 866–7 inaugurated a Viking kingdom, or indeed any sort of kingdom. The fact is that we have absolutely no information about who was ruling York in the years immediately after the Vikings captured the city; all we know is that in 872 people referred to by the sources as the 'Northumbrians' expelled the archbishop of York, Wulfhere, along with a king called Egbert I, who may have been ruling only north of the River Tyne. It is possible that it was Wulfhere who was effectively in charge of York until that time, and he may have resumed that role after the Great Army restored him in 873.[9] We have seen how that army went on to conquer Mercia in 874, and how it then split up, with one of its original leaders, Halfdan, returning to Northumbria. Then at the end of the annal for 876, which is mainly devoted to events in Wessex, the *Anglo-Saxon Chronicle* has a single sentence referring to Halfdan dividing up the land, echoed by what

[7] Richard A. Hall (1994), p. 16; see also Richard A. Hall (1988), p. 238.

[8] Sawyer (1971); see also Sawyer (1982).

[9] *Hist. Kings*, *s.a.* 872–3 (p. 110), Symeon, *Origins*, bk 2 ch. 6, Arnold (1882–5), II, p. 225, and Wendover, *Flowers*, *s.a.* 872.

may not be an independent account in the *History of St Cuthbert*.[10] From these accounts it is not clear that Halfdan was a king in any sense. If he was, his reign was transitory, for Symeon of Durham describes how he fled from the River Tyne never to return. This may have been later in the year 876 itself, if Halfdan is to be identified with the Albann, 'king of the dark heathens', who was killed, according to the *Annals of Ulster*, in Ireland in 877.[11] It is moreover only an assumption that the centre of his activities was York rather than another centre or centres.

Our information about Halfdan's immediate successors is extremely unsatisfactory, to the extent of creating a presumption that there was no settled kingship of York with a definite and continuous line of succession. According to the West Saxon chronicler Æthelweard, a king with the Scandinavian name Guthfrith was buried at York Minster in 895; he may have been the same person as the Guthred who, according to the *History of St Cuthbert*, was made king by the Community of St Cuthbert. For two succeeding Viking kings, Siefred and Cnut, we are solely dependent on the evidence of coins minted in York and bearing their names. The identity and length of reign of these kings are impossible to establish satisfactorily. A Viking leader called Sigeferð besieged the north coast of Devon in 893, and a jarl called Sichfrith caused a dissension in Dublin in the same year. If these persons were one and the same and are to be identified with the Siefred of the York coins, the information cited above is not inconsistent with a reign beginning after Guthred's death in 895; but the identification is no more than a suggestion. Similarly, later sagas describe the arrival at Scarborough around the year 900 of a Viking leader called Cnut, but whether the tradition is reliable, whether, if so, this was the Cnut of the York coins, and whether he was in any sense a king of York are all highly uncertain.[12]

We are on somewhat firmer ground with the *Anglo-Saxon Chronicle*'s account of a prefiguring of West Saxon intervention in the Viking kingship, that is the arrival in 899 of a cousin of King Edward the Elder, Æthelwold, who 'went to the Danish army in Northumbria, and they accepted him as king and gave allegiance to him'. His reign, however, was very short and he was evidently concerned with using Northumbria as a base from which to

[10] *ASC*, *s.a.* 874–6; *Hist. Cuth.*, para. 14.

[11] Symeon, *Origins*, bk 2 ch. 13, and *Annals Ulster*, *s.a.* 877; for comment, see Smyth (1977), pp. 263–4. The annal in *Hist. Kings*, *s.a.* 877 (pp. 111–12), which mentions Halfdan as having been killed by King Alfred's forces in Devon is clearly corrupt; its source evidently assigned this incident to a brother of Halfdan and Ivar, as do Asser (ch. 54) and *ASC*, *s.a.* 878.

[12] For references and discussion see *Sources*, pp. 64–5; see also Smyth (1975–9), I, pp. 47–8, and David Rollason (forthcoming). On Guthred, see below, p. 245.

gain dominance of the south, for in 901 he was accepted as king by the Vikings who were in Essex, and in 902 he was killed by King Edward the Elder. From then until the accession of Ragnall, the only hint we have as to Viking kings of Northumbria is provided by the *Anglo-Saxon Chronicle*'s account of the Battle of Tettenhall (910), at which the 'army of Northumbria' was defeated and the kings, Halfdan, Eowils and Ivar, were killed. It is quite unknown whether these persons were ruling different parts of Northumbria, or whether indeed they were kings in any real sense of the word rather than just leaders of the Viking army.[13]

The date of the accession of the next king about whom we have any information, Ragnall, may have been as early as 914 or as late as 919. If the latter, his accession was evidently violent, for the annals in the *History of the Kings* record under that year: 'King Ragnall (*Inguald*) broke into (*irrupit*) York.' This and the evidence of his coins, which have the name of York on the reverse, also serve to associate him with the city. Following Ragnall's death, we find a somewhat more consistent succession of kings, for he was succeeded by his brother Sihtric Caoch (920/1–7), who is described in the *Anglo-Saxon Chronicle*, D version, as 'king of the Northumbrians', and in whose name coins were produced at York. His successor was apparently a certain Guthfrith, who may have been another brother, and who was, according to the *Anglo-Saxon Chronicle*, E version, driven out in 927 by Athelstan, king of England (927–39), when the latter established his power over Northumbria. Although there are no coins surviving for Guthfrith, William of Malmesbury's account of how, following that king's expulsion, Athelstan razed a Viking fortress at York suggests that York was his base also.[14]

Athelstan died in 939, and in 941, according to the D version of the *Anglo-Saxon Chronicle*, the Northumbrians chose 'Olaf from Ireland as their king'. This was Olaf Guthfrithson, who, the *History of the Kings* states, 'came first to York' on assuming power. He was succeeded by Olaf Sihtricson (also known as Cuaran, 941–5), whose reign appears very confused in the historical evidence, for he seems twice to have been expelled by Edmund, king of England, and to have been expelled alongside another king, apparently of York, Ragnall Guthfrithson (943–4/5). Edmund then 'obtained the kingdom of the Northumbrians' and ruled until his death in 946, being

[13] *ASC*, *s.a.* 901–5, 910, and see *Sources*, pp. 65–6. It is possible that York had no king at all at this period, and that this accounts for the production in York of coinage in the name of St Peter but with no royal name; on this coinage, see Grierson and Blackburn (1986), pp. 322–3, 626; Pirie, Archibald and Hall (1986), pp. 34–7.

[14] *Sources*, pp. 66–7. For the texts, see *Hist. Kings*, *s.a.* 919 (p. 93), *ASC* D, *s.a.* 925, *ASC* E, *s.a.* 927, and *Sources*, pp. 165–6 (V.1.3 = William Malms., *History*, bk 2). For the coins, see North (1994), p. 113.

succeeded by his brother Eadred, who became king of England. The situation shortly afterwards becomes confused once more, with 'the councillors of the Northumbrians' accepting as king Eric (948), but expelling him in response to threats of ravaging by Eadred. In 949, Olaf Sihtricson returned to Northumbria and apparently re-established his position, but was expelled in 952 by 'the Northumbrians', who again received Eric as king until his expulsion and death in 954.[15]

In short, the sequence of these kings was broken, complex and marked by violence and instability. It is hard to believe that these were the rulers who made York 'one of the great cities of the Viking world'. Moreover, even allowing for the paucity of our evidence, it is not at all clear that these kings possessed the machinery to do anything of the kind. To take an obvious point, it is not even possible to demonstrate conclusively that they had a palace in York. The only text which mentions such an edifice is the thirteenth-century Scandinavian saga, the *Saga of Egil Skallagrimsson*. This describes how the hero was shipwrecked on the east coast of England and obliged as a result to come to York, even though his enemy Eric was king there and would have him killed. On the advice of his friend Arinbjörn, Egil went to the king in his residence in York, and the king agreed to spare his life in return for a poem in his praise. This poem is preserved under the title 'Head Ransome'; it is one of the so-called skaldic poems and is thought to be an authentic work of Egil himself. The saga itself, however, is inaccurate in detail (it refers wrongly to Athelstan as the reigning king of England at the time), late in date and clearly literary in style, so that it is extremely unreliable as a source for the Viking period.[16] It is nevertheless likely that the Viking kings did have a residence in York, as is suggested by a passage in William of Malmesbury, which describes how envoys of King Harold Fairhair of Norway were 'royally entertained' in York, and by a passage, admittedly of very doubtful reliability, in the *Life of St Catroe* which describes a visit of the saint to Eric in York. It is also perhaps suggested by a passage in the *Chronicle* of Æthelweard describing how in 894 an envoy of King Alfred 'contacted the enemy' in the city of York, where their leader was presumably resident. But it remains the case that we know nothing either of the form or of the location of the royal residence. Suggestions that King's Square in York might be the site of the palace are based only on the name which appears

[15] *Sources*, pp. 68–9. For the texts, see *ASC* D, *s.a.* 941, 948, 954, *ASC* E, *s.a.* 952, and *Hist. Kings*, *s.a.* 939, 945 (pp. 93, 94).

[16] *Sources*, pp. 167–9 (V.2.2 = Nordal (1933), pp. 176, 183, 195; trans. Pálsson and Edwards (1976), pp. 151, 153, 155–7, 163).

in an Old Norse form in early documents. The later earls of Northumbria seem to have had their residence in the St Mary's Abbey area, which an eighteenth-century antiquary called Earlsborough and which was near the church of St Olave which one of those earls founded, perhaps to have it near to his residence; but whether the Viking kings also had a residence there and what it was like are wholly unclear.[17]

Nor is it known whether these kings had effective fortifications at York. An early twelfth-century Durham text, *Chapters on the Miracles and Translations of St Cuthbert*, states that the Great Army divided into three parts, one of which (presumably that led by Halfdan) 'restored the defences of the city of York', but this is based on the *History of St Cuthbert* which simply states that this part of the army 'rebuilt the city of York'. It is likely that both references are based on no more than the inference of a later writer. Somewhat more solid is the account noted above of King Athelstan razing to the ground a Viking fortress in York. But whether this was rebuilt, where it was and what its precise character was are all wholly unknown. Certainly by the 970s, the *Life of St Oswald* was describing the 'strong walls' with which York had been built as having been 'left to the ravages of age'.[18]

Of any bureaucratic machinery at the disposal of the Viking kings, our sources tell us nothing, and the presumption must be that kings whose reigns were so disrupted and transitory did not possess such a machinery, or were unable to make use of such machinery as the kingdom of Northumbria had possessed; it is much more likely that they were basically war-leaders in charge of an army billeted in York. It should be emphasized in this connection that the geographical scope of whatever kingdom the Viking kings ruled is very uncertain. Because characteristically Viking place-names are found mostly south of the River Tees, it can be argued that that river was its northern frontier; but there was clearly nothing firm about this – we find Halfdan operating north of the river, admittedly on raiding expeditions, and we find Ragnall giving his followers Scula and Onlafball lands to the north of it in the area of what is now County Durham. As far as the southern and western boundaries go, we can only guess at them on the basis of hints in the *Anglo-Saxon Chronicle*: that Dore near Sheffield was the boundary of Mercia and thus presumably also of Northumbria; that Manchester was

[17] *Sources*, pp. 166–7 (V.2.1 = William Malms., *History*, bk 2) and p. 170 (V.2.3 = Alan Orr Anderson (1922), I, p. 441); and Æthelweard, p. 51. On King's Square, see Palliser (1978), p. 8 (Coney Garth); on Earlsborough, see *Sources*, pp. 171, 175, and Richard A. Hall (1988), p. 235. A contemporary account of the foundation of St Mary's Abbey refers to this area as 'burgum ... extra ciuitatem, iuxta ipsam ciuitatem' (Dugdale (1817–30), III, p. 546. I owe this reference to Christopher Norton).

[18] *Sources*, p. 165 (V.1.2 = Arnold (1882), I, p. 229), and pp. 171–2 (V.3.1 = Raine (1879–94), I, p. 454).

'in Northumbria' and that Edward the Elder built a fortification (*burh*) there and at nearby Thelwall as if he were fortifying a frontier; that Edward the Elder built a *burh* and received the homage of various northern rulers, including Ragnall, at Bakewell in 920, suggesting that this may have been a frontier location (map 9). There was, however, clearly very considerable instability. Although the Viking Five Boroughs (Nottingham, Lincoln, Derby, Stamford, Leicester) were under different control from that of York at the beginning of the Viking period and therefore constituted a frontier to the south, it appears that they were for a time, even if only a short time, under the control of the Viking kings of York, for a piece of alliterative verse preserved as the annal for 942 in the *Anglo-Saxon Chronicle*, A version, describes the liberation of the Danes of these boroughs by Edmund, king of England, from forced subjection to the heathen 'Norsemen'. These were presumably the Vikings of York, whose king Olaf Sihtricson was one of the Norse of Dublin. In fact, the ambitions and interests of these kings were by no means always focused on York. Several of them were as concerned with events in Ireland and with the kingdom of Dublin as they were with York, and it is unclear whether they thought of York as a settled kingdom or as another base from which to conduct raids.[19]

The view that the Viking kings were sophisticated rulers who had a major impact on Northumbria south of the Tees is not based, however, on what we know about their individual reigns, but rather on two main types of evidence: first, the archaeological and topographical evidence for the commercial and industrial development of York; secondly, the coins produced in York before and during their reigns. Archaeological excavations, conducted in the area between the Roman fortress and the River Foss, have revealed that from around 900 York was importing a wide range of commodities, including honestones from Scandinavia, pottery (perhaps with wine in it) from the Rhineland and even silk from the Far East; and that there was vigorous activity in working of leather, metal, wood, glass and bone to produce, apparently on a commercial scale, objects such as jewellery, shoes, turned wooden bowls and even bone skates. The evidence suggests that at least in this area of the city all this activity was new. The fullest excavations, those alongside the street called Coppergate, showed that the site was

[19] In general, see Smyth (1977, 1975–9). For place-names, see David Hill (1981), no. 68; for the five boroughs, see Richard A. Hall (1989); and on frontiers, see above, pp. 26–7. For the texts, see *ASC* A, *s.a.* 924, recte 920 (Bakewell), 942 (for comment, Mawer (1923)), and *Hist. Cuth.*, para. 23 (Scula and Onlafball).

a 'green-field' one since the bones of animals characteristic of fields and open spaces were found there. At the end of the ninth century, however, industrial activity commenced on the site in the form of a glass works, and *c.* 910 there was intensive and closely packed development, comprising craft workshops and housing. The activities on the Coppergate site parallel those elsewhere in this area of the city, notably along what is now Parliament Street.[20]

The Coppergate excavation also suggested that development was organized rather than haphazard. Crucial to this was the recovery of property boundaries, and the archaeological proof that these were aligned on Coppergate as it exists today (fig. 17). Although the street itself has not been excavated, this evidence creates a strong presumption that it is in precisely the same position as it was in the early tenth century when the property boundaries were laid out. The implications of this are much wider than appears at first sight, for the street-plan of this part of York shows that Coppergate is only one of a group of streets focused on Pavement and forming a grid pattern such as is often associated with newly laid out towns – or quarters of towns – in the middle ages (map 8). The conclusion is inescapable that the streets in question represent the laying out of a new quarter of York in the Viking period.[21] Now, as far as is known, the only Roman crossing of the Ouse at York was the bridge, now disappeared, which led from the civil settlement (the *colonia*) to the main gate of the legionary fortress on the other side of the river. The pronounced curve in the line of the main street of the *colonia* area, Micklegate, may result from the fact that this street originally provided access to the Roman bridge, but that at some time it was diverted to lead to what is now Ouse Bridge. This latter, however, is aligned on Pavement and the grid of streets of which Coppergate forms part. In short, the street-plan and the evidence from Coppergate suggest that the laying out of the buildings on the Coppergate site was part of a much wider scheme of 'town planning' in York – the creation of a whole new quarter and the diversion of the main street of the former *colonia* and of the river crossing to serve it. That new quarter, to judge from the evidence of archaeological finds, was the principal industrial and commercial quarter of the city in the Viking period, so its laying out was a matter of great importance to the development of York. Part of this development may

[20] See Richard A. Hall (forthcoming); for surveys, see Richard A. Hall (1984b, 1994); see also Waterman (1959).
[21] Richard A. Hall (1984b), pp. 43–51.

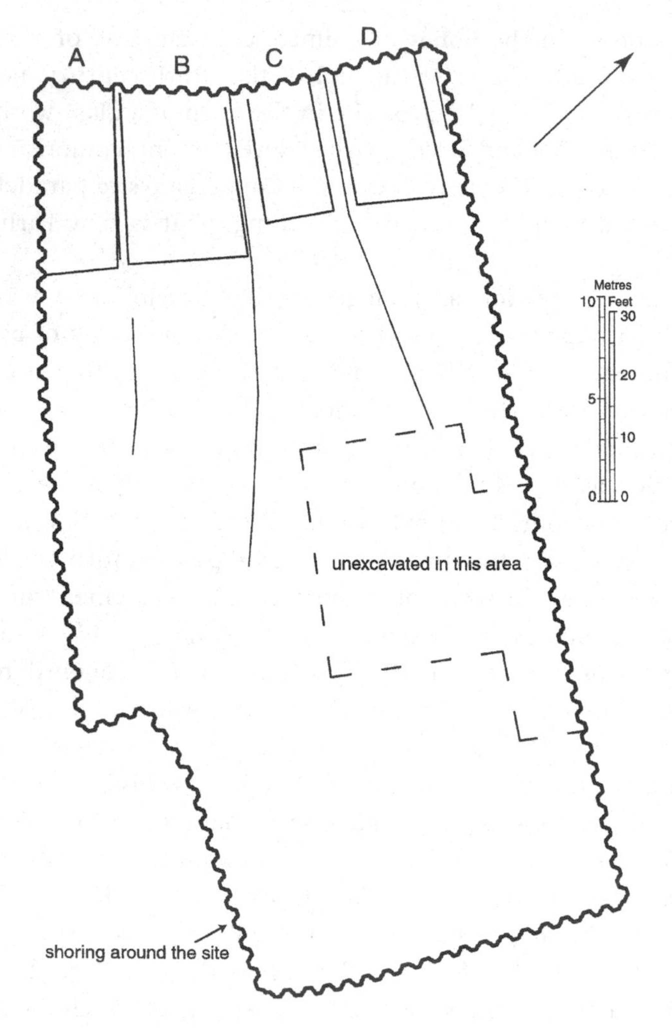

FIGURE 17 York, 16–22 Coppergate, Phase 4B (*c.* 930/5–*c.* 975)
At this period, the area under excavation was divided into four tenements (A–D), divided from each other by wattle fences as shown on the plan. Each tenement contained a building with its narrow end facing the street of Coppergate at the top of the plan. The alignment of the tenements shows that the street Coppergate must have been in existence at the same time (Richard A. Hall (1992)).

have been the church of St Mary Castlegate, whose Viking founders Grim and Æse are named on the dedication stone still preserved in the church.[22] Another important development was the removal of the south and east walls of the Roman fortress, so that the city was opened out on the east side and extended towards Pavement and Ouse Bridge. The Scandinavian names of streets leading in that direction from the site of the former Roman fortress tend to suggest that this development also belonged to the Viking period

[22] Richard A. Hall (1978, 1994), pp. 34–7; for St Mary Castlegate, see *Sources*, pp. 176–7 (V.4.6) and Okasha (1971), no. 146.

and was more or less contemporary with the laying out of the Coppergate area.[23]

The development of York in the period of the Viking kings may therefore have been rapid and spectacular. By the end of the tenth century, Byrhtferth in his *Life of St Oswald*, archbishop of York, asserted that 'the city rejoices in the multitude of its population, which, counting men and women but not infants and children, is numbered at not less than 30,000'. The figure given for the population seems implausible in view of the approximate figure of 9,000 derived from *Domesday Book*'s information about the city, but at least it is clear that Byrhtferth regarded the city as flourishing and populous.[24] Since the archaeological evidence suggests that the development of York began on a considerable scale soon after the Vikings' capture of the city, it is natural to attribute it to them. Their role in it would thus indicate that they really were sophisticated and powerful despite our strictures on their authority set out above. But is it certain that they were in any sense responsible? First, although Byrhtferth highlighted Danish merchants as an important element in the city's life, York's commercial and industrial activity, even at the beginning of the tenth century, was not by any means wholly Viking in character. To be sure, some of its imports were from Scandinavia, such as the honestones and amber recovered on a number of sites, but many, including fine pottery and textiles, were from the Rhineland, and two very important objects recovered from the Coppergate excavations (a silk headdress from the area of Samarkand and an Arabic dinar) point to connections with the Muslim world. It is true of course that Vikings are known to have had trading contacts with Muslim lands, but it is not certain that these objects came to York through a Viking trading nexus. Nor was there anything characteristically Viking about the laying out of the city around Coppergate. Excavations have shown that Winchester was laid out on a grid pattern at more or less the same time as is proposed for the Coppergate area, and Southampton and Ipswich had been similarly laid out a century or more before. The house types themselves which have been recovered at Coppergate are not characteristically Viking either, and significantly they have no affinity with the timber buildings recovered from Dublin in the Viking period, so there was clearly no 'Viking' way of building which would prove to us

[23] Royal Commission on Historical Monuments (England) (1972), pp. 111–16 and *passim*; for the street-names, Palliser (1978), under Coney Street, Feasegate and Coney Garth, and p. 3 on 'gate' names in York generally.

[24] *Sources*, pp. 171–2 (V.3.1 = Raine (1879–94), I, p. 454).

that York's development was the work of Scandinavian incomers. In short, development need not have been the result of a Scandinavian influx into the city, although any such may have increased the size of the population.[25] Secondly, impressive as York's commercial and industrial activity was in the Viking period, the archaeological evidence of an eighth-century trading settlement at Fishergate and the references in the late eighth-century *Life of St Liudger* to Frisian merchants at York suggest that it had begun earlier and was not the result of the activity of the Viking kings. The Viking period may have seen a relocation of commercial and industrial activity from outside the city (perhaps Fishergate itself) to an area within it (the Pavement area), but such a development was typical of town evolution in the tenth century elsewhere, for example at London. We need not therefore assume that the development of York was the work of the Viking kings, and attribute power and sophistication to them on the basis of it.[26]

The second type of evidence on which the reputation of the Viking kings rests comprises the coins from York in their time. The coins of ninth-century Northumbria before the Viking capture of York consisted, as we have seen, of the so-called *stycas*, coins so debased that they were made of little more than bronze, and so diminutive and poorly designed as to make them a miserable contrast with the fine silver pennies of the English kings in the south, or of the Carolingian kings on the continent (ill. 23). Very soon after the Viking capture of York this changed radically. The hoard of coins deposited at Cuerdale on the banks of the River Ribble in Lancashire around 905 contains many fine silver pennies minted in York in the names of the Viking kings Siefred and Cnut (ill. 26). These royal coins were succeeded by the so-called 'St Peter's coins', which are also silver pennies and have on them some version of 'Money [or mint] of St Peter' (*Sci Petri Mo*). They begin around 905 and continue, with one interruption, until King Athelstan of England seized control of York in 927. The interruption appears to have been the reign of King Ragnall (914 × 919–920/1), during which such coins may have been minted, but the more frequent issue consisted of coins with the king's name and an open hand (perhaps the hand of God), or a Charlemagne monogram, or a hammer, symbol of the pagan Scandinavian god Thor. After Athelstan's reign in York, during which his own English pennies were minted, the reign of the next Viking king, Olaf Guthfrithson (939–41), saw the minting at York of pennies with a design featuring a raven, symbol

[25] Richard A. Hall (1984b), esp. pp. 49–116, and (1994), esp. pp. 83–8; on buildings, see Richard A. Hall (1984a). On planned urban layouts elsewhere, see Biddle (1975).

[26] See above, pp. 178–80.

ILLUSTRATION 26 Penny of King Cnut, obverse and reverse (Cambridge, Fitzwilliam Museum)
The obverse of this silver coin has the name CNVT REX around a cross; the reverse has the words MIRABILIA FECIT (He has worked wonders); see Grierson and Blackburn (1986), no. 1444, and North (1994), no. 497.

of the pagan Scandinavian god Odin. Later kings down to Eric had coins with a cross on them minted in their name.[27]

All the coins from York of the period of the Viking kings were silver pennies, and it seems very likely that those kings had been responsible for an injection of silver bullion from their plundering to make such a coinage possible, even if it was then sustained by York's commercial and industrial development. Whether they were also responsible for the design and production of the coins is less clear, and any assessment must take into account the coins showing Christian influence. Those of Siefred and Cnut have markedly Christian inscriptions, which read 'He has worked wonders' (*mirabilia fecit*), a quotation from the book of Psalms, and 'The Lord God is king' (*dominus deus rex*). On some coins these are combined with the name of one of the kings; in others, only these legends occur, although the same die has been used as in the coins with the king's name on them so it is clear that these coins belong to the same phase of minting. The inscriptions have to be seen in conjunction with the designs on the coins, which are chiefly crosses, including a cross with steps at the base, which is thought to reflect images of Christ's actual cross at Golgotha. In rare cases the coins

[27] The most up-to-date discussion of the coinage of Viking York is Blackburn (forthcoming), where arguments for the sophistication and dominance of the Viking kings are advanced. See also Grierson and Blackburn (1986), pp. 316–25, and Dolley (1965); and, for a concise catalogue, North (1994), pp. 109–16. On the date of the Cuerdale Hoard, see Archibald (1992).

ILLUSTRATION 27 St Peter coin with sword, obverse and reverse (Cambridge, Fitzwilliam Museum)
The obverse of this silver coin has the words SCI IECII IIO (a blundered version of SANCTI PETRI MONETA, 'Money of St Peter') and a sword; the reverse has the meaningless word REVITII and a hammer which is the attribute of the pagan god Thor; see Grierson and Blackburn (1986), no. 1449, and North (1994), no. 556.

have the monogram of the great Christian emperor Charlemagne (d. 814). It can be argued that the Christian connotations of these coins and their visual reference to Charlemagne show that the Viking kings were being rapidly assimilated to Christian culture and were so sophisticated that they were using the coins to make a claim to a place in the wider Christian culture of Europe. A similar argument can be advanced for the St Peter's coins, which formed the exclusive coinage of York from *c.* 905 to 927, aside from the reign of Ragnall (914 × 919–920/1). According to this, the Viking kings were continuing to behave in a very sophisticated way, producing coins deliberately intended to assert for their kingdom a Christian identity, and for themselves full membership of Christendom under the tutelage of St Peter. This argument asserts further that some compromise occurred to make this all the more sophisticated, for after Ragnall's reign the image of a sword was added to the St Peter's coins, arguably some sort of a pagan symbol, which would have appealed to the pagan Vikings just as the St Peter legend did to Christendom at large (ill. 27). Similarly, the raven on the coins of Olaf Guthfrithson can be held to reflect the appearance of a raven in the hagiography of the Christian saint-king Oswald (according to his twelfth-century *Life* a raven miraculously retrieved his right arm from the stake on which it had been impaled), as well as the cult object of Odin (ill. 28). The

ILLUSTRATION 28 Penny of King Olaf Guthfrithson (939–41), obverse (London, British Museum)
The obverse of this silver coin has the name ANLAF CUNUNC, the second word standing for *cununc*, Old Norse for 'king'. The reverse has the name of the moneyer, using the Old Norse word *minetre*, around a cross. Cf. Grierson and Blackburn (1986), no. 1119, and North (1994), no. 537.

later coins with crosses also reflect, according to this argument, the Viking kings' desire to be accepted as full members of Christendom.[28]

Doubts can be raised about this argument. Apart from Guthred (Guthfrith) who was buried in York Minster in 895, we have no reason to suppose that the Viking kings were Christian. Sihtric Caoch certainly was not, until he underwent a short-lived conversion to Christianity as part of a diplomatic agreement with King Athelstan in 926. Eric's queen is presented as Christian in the *Life of St Catroe*, but the source is manifestly ill-informed on a number of points and we have no reason to accept this particular piece of information from it.[29] Moreover, the arguments presented above make it unlikely that the kings themselves had the administrative capability to produce coins of this quality. Instead of being a royally inspired and controlled

[28] Blackburn (forthcoming); see also Blackburn (2001), pp. 135–6.

[29] On Guthfrith's burial, see *Sources*, pp. 173 (V.4.1 = Æthelweard, p. 51); for Sihtric Caoch's conversion the source is admittedly only the thirteenth-century Wendover, *Flowers*, *s.a.* 925, but with partial corroboration from *ASCD*, *s.a.* 926; for the *Life of St Catroe*, see *Sources*, p. 170 (V.2.3).

coinage, it is possible that the coins are evidence of the varying degrees of control exercised by the archbishops over York and over the Viking kings themselves. The St Peter's coins mark their ascendancy with the name of the patron saint of their cathedral which is the only name on the coins. The use of Christian inscriptions on the coinage of Siefred and Cnut is further evidence that minting was in their hands. It seems much more likely that they would have introduced these inscriptions than that the unstable and violent Viking kings could have done so. The coinage of Ragnall then represents a partial set-back for the archbishop, with a greater degree of power and influence in the hands of the king; when the St Peter coinage returns after his reign, the archbishop has had to make some compromise and introduce the sword with its possible pagan overtones. The crosses on the coins of the later Viking kings represent the re-assertion of their influence, although not to the extent that they were able to discard the royal name.

The development of York and the coinage of York do not require us to accept that the Viking kings were the predominant force in the city, and thus open the way to the alternative hypothesis that it was the archbishops who wielded the real power, using the military capabilities of the Viking kings when it suited them to do so against, in particular, the military and political ambitions of the kings of England to the south, in much the same way that late Roman rulers used the military power of barbarian federates. We have seen the political role of the archbishops in the pre-Viking period, not least in the councils which expelled and elected kings. With the elimination of the native kings in the last defence of York in 867, the position of the archbishops may have been enhanced. In 872 the revolt against Viking dominance also involved the expulsion from York of Archbishop Wulfhere. The inference must be that he had somehow been ruling York in association with the Viking conquerors (with King Egbert I ruling north of the Tyne) and that he was already in alliance with the Viking kings and dependent on their military power. Nothing is known of the political activities of the archbishops between Wulfhere and Wulfstan I (930/1–52, possibly 950/1), but we find the latter in company with King Olaf Guthfrithson when, sometime between 940 and 943, that king, after sacking the English royal centre of Tamworth, was besieged by Edmund, king of England, in Leicester. It is hard to escape the conclusion that Wulfstan was collaborating with Olaf at this time in a military raid on territory controlled by the king of England. Again in 947, we find the *Anglo-Saxon Chronicle*, D version, specifically naming Wulfstan alongside 'the councillors (*witan*) of the Northumbrians' in pledging loyalty to King Eadred, a pledge to which, the *Chronicle* tells us, they were afterwards false – that is in accepting Eric as their king. That Wulfstan had played a role in bringing Eric to power is corroborated by the fact that he was arrested and imprisoned by King Eadred in 952, because 'accusations had often been made to the king about him'.[30] Two years later, Eric was once more expelled and this time killed, and it is tempting to conjecture that his fall was hastened by the fact that Wulfstan was no longer in York to support him.

Apart from Wulfstan I, the archbishops are shadowy figures and we can do nothing to fill out this image of 'prince-archbishops' which we are assigning to them. General trends in the history of the archiepiscopal see of York at this time, which we can infer with some confidence, tend nevertheless to support it. Admittedly, the see of York lost substantial estates in the period of the Viking kings, information which we learn from one of our few documents referring to land that Archbishop Oswald was trying forlornly to retrieve. But there are indications that the see was amply compensated for this by an extension and consolidation of its jurisdiction. Its southern boundary makes a loop south into Nottinghamshire, far beyond what is ever likely to have been its original course, to take in the rich abbey of Southwell where, we learn from a surviving charter, Archbishop Oscytel received a grant of twenty hides from King Eadred in 956. Often regarded as the foundation charter of the abbey, this grant may be rather a further stage in the process, begun perhaps in the period of the Viking kings, by which the archbishops were expanding their power. A similar expansion of power may have been involved in the grant of the vast area of Lancashire called Amounderness to the archbishops of York by King Athelstan, although in this case the archbishops had lost control of it before the Norman Conquest, and it was in any case in an area of Viking settlement from which the archbishops may not have been able to benefit. It may have been at this time that the church of York claimed Whithorn in Galloway as part of its archiepiscopal province; the claim is made in a Latin text of c. 1200 appended to a copy of Bede's *Ecclesiastical History*, but it refers to its source being an Old English text which may have been composed in the period we are considering. Also to this period may belong the process by which the archbishops assigned to the ancient churches of Ripon, Beverley and Southwell, which were served by communities of secular canons, the role of informal 'subsidiary cathedrals' which we see them exercising in the

[30] On Wulfhere, see above, p. 215; on Wulfstan, see *ASC* D, *s.a.* 940–3, 947, 952.

documentation of the time of Archbishop Thurstan (1114–40). If our evidence is interpeted in this way, there would appear to have been continuity in the position and power of the archbishops of York between the Viking period and the Northumbrian past. The Viking kings may have been titular heads of Northumbria south of the Tees; their military power may have underwritten the preservation of the independence of the area from the English kings; but real power lay with a church which was organized and acted much as it had done in the pre-Viking period, and maintained a similar ascendancy.[31]

The archbishops, however, must surely have shared any position of political dominance with the Northumbrian aristocracy, which as a body makes several appearances in the *Anglo-Saxon Chronicle* as the 'councillors of the Northumbrians' or simply as 'the Northumbrians', and was in a position, as in the case of Eric, to accept and expel kings. That aristocracy was evidently capable of acting in concert in a political way, and may have been as important under the Viking kings as it had been under their predecessors. As we shall see, it is clear that its ethnic composition – or at least its culture – changed as a result of a Viking contribution to it; but the evidence we have, slight as it is, suggests that its position in the Northumbrian polity was much as it had been before the arrival of the Great Army.[32]

It is arguable then that the political discontinuity produced between the River Tees and the Humber by the Viking invasion was much less than appears at first sight. Despite the elimination of the native kings, the Viking kingdom of York had recognizable political elements from the former kingdom of Northumbria: namely the predominant position of the archbishops and the aristocracy. Moreover, although the kingdom was fractured broadly along the line of the River Tees, that fracturing did not affect the heartland south of the Tees, which we identified in an earlier chapter as crucial to the integrity of what was originally the kingdom of Deira.[33] This heartland remained intact as it was in the earlier period.

[31] The Oswald document is Robertson (1956), no. 54; trans. also in *EHD I*, no. 114; for Southwell, see above, p. 27, and Sawyer (1968), no. 659; for Amounderness, *EHD I*, no. 104 (Sawyer (1968), no. 407); see also Kenyon (1991), pp. 112–14; for Whithorn, Howlett (2000), pp. 185–7, Brentano (1953); and, on Ripon, Beverley and Southwell, Nicholl (1964), pp. 123–7.

[32] *ASC*, *s.a.* 924 (A – 'ge ces þa to fæder 7 to hlaforde . . . ealle þa þe on Norþhymbrum bugeaþ'), 941 (D – 'Her Norðhymbra alugon hira getreowaða 7 Anlaf of Yrlande him to cinge gecuron'), 947 (D – 'ealle Norðhymbra witan wið þone cyning hi getreowsoden'), 948 (D – 'þa Norðhymbra witan þæt ongeaton, þa forlæton hi Hyric 7 wið Eadred cyning gebeton þæt dæde'), 952/4 (E/D – 'Her Norðhymbre fordrifan Anlaf cyning, 7 under fengon Yric Haroldes sunu').

[33] See above, pp. 45–8.

The Viking kingdom of York: ethnic transformation?

How far was the ethnic make-up of the Viking kingdom of York different from that of the former kingdom of Northumbria? Specifically, was the establishment of the Viking kingdom of York accompanied by extensive Scandinavian settlement, displacing or reinforcing the existing population, or was there rather a take-over of the lands of English landlords by an incoming Viking elite? The relevant evidence is chiefly the annal for 876 in the *Anglo-Saxon Chronicle*, Scandinavian personal names recorded in *Domesday Book* and other sources, a rather small amount of archaeological evidence, and place-names derived wholly or partly from the Vikings' language (Old Norse). The annal reads: 'And that year Halfdan shared out the lands of the Northumbrians and they proceeded to plough and to support themselves.' The *History of St Cuthbert* provides some corroboration for this in stating that the Viking army 'rebuilt the city of York, cultivated the land around it, and remained there'. The implication of both texts is that the Great Army, or at least that part of it which Halfdan had led to Northumbria, took possession of the land of that kingdom and its members settled on it. There is, however, room for doubt as to how well informed the *Anglo-Saxon Chronicle*, written in Wessex, really was about events in Northumbria, and how far the author of the *History of St Cuthbert*, probably writing in the mid-eleventh century at the earliest, had reliable information to draw on. Moreover, in a different section, that author represented a later Viking leader, King Ragnall (914 × 919–920/1), as dividing up vills in what was to be County Durham between two of his – presumably aristocratic – followers, Scula and Onlafball; and it seems possible that what Halfdan too was doing was enfeoffing his men rather than literally handing over land for settlement.[34]

Given the limited and problematic nature of the texts, the evidence of personal names and place-names is crucial. *Domesday Book* for Yorkshire indicates a substantial element in the population bearing Old Norse names, although we cannot tell when such naming practices originated or whether they were the result of in-depth settlement or the dominance of a Viking elite leading to a change of fashions in naming in what remained an English population.[35] Likewise, place-names derived from the Vikings' language,

[34] For the texts, *ASC*, *s.a.* 876, and *Hist. Cuth.*, paras. 14, 23. Note that *ASE* E has *hergende* ('harrying') in place of *ASC* A's *ergende* ('ploughing'), dismissed by Swanton (1996), p. 75 n. 14, as 'an ironic slip'. See also Plummer (1892–9), II, p. 91.

[35] Feilitzen (1937).

which are widespread but not predominant in Yorkshire, are unlikely to allow us to assess the level of Viking settlement in any simple way, for, in the light of recent research, it is no longer possible to regard a village with an Old Norse name as necessarily one settled by Vikings. There are in Yorkshire forty-two names made up of an Old Norse personal name and the English suffix *tun* meaning 'homestead' or 'village'. Most of these so-called 'Grimston hybrids' have been interpreted as established English villages, which were taken over by Viking landlords who gave their own names to them. Almost three hundred names in Yorkshire have the Old Norse ending *by*, also meaning 'homestead' or 'village'. The occurrence of English personal names combined with this ending, and the fact that some of the villages concerned are located on good land which must have been occupied in the pre-Viking period, suggest that these names may have originated when a large, formerly English-held estate was split up into smaller units. The *by* element in their names may indicate that they were owned by incoming Viking landlords; or it may indicate that Viking dominance perhaps of the old estate centre had led to imposition of this Old Norse element even if the new villages were entirely English. Names with the Old Norse element *thorp*, meaning 'smaller village', may also represent renamings of English villages rather than the incursion of Viking settlers; and only names with the Old Norse element *thveit*, meaning 'meadow' or 'piece of land', seem to indicate new colonization of land by Danish incomers.[36]

These studies of place-names are predicated on the assumption that there was a broad continuity across the Viking period in rural organization, and that whatever settlement took place did not radically disrupt that. As discussed earlier, the pattern of head vills with their subordinate vills ('berewicks' or 'sokelands') making up sokes or 'multiple estates' is found extensively in *Domesday Book* for Yorkshire, and is also evident north of the Tees, where groups of vills are called 'shires' and are especially prominent in the twelfth-century survey called *Boldon Book*, but also in some of the documents preserved in the *History of St Cuthbert*. This pattern is also perceptible in southern Scotland, Kent and East Anglia. There is nothing characteristically Viking about it and much to suggest that it antedates the Viking period. The incidence of Old Norse settlement-names, as we noted above, seems to point to the splitting up of these units but not to fundamental change. Although research has not yet been undertaken systematically, it appears often to be the case that the head vill of a soke had an Old English name, while

[36] For a summary of research with tentative conclusions, see Fellows-Jensen (1995).

some at least of its subordinate vills had Old Norse names. A classic example is that of Pickering (Yorkshire North Riding), an unimpeachably Old English name. Some of its dependencies are equally English in nomenclature: Barton, Newton, Ebbertston, Allerston, Wilton, Levisham, Middleton and so on. But several have the characteristic *by* or *thorp* endings of Old Norse names, as in Blandsby, Eastthorpe, Roxby and Kingthorpe. This pattern could be interpreted as meaning that the process of establishing vills on the territory of Pickering happened partly before the Viking period (hence the English names) and partly during it (hence the Old Norse names). Perhaps in that latter stage a Viking landlord owned Pickering; perhaps he was enfeoffing his followers in the newly created vills – we have no way of telling. At one of the vills with an English name, Middleton, there is the further complication that its church has in it several grave-stones apparently representing Viking chieftains with characteristic weapons and headgear, and decorated with animal motifs in Viking style (ill. 29). The English name combined with the Viking-style sculpture underlines the complexity of the situation: was the vill held by a Viking who retained its English name, or is the sculpture simply evidence of the spread of Viking taste among the English promoted by a dominant Viking elite? Further, the splitting up of estates at this time was part of the mainstream of northern European development as was the creation of new villages, so it may be purely coincidence that some of this took place while Northumbria south of the Tees was under Viking dominance. Whether that splitting up was accompanied by Viking settlement, we cannot tell.[37] Aside from place-names, some aspects of the terminology of territorial organization changed. We find, for example, territorial units called by the Old Norse term 'wapentake', but there is no reason to suppose that the structures it was describing were anything other than the units called 'hundreds' found in the south of England and in eastern Yorkshire. What archaeological evidence we have suggests that the villages themselves were not affected by Viking influence or settlement. The extensively excavated site of Wharram Percy in the Yorkshire Wolds seems to continue as a settlement unchanged. The small settlement of Simy Folds at the head of Teesdale continued through the Viking period; it was in the mainstream of rural settlement of the period and has nothing diagnostically Viking about it.[38]

[37] On estates, see above, pp. 91–3; for *by* names and subordinate vills, see Kapelle (1979), pp. 65–6, and Fellows-Jensen (1995); for Pickering, see *Domesday Yorks.*, 1Y4; and for the sculpture Lang (1991), pp. 181–7.

[38] On Wharram Percy, see Beresford and Hurst (1991); on Simy Folds, Coggins, Fairless and Batey (1983); for a useful summary, see Richards (2000b), pp. 56–7. For hundreds and wapentakes, see Hadley (2000), pp. 101–7, and *passim* for the general argument of this paragraph.

ILLUSTRATION 29 Middleton (Yorkshire North Riding), St Andrew's Church, cross
Dated after *c.* 920 because of the ring-headed form of the cross, the ornament on this cross (there is a beast on the opposite side) has been characterized as 'thoroughly Anglo-Scandinavian'. The human figure is an armed man with conical helmet, knife in sheath at the belt and spear, sword and axe at the sides. Although it seems unlikely that it represents a pagan burial, it does underline the likely patronage of this cross by a layman of the warrior class in the Viking period. See Lang (1991), pp. 182–6 and references therein.

In assessing possible Viking impact on Northumbria, we have no way of knowing how many Vikings came to England, or even whether the Great Army which captured York in 866–7 was a force of thousands, involving men of peasant class, perhaps accompanied by their women folk, or an elite force, numbered rather in hundreds, and composed of aristocratic adventurers. The sizes of Viking forces are often indicated in the *Anglo-Saxon Chronicle*, but doubt has been cast on its reliability in this respect, on the grounds that contemporary chroniclers, who were of course Christian, were anxious to exaggerate the scale of the Viking attacks in order to magnify the menace which they, as pagans, represented for the church. There is, however, no escaping the fact that the *Anglo-Saxon Chronicle* was so impressed with the scale of the army which captured York that it specifically referred to it as the Great Army; and nor is there any escaping the astonishing military success of that army in effectively liquidating not only Northumbria but also the kingdoms of East Anglia and Mercia. In the case of Northumbria, the *Anglo-Saxon Chronicle* seems to be attempting an explanation for the success of the Viking invasion by explaining that 'there was great civil strife going on in that people, and they had deposed their king Osberht and taken a king with no hereditary right, Ælle'. The compiler may have exaggerated the political instability of Northumbria at the time of the Viking attack. The *History of St Cuthbert* represents Ælle as Osberht's brother rather than a 'king without hereditary right', while numismatic evidence has suggested that Ælle's reign was limited to 867, with Osberht reigning from *c.* 852 to 867, so that any political disturbance in Northumbria was of very recent origin, and may indeed have been precipitated by the first Viking capture of York on 1 November 866, rather than being a cause of it. At all events, there is no reason to suppose that the Northumbrians were not in a position to mount a major resistance to the Vikings, and the accounts of their counter-attack can be read as showing that they did. There is every reason therefore to think that the Great Army was an extremely formidable force, one that remained capable of subjugating Northumbria south of the Tees even when it had split into three with only the part led by Halfdan returning to that kingdom. But it remains impossible to estimate how large it was, or to establish the social class of its members. F. M. Stenton argued that the Viking armies were made up primarily of men of peasant class in large numbers and that these men settled down in England, so that their descendants became the sokemen and freemen recorded in *Domesday Book*; but this interpretation has seemed increasingly unlikely, particularly now that sokes (multiple estates), on which much of it was based, appear not to be Viking creations, but to have had their origins in the pre-Viking period, as a common development across western Europe. Sokemen cannot therefore be equated with incoming Viking settlers, as is required by Stenton's interpretation.[39]

There can be no doubt, however, that some Vikings did settle. There is evidence by the time of *Domesday Book* for the presence – or at least the influence – of Vikings in the upper echelons of Northumbrian society between the River Tees and the Humber. In York, the moneyers producing coins at the mint in the reign of William I had Old Norse names. Moneyers were an elite group in the early middle ages and this evidence therefore suggests that at least the upper echelons of York society were either Viking in origin or heavily influenced by Viking settlers. Several known moneyers – and other men with Old Norse names – appear as witnesses to a survey of the *Rights and Laws of the Archbishop in York* made in the late eleventh century, and this tends to confirm our conclusion. The founders of the York church of St Mary Castlegate, Grim and Æse, recorded in the dedication stone preserved in the church, were evidently also Vikings of high status associated with York, as also was Ulf, who presented lands to the Minster together with an ivory horn, now in the Minster Treasury, which was presumably a token of the gift of those lands. His son was Styr, who gave Darlington (County Durham) to the church of St Cuthbert at Chester-le-Street, and whose daughter married the English earl of Northumbria, Uhtred, on condition that the latter would kill his enemy Thurbrand, also apparently an inhabitant of York and one with an Old Norse name. Such names do not prove that the men who bore them were descended from Viking warriors who came with Halfdan, as there was no doubt immigration from Scandinavia into Northumbria subsequent to the capture of York in 866–7. They do not prove that the men who bore them were of Scandinavian extraction at all, but they create a strong presumption that they were, and it would be perverse to argue other than that by the time of the Norman Conquest at any rate the upper levels of the population of York had been heavily Scandinavianized, either because Vikings had been introduced into it or because there was strong Viking influence on it.[40]

[39] For the numbers of Vikings, see Sawyer (1971), pp. 120–31, and, against, Brooks (1979); for the capture of York, see above, p. 212, and specifically *ASC*, *s.a.* 867, and *Hist. Cuth*, para. 12; for the length of Ælle's reign, see above, p. 196; for a broad discussion of earlier scholarship, see Hadley (2000), pp. 1–41, 180–9.

[40] For *Rights and Laws*, see *Sources*, pp. 192–3, 212–13 (N.5.1); for the street-names, see Palliser (1978). For Grim and Æse, see Okasha (1971), no. 146; for Ulf, R. M. T. Hill and Brooke (1977), p. 39, pl. 2 and note; for Styr, Fletcher (2002), pp. 52–3, 76–7 (for the texts, *Hist. Cuth.*, para. 67, and *Siege of Durham*).

The Viking kingdom of York: cultural transformation?

What did the Viking kingdom of York owe to the culture of the former kingdom of Northumbria? Like the English of an earlier period, the Vikings were pagans, and we might expect considerable discontinuity in cultural development. In fact, however, there are very few indications that this was the case. If we begin with York itself, our evidence suggests that the Viking incomers were relatively quickly converted to Christianity and assimilated to Christian culture. That evidence is in part of a negative character: the absence of evidence for paganism in the city, aside from the coins of Olaf Guthfrithson referred to above, and maybe also the coins of Sihtric Caoch which have on them a sword, possibly a symbol of the pagan Vikings of Dublin; and aside also from two burials from the church of St Mary Bishophill Junior which, since they have grave-goods, may be of recently arrived Vikings. But the fact that the deceased were buried in a churchyard suggests that their families were already in the process of assimilation to Christian culture. In the early eleventh century, the *Northumbrian Priests' Law* had clauses against pagan practices – sacrifice, divination, witchcraft, worship of idols, having a sanctuary round 'a stone or a tree or a well or any such nonsense' – but it is not clear that this applied to York in particular, and in any case the overall impression of this legislation is that it was directed at an essentially Christian society, with some relict pagan practices. What is striking is not the evidence for paganism, but the much stronger evidence for the Christian character of York continuing after the Viking capture of 866/7. One of the earliest Viking kings associated with York, Guthred, was apparently a Christian, for not only was he supposed to have owed his throne to the religious Community of St Cuthbert at Chester-le-Street, but he was also buried at York Minster. Indeed, the cemetery excavated under the Minster, in which his grave may have lain, is itself strong evidence for Christianity in the city in the Viking period. Although the Minster of that period has not yet been located archaeologically, there is little doubt that the cemetery under the present Minster was attached to it, and there is no doubt at all that this was a Christian cemetery, where high-status citizens of York were buried. The changing styles of the grave-stones point to Scandinavian influence, the influence of the Viking incomers presumably; but the burial practices and the location of the cemetery show uninterrupted Christianity at the heart of the city.[41]

[41] For a survey and discussion of the rapidity or otherwise of the conversion of the Vikings, see Abrams (2001); for the coins, see above, p. 224; for St Mary Bishophill Junior, see Wenham, Hall, Briden et al. (1987), pp. 80, 83. The *Northumbrian Priests' Law* is trans. *EHD I*, no. 53 (text, Liebermann (1903–16), I, pp. 380–5, and Tenhaken (1979), dating, Wormald (1998), pp. 396–7). For Guthred, see

Other churches in the city similarly appear to have functioned without interruption. It is true that York did not feel the effects of the tenth-century reformation of the monasteries, by which religious foundations in the south and the midlands were brought into line with the Rule of St Benedict and the precepts of continental centres of reform such as Fleury on the Loire, so there were no monastic houses as such in pre-Conquest York, and probably only a couple of houses of secular canons. These were St Olave's, founded by the earl of Northumbria, Siward, a century after the death of Eric, and the church which later became Holy Trinity Micklegate, the eleventh-century foundation charter of which refers to its having been originally founded in ancient times by canons (the exact period is unknown). Many of the numerous parish churches in existence in late medieval York, however, are likely to have been founded in the Viking period, or at least in the pre-Conquest period, perhaps as private (proprietary) churches. Admittedly, this can only be asserted by reference to similar developments in other cities such as London, since *Domesday Book* only documents eight churches in York. It is, however, notoriously patchy in its recording of churches, and these eight seem to be mentioned not for their own sake but because a change of ownership was involved.[42]

Such steady progress of Christianity seems to have affected the whole of the Viking kingdom of York. The great Christian achievements of the eighth century admittedly found no echo in the Viking period; and a number of churches which had been central to those achievements were much reduced in status by the time of *Domesday Book*. The great Northumbrian churches of Whitby and Ripon figure in that record respectively as a ruined site and as a house of canons; while Lastingham and Beverley, monasteries founded in the seventh and eighth centuries, seem to have become respectively a parish church and a house of secular canons. We cannot be sure, however, when the degradation of these institutions took place. Whitby's condition may have been the result of William the Conqueror's harrying of the north, for the site of the former monastery was still known as Presteby ('settlement of priests') in the late eleventh century, and 'almost forty monasteries or oratories' were still recognizable on the site, although ruined; Ripon's conversion to a house of canons may have been the result of a sack of the church by Eadred,

below, p. 245; for the cemetery under the Minster, see Phillips, Heywood and Carver (1995), pp. 75–92, 438–63.

[42] On the reformation, see Robinson (1923) and Parsons (1975); on St Olave's and Holy Trinity Micklegate, see *Sources*, pp. 175 (V.4.4) and 204–5 (N.4.14); on York parish churches, see *Sources*, pp. 207–10 (N.4.17–29); on London, Christopher N. L. Brooke and Keir (1975), pp. 131–43; for further discussion, David Rollason (forthcoming).

king of England, in 948. In short, there is nothing to suggest that these churches were fatally disrupted by Viking activity. It may be indeed that we are underestimating the cultural importance of York Minster in the Viking period as a result of the destruction of its archive and library probably in the great fire started by the Norman garrison of the city in 1069.[43]

The decline in the status of the major churches in the Viking kingdom of York, even if it really occurred in the Viking period, was more than balanced by an increase in the number of rural churches. The principal evidence for this is the distribution of Christian sculpture in a style which makes it possible to date it to the period of Viking influence. Thus crosses with heads set in a ring (the so-called ring-head crosses (ill. 30)), and carving with particular types of plait-work or animal decoration derived from Viking metalwork in Scandinavia (especially so-called Jellinge-style ornamentation comprising ribbon-like intertwining animals with long jaws (ill. 31)) can with reasonable confidence be assigned to the Viking period. Although much of this sculpture has been recovered from the fabric of later churches, there is every reason to suppose that it has always been on the sites of those churches and was associated with their predecessors. It is apparent that such sculpture was amazingly common, so that fragments of it have been found in a large proportion of churches in the area. Moreover, it is generally the case that where a church has pre-Viking sculpture, it also has Viking-age sculpture, whereas there are many churches which have the latter but not the former. The inference is clear: the Viking age saw much increased patronage of sculpture for churches and churchyards, and many more churches for which it could be made. Proliferation of parish churches, which was part of a much wider trend in the ninth and tenth centuries, suggests that Viking assimilation to Northumbrian culture and society was such that the progress of the Christian church was unaffected. It is significant that, where we find Scandinavian pagan images in the area, as we do in a few cases as at Leeds where there is a representation of the smith Weyland and his flying outfit, it is in the context of sculpture associated with churches and seems to indicate the church's efforts to gather previously pagan Vikings into its fold by relating pagan stories to Christian ones.[44]

[43] Hadley (2000), pp. 233–6, 239–41, 245–7; for the ruins at Whitby, see Burton (1999), p. 33. For the burning of York, see *Sources*, pp. 184–5 (N.1.7 = *Hist. Kings, s.a.* 1069 (pp. 187–8)); see also p. 188 (N.3.1 = *ASC* D, *s.a.* 1069).

[44] The most useful study is Bailey (1980): see esp. on dating pp. 45–75, on distribution of sculpture, see pp. 80–1, and, on pagan images, pp. 101–24 (esp. for Leeds, p. 104). See also Lang (1991), figs. 1, 4, pp. 16–18, 26–7 (distribution), and, on the wider pattern of ecclesiastical development, Richard Morris (1989), pp. 140–67.

ILLUSTRATION 30 Gosforth (Cumberland), St Mary's Church, cross
The ring-head shows that the cross belongs to the Viking period, as also does the ring-chain decoration below it. This is in the style called Borre after a group of objects from Borre in Norway, on which it is found. It has been dated to the first half of the tenth century (Bailey and Cramp (1988), pp. 100–4). For a detail of the lower part of the cross, see ill. 35.

ILLUSTRATION 31 York Minster, part of a cross-shaft excavated from the foundations of the eleventh-century nave, and dated to the late ninth or early tenth century The decoration of this and the other narrow face consists of a chain of beasts linked by the neck. Note the double outline and spiral joints which reflect Viking Jellinge style. Although they could have been derived from Insular art, their appearance here 'can be attributed to colonial Viking taste'. See Lang (1991), pp. 54–5 and references therein.

ILLUSTRATION 32 Brompton-in-Allerton (Yorkshire North Riding), All Saints Church, hogback stones with beasts clasping the stone at each end
For description, see Lang (2001), pp. 73–9 (ills. 79–108).

In other respects, too, the evidence of Christian sculpture points to assimilation of the Vikings rather than to the dominance of their own culture. Stone sculpture was not part of Viking art in Scandinavia, as it was of Christian art in pre-Viking Northumbria; the fact that it continued to be important between the River Tees and the Humber in the Viking period suggests the continued dominance of native cultural traditions and the assimilation of the Vikings to them. The same applies to the style in which that sculpture was carved. Jellinge style, although certainly of Scandinavian origin, is not far removed from the animal decoration of Northumbrian art of the pre-Viking period, as found in the Lindisfarne Gospels, for example, and may well have been influenced by it. The 'ring-headed crosses' are associated with the Viking period, but in fact they seem to have been an idea derived from Irish sculpture, perhaps brought to Northumbria by Vikings coming across from Ireland. More intriguing still is another characteristic form of Viking period sculpture, the so-called hogback stones, consisting of elongated rounded stones more or less in the shape of a house, sometimes with beasts clasping the ends, sometimes without (ills. 32–3). The function of these stones is unclear, although they may well have been grave-stones; but they have a clear resemblance to the so-called house-shrines of Irish

ILLUSTRATION 33 Gosforth (Cumberland), St Mary's Church, part of a hogback stone called the Warrior's Tomb
Dated to the first half of the tenth century, this hogback stone is of the type without clasping beasts (compare ill. 32), and is more obviously house-shaped. The procession of at least fifteen warriors which is the principal decoration of the stone underlines the probability that it was made under aristocratic lay patronage (compare ill. 29). See Bailey and Cramp (1988), pp. 105–6.

art, and to the eighth-century Hedda Stone now in Peterborough Cathedral which may itself have been the cover for the grave of a saint. Here too then we may be seeing Viking assimilation and adaptation to native traditions rather than any dominance of Viking culture.[45]

The influence of the Vikings on the language of the area is a much debated issue, and, although the evidence of the place-names suggests that Scandinavian language was spoken there, we do not know how long this persisted. That language had certainly disappeared by the Norman period, and it may be that it had been largely assimilated into English long before that. It is striking that two early eleventh-century legal compilations, the

[45] Bailey (1980), pp. 53–72 (style), 85–100 (hogbacks); on hogbacks, see further Lang (1984).

Law of the North People and the *Northumbrian Priests' Law* were both written in Old English and presumably intended to be understood by the inhabitants of Northumbria in general. Despite the destruction of the Northumbrian royal line, therefore, in political, ethnic and cultural terms Northumbria between the River Tees and the Humber in the Viking period can be seen as a continuation of what had come before, with the emergence of the Viking kingdom of York as nothing more than the splitting of the former kingdom of Northumbria along the ancient divide between Bernicia and Deira that we examined in chapter 2.[46]

North of the River Tees: the 'liberty' of the Community of St Cuthbert and the earls of Bamburgh

The evidence of place-names suggests that there was very little Viking settlement or influence north of the River Tees. Aside from a group of place-names just to the north of the river, around the wapentake of Sadberge (note the Old Norse term), there are very few Scandinavian place-names of any type, and the contrast with the dense occurrence of such names south of the river is striking. It may be of course that there was nevertheless a Viking elite north of the Tees which did not impose a new nomenclature on the names of settlements – Ragnall's followers, Scula and Onlafball, were both granted vills with no apparent name-changes, as in the case of South Eden (County Durham). But the *History of St Cuthbert* suggests that the tenure of these vills by Onlafball at least was brief, for he was supposed to have been miraculously struck down for slighting St Cuthbert and his bishop. That there was no settled Viking elite north of the Tees is suggested by a later passage in the same text, which shows the bishops of the Community of St Cuthbert in the late tenth and eleventh centuries surrounded by men who were clearly aristocratic vassals and had names of English derivation.[47]

The political organization of the area between Tyne and Tees in the Viking period consisted of the 'liberty' of the religious Community of St Cuthbert, based at Chester-le-Street from 883 and from 995 at Durham. That community claimed to be the direct successor of the monastic community founded in 635 on the island of Lindisfarne by Bishop Aidan, but it is hard to establish how it came to move from Lindisfarne and how it came to acquire its

[46] In general, see Parsons (2001) and Hines (1991); for the texts see above, p. 170 n. 130, and Liebermann (1903–16), I, pp. 456–69, and *EHD I*, no. 52B (text Liebermann (1903–16), I, pp. 442–4).

[47] For the place-names, see Watts (1995); for the texts, see *Hist. Cuth.*, paras. 23–4. For discussion, see Christopher D. Morris (1977, 1981).

'liberty', because of the hagiographical character of the sources. The monastery of Lindisfarne had been attacked by the Vikings in 793, but the *History of St Cuthbert* suggests that it was flourishing in the early ninth century, when Bishop Ecgred of Lindisfarne (830–45) was active in acquiring property in the area between Tyne and Tees, notably Gainford-on-Tees (County Durham) with all its subordinate vills, and Billingham (County Durham), where he built a church. The *History of St Cuthbert* then recounts how, threatened with further Viking attacks (presumably the return of Halfdan to Northumbria with his portion of the Great Army), the religious community of Lindisfarne set out in 875 on a period of wandering across northern England which took it to the Cumbrian coast, to Crayke near York and in 883 to Chester-le-Street, on the River Wear (between Tyne and Tees, map 9), where it established the church, the bishop's see and the shrine of its principal saint, Cuthbert (bishop of Lindisfarne, 685–7). As regards how the community acquired its 'liberty', the *History of St Cuthbert* describes how, in the course of the community's wanderings, St Cuthbert appeared to one of its leaders, Abbot Eadred, and instructed him as follows:

> Go over the Tyne to the army of the Danes, and tell them that if they wish to be obedient to me, they should show you a certain young man named Guthred son of Hardacnut, the slave of a certain widow. In the early morning you and the whole army should offer the widow the price for him, and at the third hour [take him] in exchange for the price; then at the sixth hour lead him before the whole multitude so that they may elect him king, and at the ninth hour lead him with the whole army upon the hill called *Oswigesdune* and there place on his right arm a gold armlet, and thus shall all constitute him king. Tell him also, after he has been made king, to give to me all the land between the Tyne and the Wear, and [to grant that] whoever shall flee to me, whether for homicide or for any other necessity is to have peace for thirty-seven days and nights.

Abbot Eadred did as he was instructed, and then 'brought to that host and to that hill the body of St Cuthbert, over which the king himself and the whole host swore peace and fidelity as long as they might live, and this oath they faithfully observed'.

This is clearly a miracle-story, the primary purpose of which in the Durham traditions was to emphasize the power of St Cuthbert; but it may nevertheless cast light, however dim, on the political circumstances of the 880s. The Viking king in question was almost certainly the Guthfrith buried at York Minster in 895, so the *History of St Cuthbert*'s account that he was sympathetic to a Christian community and began to reign in around 883 is not unconvincing. The reference to the gold arm-ring is also a convincing feature, since something rather like it is found in the *Anglo-Saxon Chronicle*'s account of King Alfred making peace in the south with Vikings who 'swore him oaths on the sacred ring'. So too is the account of the sanctuary rights given to the community to the extent at least that the church of Durham possessed such rights at a later date, and the idea that they originated in the late ninth century with a gift such as this is not implausible, especially as the rights specified echo the arrangements for sanctuary made by King Alfred in his law-code, and somewhat later Durham traditions claimed that Alfred himself was involved in the grant of them.[48]

In short, setting aside the miraculous elements of the *History of St Cuthbert*'s story, it is not impossible to accept that the community really had come to an arrangement with the Viking army on the Tyne, and had somehow been of service to a future Viking king called Guthred. It is reasonably certain (although not explicit in the *History of St Cuthbert*'s account) that the gift of land between the Rivers Tyne and Wear resulted in the community's establishment at Chester-le-Street. Since Bishop Ecgred had built up extensive estates between the Rivers Wear and Tees in the earlier ninth century, this meant that the community's hold on the area from Tyne to Tees must have been considerable. The significance of the grant of sanctuary rights is not clear; but it may have conferred on the community some form of jurisdiction over its lands and thus have put it in a position of political and juridical as well as territorial dominance. In short, the community's 'liberty' chiefly in the area of what was to be County Durham can be seen as emerging in the wake of the Viking invasions, the core of the palatinate of the later bishops of Durham. In addition, our evidence suggests that the community retained its hold over part at least of the territory of the church of Lindisfarne north of the Tyne, including the area which was to be known in the later middle ages as Islandshire and Norhamshire or, alternatively, as North Durham.[49]

This 'liberty' of the Community of St Cuthbert recalled in certain important respects the former kingdom of Northumbria, and to that extent showed elements of continuity with it. The pattern of the lands of the church of St Cuthbert reflected the territories of the former kingdom of Bernicia, south and north of the Tyne; the rights of sanctuary, although reinforced in

[48] For the attack on Lindisfarne, see Alcuin, *Letters*, nos. 19–22; Symeon, *Origins*, bk 2 ch. 5 (pp. 86–91). For the text, see *Hist. Cuth.*, paras. 9, 13, 20. For discussion, see Johnson South (2002), Aird (1998b), pp. 9–59, and Craster (1954). For the ring, see *ASC*, *s.a.* 876 (comments Swanton (1996), p. 74 n. 3), and, on sanctuary, David Hall (1989), pp. 430–1.

[49] Craster (1954), pp. 178, and Raine (1852).

the ninth century, went back to those attaching to the tomb of St Cuthbert itself, mentioned by Bede in his account of the saint. Moreover, although the texts emanating from the community describe its history as one of disruption and flight in the face of the threat of Viking attack, it is tempting to conjecture that, in view of Bishop Ecgred's earlier acquisitions in the future County Durham, the community was really moving its centre of operations southwards as its territorial base had grown in the same area. In other words, it was engaged in consolidating and enhancing its position, rather than suffering the flight into the wilderness of an exiled community which its traditions presented it as. Moreover, sometime in the second quarter of the ninth century the see of Hexham had come to an end in circumstances which are utterly obscure but unlikely to have had anything to do with Viking activity. Chester-le-Street had been within the diocese of Hexham, so the move of the Community of St Cuthbert to that centre was possible because Hexham had ceased to function as an episcopal see earlier in the ninth century. We may well suspect that the two things were related to the extent that the Community of St Cuthbert's southward move was designed to take advantage of Hexham's demise.

In view of this, Lindisfarne's history in the former kingdom of Northumbria became of crucial importance to the Community of St Cuthbert's legitimacy. When the community removed in 995 with its saint and the see to the rocky peninsula of Durham some way upstream along the River Wear, Symeon of Durham relates that this site was chosen by the miraculous intervention of St Cuthbert, whose body, while being carried back to Chester-le-Street after a temporary absence, became too heavy to move until the community selected Durham as its new home. The site might be new, but the links with the Northumbrian past had thus to be forged. Other saints associated with St Cuthbert served to emphasize those links. The head of the martyred Northumbrian king Oswald was kept in St Cuthbert's coffin; and in the 1030s the sacrist of the community, Elfred, son of Westou, collected the relics of other saints from the golden age of Northumbria (including those of Bede, Balthere, Billfrith, Acca, Alchmund, Oswine and Boisil) as if to emphasize the Northumbrian past of the community.[50] The community seemingly went out of its way to emphasize its roots, commissioning sculpture which was a conscious revival of the styles of the former kingdom of Northumbria, for example a splendid grave-stone at Durham, decorated with carved interlace patterns which represent a conscious revival of the style of

[50] David Rollason (1987, 1995a); on Elfred, see Symeon, *Origins*, bk 3 ch. 7.

ILLUSTRATION 34 Gainford-on-Tees (County Durham), St Mary's Church, part of a cross-shaft
Note the four-footed beast above, partially broken away, with a similar beast below, its head facing to the left and its jaws biting at the interlace which links it to the beast above. The hooked feet and spiral hip-joints of the animals reflect carving in the Danelaw and characterize this cross as influenced by Viking taste, although it lies on the north bank of the River Tees. It is dated to the first half of the tenth century.

the Lindisfarne Gospels. Nevertheless, some elements of the style associated with the Viking period in Yorkshire are found between Tyne and Tees, showing that the lands of the community were not immune from Viking influence or at least from influence emanating south of the Tees: the Jellinge-style beasts on the crosses at Gainford-on-Tees and Aycliffe, for example, and the hogbacks at Sockburn just to the north of the Tees (ill. 34).

North of the Tyne, the rule of the earls of Bamburgh seems in some respects to hark back to the period of the Northumbrian kings. Shortly after 867, it appears that a king called Egbert ruled the area, presumably in principle as far north as the frontier of the former kingdom on the Firth of Forth. He was succeeded by a king called Ricsige who was in turn succeeded by another king called Egbert. We hear of no more kings in the area after the second Egbert, but early in the tenth century we find a dynasty of earls, ruling – surely significantly – from the ancient Northumbrian royal palace of Bamburgh. It is not impossible that they were scions of the Northumbrian royal house, or maybe of the families of its chief officers, who were in some form carrying on a rule which approximated to that of the Northumbrian kings of yore. The first of the earls, Eadwulf, was in fact called 'king of the north Saxons' by the contemporary *Annals of Ulster*.[51] Very different then as was the political organization of the lands north of the Tees, their ethnic and cultural character were not radically altered. Even certain aspects of their political organization reflected, sometimes consciously, the former kingdom of Northumbria.

Cumbria

Consideration of the degree of political, ethnic and cultural transformation in Cumbria in the Viking period is complicated by the question of the kings who emerged in this area in the tenth century, described by our sources as 'kings of the Cumbrians'. These kings begin with the Owain who was probably the ruler defeated along with others by King Athelstan at the Battle of *Brunanburh* in 937; they continue with Dunmail (Donald), expelled in 945 by King Edmund of England who also blinded his sons; and further with Malcolm (d. 997), who was also known as 'king of the Britons of the North'. It appears then that these kings began to rule the north-west at least as early as the early tenth century; to judge from sources for the somewhat later history of the region, their southern frontier was as far south as the monument known as the Rey Cross (or Rere Cross) on Stainmore (Yorkshire North Riding) on the line of the modern A66 road from Scotch Corner to Brough (map 10). Who they were and how they came to power are much more problematic. Following the Viking sack in 889 of Dumbarton, the royal centre of the kingdom of Strathclyde, the native British dynasty of that kingdom

[51] See above, pp. 213, 215; on Uhtred and the house of Bamburgh, see Kapelle (1979), pp. 15–22, and now the excellent and readable account by Fletcher (2002).

became extinct, and Strathclyde was annexed by King Donald II of Scots. It was until recently accepted that what happened in north-west England after this was that the kings of Scots, having installed junior members of their dynasty as kings of Strathclyde, used that kingdom as a springboard to expand their power southwards into north-west England. In this scenario, the term 'king of the Cumbrians' referred to these Scottish kings of Strathclyde, whose rule in Cumbria thus represented an extension of Scottish power. The justification for this was, first, the possibility of identifying in written sources kings of the Cumbrians with kings of Strathclyde; and, secondly, the presence of British place-names in the north-west (Strathclyde was, as we have seen, a British kingdom) which are held to have been introduced as a result of this supposed invasion from Strathclyde, albeit led by Scottish kings. According to this line of argument, rulers of Strathclyde continued to rule the north-west until the death of the last king of Strathclyde, Owain the Bald, in or around 1018, after which the kings of Scots ruled his former kingdom directly. There was then no real continuity with Northumbria in the north-west: an invasion from Strathclyde effectively introduced a new political order to the area.

Another intepretation, however, has been advanced by Charles Phythian-Adams, and we have already seen the part of it relating to the pre-Viking period. According to this, this area of Northumbria had retained a marked British character, and was in some sense the successor of the British kingdom of Rheged, albeit under the control of a branch of the Northumbrian royal dynasty. What happened in the tenth century was that, with the extinction of the Northumbrian kings, the people of the north-west re-asserted their British identity by choosing the 'kings of the Cumbrians' noted above. Phythian-Adams argues that this term and 'king of the North British' as applied to Malcolm were not synonyms for 'king of Strathclyde' as the earlier interpretation maintained. Moreover, they not only indicated that these kings were not identical with the kings of Strathclyde, but also that Cumbria had retained its British identity (the name Cumbria is of course derived from the British word *Cymry*, which is still used for the Welsh). In this scenario, the British place-names represent survivals from the pre-English period and not evidence for an invasion from Strathclyde in the tenth century, which might have been expected to result in Brish place-names being attached to more strategic places than is in fact the case. It is impossible to resolve this dispute until the place-name evidence is re-evaluated; but there seems on the face of it much in favour of Phythian-Adams's interpretation. If it is correct, then we can perceive a broadly similar pattern to that of the continuity which we have identified in the east. Just as the north-west was a region of somewhat greater British continuity in the period of Northumbrian rule, so the fragmentation of Northumbria to some extent enhanced that characteristic and led to the emergence of a line of native British kings. It is an attractive hypothesis even if it is hard, given the nature and date of our sources, to make it more than that.[52]

In terms of the ethnic make-up of the area, Phythian-Adams's interpretation implies that there was considerable ethnic and cultural continuity from the Northumbrian period, the area being one preserving a substantial British population and culture then as in the Viking period. The place-name evidence, however, suggests that there was also a considerable Viking influence on the area, although it differs from that between the Tees and the Humber in that many of the Old Norse names are derived from the Norwegian (western Scandinavian) rather than the Danish (eastern Scandinavian) language. This suggests that the Vikings who either settled in or dominated the area were Norwegians, possibly from colonies in Ireland, although some of the personal names forming part of the place-names suggest Scottish (Gaelic) influence and the possibility that the Vikings in question might have come from colonies elsewhere in Scotland. There are in addition some place-names ending in *by*, for example, Appleby in the Eden Valley, and Bysbie in the Machars of Galloway, which suggest Danish influence from across the Pennines. In other respects, the type and distribution of Scandinavian place-names and their relationship to English names are not dissimilar from that found in the Viking kingdom of York and they present similar problems of interpretation. Do the place-names, in other words, show Viking settlement, political dominance or merely influence? Names in the uplands, for example in the mountainous upper reaches of the Derwent Valley, may represent new colonization; those in more developed regions such as the Eden Valley or the Machars of Galloway suggest dominance or displacement of existing populations; but certainty in any of these cases is impossible. Archaeological evidence is very limited. There are some apparently Viking burials from the area, and a settlement of the Viking period at Bryant's Gill in Cumberland has been excavated; but results are not very informative. Excavations at Whithorn have shown the replacement of the ecclesiastical

[52] Phythian-Adams (1996), pp. 77–87, 110–22, and Kirby (1962). See also, above, pp. 63–4. For the contrary interpretation, see Jackson (1963a), Stenton (1970b) and, for arguments against viewing Cumbria as distinct from Strathclyde, P. A. Wilson (1966) and Duncan (1975), pp. 93–4.

community there by a Viking trading settlement, which was probably related to the spread of Scandinavian place-names in the Machars of Galloway in which Whithorn is located. None of this evidence, however, suggests radical change in the ethnic character of the area, although there was arguably quite extensive settlement unevenly distributed. As for the possibility that we are dealing with the take-over of existing estates by incoming Vikings rather than with Viking settlement in depth, the evidence in this area too is restricted to brief and enigmatic references – indeed we do not here even have accounts of Viking invasions as we do in the east. Thus the *History of St Cuthbert* refers to a certain Elfred son of Brihtwulf who 'came over the mountains in the west... fleeing pirates' and asked the bishop of the Community of St Cuthbert to grant him some land. We can speculate that he had been expropriated by Vikings. The name of the estate which King Athelstan granted to the church of York, Amounderness, may be a record of just such an expropriation, if it embodies the name of *Agmund hold*, perhaps its original proprietor.[53]

As for the area's culture, we seem to see a similar pattern as in the area between the River Tees and the Humber. Monasteries such as Dacre, Whithorn, Hoddom and Heversham which flourished in the pre-Viking period did not survive as monasteries. We cannot be sure that this was the result of Viking activity, although it seems likely. The abbot of Heversham is recorded in the time of King Edward the Elder at the beginning of the tenth century giving land to Norham on the River Tweed 'so that he might be abbot there'. We cannot prove that his move was prompted by the destruction of Heversham by Vikings, but it is a plausible speculation. In the case of Whithorn the archaeological evidence shows the replacement of the Christian complex by the Viking trading settlement, even if it cannot prove that the Vikings were the cause of the change.[54]

Nevertheless, as in the area between the River Tees and the Humber, the denser distribution of post- than pre-Viking Christian sculpture points to rapid assimilation of the incoming Vikings to Christian culture. Moreover, we find rather more – and more striking – examples of the juxtaposition of pagan Viking and Christian images as if to emphasize the process of assimilation. There are possible examples at Lowther and Kirkby Stephen

[53] The fullest study of north-west England place-names is Fellows-Jensen (1985b). For useful summaries of scholarship relating to England and Galloway, see Fellows-Jensen (1985a, 1992) and, relating to south-west Scotland only, Oram (1995). For Bryant's Gill, see Dickinson (1985), and for Whithorn, see Peter Hill (1997). On Lancashire, see Fellows-Jensen (1983). For the text, see *Hist. Cuth.*, para. 22.

[54] *Hist. Cuth.*, para. 21; Peter Hill (1997).

ILLUSTRATION 35 Gosforth (Cumberland), St Mary's Church, cross, Loki bound
As with other scenes on the cross, these images can be interpreted as aspects of the Scandinavian pagan legend of the Ragnarok (end of the gods) as described by Snorri Sturlusson in the thirteenth century. The figure with the horn is probably Heimdallr, the watchman god who will rouse the gods to the struggle with their enemies at the Ragnarok. The horseman is unidentified, but the bound figure below is interpreted as the god Loki, who was bound in this way for his part in the killing of the god Baldr. He was bound so that a serpent would drip venom on his head, but his wife collected it in a bowl; both the wife and the serpent are represented in the carving (Bailey and Cramp (1988) and Bailey (1980)).

ILLUSTRATION 36 Gosforth (Cumberland), St Mary's Church, the 'Fishing Stone' Considered to be by the same sculptor as the Gosforth Cross (figs. 30, 35), this carved slab shows a scene convincingly interpreted as the god Thor and the giant Hymir out fishing. Thor used the head of one of the giant's oxen (shown at the bottom) to catch the serpent which encircled the world. Above this scene is another representing the struggle of a serpent and a hart, and possibly drawing a Christian parallel. See Bailey and Cramp (1988), pp. 108–9, and Bailey (1980), pp. 131–2.

(Westmorland), although we cannot always be sure that we are seeing actual pagan images as opposed to images common to Christian artistic tradition. In the famous and elaborate cross at Gosforth in Cumberland, however, there is no doubt. The sculptured scenes combine a representation of the Crucifixion of Christ with scenes from the end of the gods – the Ragnarok – in Scandinavian mythology (ill. 35). The precise significance of this juxtaposition of scenes has been much debated, but it seems likely that the theme was either the transition of pagans to become Christians, or the end of the pagan order to be replaced by that of Christ represented by the Crucifixion. The church of Gosforth also preserves a sculptured stone, the so-called Thor's Fishing Stone, which in representing the god Thor fishing for the World Serpent is a definite allusion to Scandinavian paganism and was also, like the cross, presumably an attempt at assimilation and acculturation (ill. 36). Other representations of pagan mythology on ecclesiastical sculpture are less detailed than those at Gosforth, but it seems likely that they are to do with the same themes of conversion and integration.[55]

What then did the emergence of the 'successor states' mean for the Viking kingdom of Northumbria? In political terms, the heartlands which we identified in chapter 2 were left intact, and in several respects the 'successor states' resembled the former kingdom of Northumbria, which, arguably, was changed but not destroyed by their emergence. As for the ethnic make-up of 'successor states', the importance of Viking settlement or at least influence is clear apart from in most of the lands between the Tyne and Forth; but its impact seems to have been much less than that of the English in earlier centuries, and the assimilation of Viking elements to the existing English ethnic make-up, as well as to the dominant Christian culture, was the most prominent process. In short, the history of the 'successor states' provides us with an object lesson in how the essential characteristics of a kingdom were not destroyed just because its political unity was fractured, since the Viking kingdom of York, the 'liberty' of the Community of St Cuthbert, the earldom of Bamburgh and the kingdom of Cumbria preserved important aspects of the former kingdom. By 1100, however, they too had been absorbed by the kingdoms of England and Scotland. To the questions of how and when that absorption took place, and the extent to which the aspects of the former kingdom of Northumbria which had helped to shape the 'successor states' went on to influence and shape English and Scottish rule over them, we must now turn.

[55] Bailey and Cramp (1988), figs. 2–3, pp. 10–11, 24–6 (distribution), pp. 120–1 (Kirkby Stephen), p. 130 (Lowther), and pp. 100–9 (Gosforth). For some caveats about the pagan character of some of the motifs, see Bailey (1985). For sculpture in Galloway, see Craig (1991).

CHAPTER 7

The English and Scottish impact: partition and absorption of the Northumbrian 'successor states'

The 'successor states' of the former kingdom of Northumbria (the Viking kingdom of York, the 'liberty' of the Community of St Cuthbert, the earldom of Bamburgh and the kingdom of Cumbria) were quite short-lived. Following the expulsion of King Eric from York and his death in 954, the rule of the Viking kings of York came to an end and the lands which their kingdom had comprised passed definitively into the hands of the emerging kings of England, who were then drawn from the dynasty of Wessex. The lands of the earls of Bamburgh were truncated with the handing over of the lands between the River Tweed and the Firth of Forth (known as Lothian) to the king of Scotland in the late tenth or early eleventh century. The lands of the kings of the Cumbrians also came to be amalgamated into the kingdom of Scotland, probably not much later than the death in battle of King Owain the Bald in 1018, although for a short time in the 1040s they were partitioned with Siward, then governing from the east as earl of Northumbria. That arrangement too was radically altered by the northward advance of King William II (Rufus) and the establishment of Norman power at Carlisle in 1092. Only the 'liberty' of the Community of St Cuthbert had a longer existence, although it too was increasingly integrated into the kingdom of England.

We examined in the previous chapter evidence to suggest that these 'successor states' preserved important aspects of the former kingdom of Northumbria, so that the lands of that kingdom were less radically reshaped in the Viking period than might appear at first glance. The purpose of the present chapter is to explore what impact the advance of West Saxon (then

English) power on the one hand, and Scottish power on the other, had on those relict aspects of the kingdom of Northumbria. How deep, in other words, were the changes to the 'successor states' brought about by their division between England and Scotland and their absorption by one or other of those two rising kingdoms? We shall consider in turn the impact on the political, ethnic and cultural make-up of those 'successor states' produced by the kings of Wessex and later the kings of England as they expanded the power of the English crown from the south; the impact of the Scottish kings and especially the establishment of the Scottish Border; and, finally, the impact – perhaps more definitive than the others – of the Norman kings of England.

The West Saxon kings and the kings of England

The process by which the power of the West Saxon kings and later the kings of England was extended over the 'successor states' south of what was to become the Scottish Border is itself instructive in assessing the degree of their impact on the lands in question. That process began with the military successes of Alfred, king of the West Saxons (871–99), which led to the restriction of Viking political and military dominance to a zone often known as the 'Danelaw', consisting of Northumbria south of the Tees, the north-east midlands and eastern England. Beginning under Alfred but gaining momentum in the reign of Edward the Elder (899–925), the West Saxon kings conducted an unremitting military campaign to absorb these Viking-held lands into their kingdom. This campaign could be presented as a 'liberation' of subject peoples (as when the *Anglo-Saxon Chronicle* reported that King Alfred had captured London from the Danes and 'all the English people who were not under subjection to the Danes submitted to him'); but in reality it was a campaign of military and political expansion by what was becoming one of the most highly organized and militarized states which Europe had seen since the Roman period.[1] The campaign was systematic and steady in the midlands. Just as West Saxon resistance to the Vikings in Wessex itself had been based on the construction of *burhs*, that is fortifications which were strongpoints and refuges as well as nascent urban centres, so West Saxon power was extended north-eastwards in the midlands by the steady construction of such *burhs*, which also formed the foci of an administrative system. This

[1] Stenton (1971), pp. 260–1, 320–63; for more modern accounts of this area, sometimes known as the Danelaw, see Hart (1992) and Hadley (2000).

system was based on shires, with principal *burhs* serving as their centres, as, for example, Warwick and Warwickshire or Stafford and Staffordshire. Elaborate arrangements were made for *burhs* to be maintained and garrisoned by levies from the surrounding shire, and, at least in Wessex, there existed a numerical ratio for the number of hides of lands responsible for so many feet of the defences of a *burh*. The West Saxon advance, which was greatly assisted by the close alliance between Wessex and Mercia (Edward the Elder's sister Æthelflæd effectively ruled the English part of the former kingdom until her death in 918), meant that West Saxon power was established with some solidity throughout the midlands, East Anglia and Lincolnshire, with a close network of *burhs* in the first of these and a sufficiency in the other two areas.[2]

The progress of West Saxon power northwards into what had been the kingdom of Northumbria, however, was much slower and much less systematic, and the kings seemed to lack either the ability or the determination to impose West Saxon political and military organization such as characterized their activity in the midlands. There was some building of *burhs*, but this was restricted to the very south-west of the Northumbrian 'successor states' and there was no parallel for the systematic construction of *burhs* which marked the West Saxon advance into the midlands (map 9). Æthelflæd 'restored' the Roman fortress of Chester in 907, which presumably meant that she converted it into a *burh*. In 914, a pre-Roman hill-fort was refurbished as a *burh* at Eddisbury (Cheshire), and in 915 a *burh* was constructed on a promontory at Runcorn (Cheshire). The reason for the construction of these *burhs* appears different from that for the construction of those in the midlands. Aside from Chester, there is no real indication that they had the sort of administrative and urban functions which more southerly *burhs* exercised. Moreover, all three look from their position to have had less to do with the advance of West Saxon power into the Northumbrian 'successor states' than with defending north-east Mercia against Viking incursions across the Irish Sea such as had occurred in the early tenth century when the Wirral was invaded by the Viking leader Ingimund, or playing a part in West Saxon military pressure on the Welsh. The two *burhs* constructed in 922 at Thelwall (Cheshire) and Manchester may also have been intended to secure areas to the south of them rather than to form a basis for a northerly

[2] On *burhs*, see David Hill (1981), pp. 85–6 and map 233; see also David Hill (1969), and David Hill and Rumble (1996). On Æthelflæd, see Wainwright (1975), pp. 305–24.

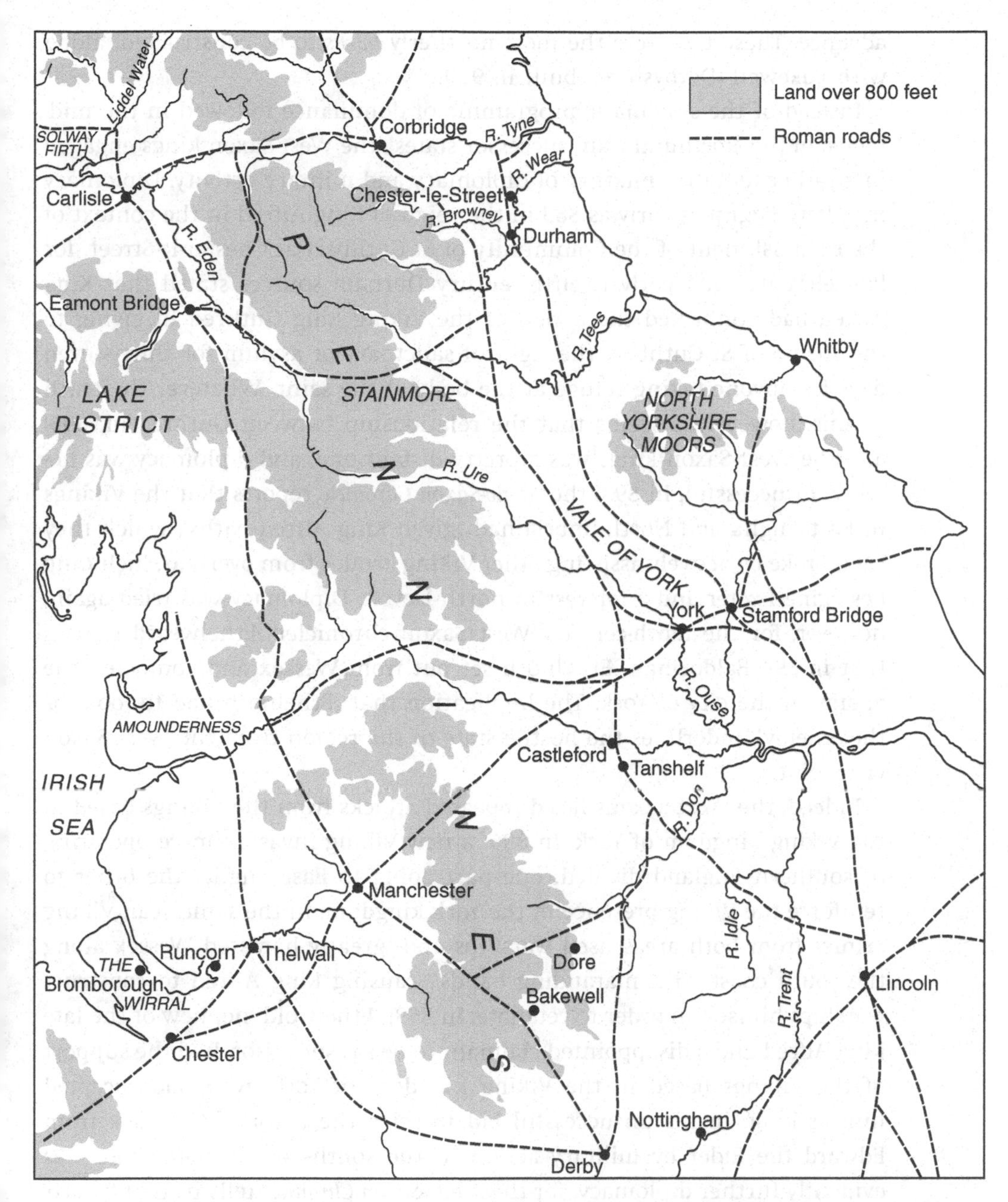

MAP 9 Map showing location of places relevant to the dealings of the kings of England with Northumbria

advance. These two were the most northerly *burhs* to be constructed, along with Bakewell (Derbyshire), built in 920.[3]

Instead of the systematic programme of dominance followed in the midlands, in the Northumbrian 'successor states', the West Saxon kings engaged in a rather tentative mixture of diplomacy and military activity. Diplomacy may have begun as early as 883 in the reign of King Alfred in the context of the establishment of the Community of St Cuthbert at Chester-le-Street, for late eleventh- and early twelfth-century Durham sources stated that King Alfred had confirmed the action of the Viking king Guthred in giving to the shrine of St Cuthbert the right of sanctuary or asylum for thirty-seven days for anyone taking refuge at the body of the saint. Whatever the truth of this, however, it seems that the relationship between Guthred (d. 895) and the West Saxon kings was a pretty distant one, and diplomacy was notably unsuccessful. In 893, the *Anglo-Saxon Chronicle* reports that the Vikings in East Anglia and Northumbria had 'given King Alfred oaths', which they then broke by actively assisting other Viking armies from overseas, including besieging Exeter and a fortress in north Devon. Diplomacy was tried again, however, for the tenth-century West Saxon chronicler Æthelweard reports that in 894 Ealdorman Æthelnoth set out from Wessex and contacted the enemy in the city of York. The implication that Æthelnoth had to look for the 'enemy' underlines the hostile state of the region from the West Saxon viewpoint.[4]

Indeed, the West Saxons faced repeated attacks from the Vikings based in the Viking kingdom of York. In 896, a new Viking invasion force operating in southern England divided, one part going to East Anglia, the other to reinforce the Viking presence in the York kingdom. In the same year, Viking armies from both areas used warships and 'greatly harassed Wessex along the south coast with marauding bands', causing King Alfred to construct warships himself in order to retaliate. In 899, Æthelwold, nephew of the late King Alfred and a disappointed claimant to be his successor, had the support of the Vikings based in the Viking kingdom of York, who had accepted him as king, in his unsuccessful bid to seize the throne of Wessex from Edward the Elder by military action in the south-east. In 906, there was evidently further diplomacy, for the *Anglo-Saxon Chronicle* tells us that 'peace'

[3] *ASC* C, *s.a.* 907, 914, 915; *ASC* A, *s.a.* 923 (*recte* 922), 924 (*recte* 920); on the sites and the difficulties of identifying them, see Griffiths (2001), pp. 174–8; on Ingimund's invasion, see Wainwright (1975), pp. 131–61; and, for emphasis rather on the importance of the Welsh for West Saxon strategy, see Griffiths (2001), pp. 179–81.

[4] For the grant of sanctuary, see above, p. 246; for the oath-breaking, *ASC*, *s.a.* 894 (*recte* 893); for Æthelnoth's journey, see *Sources*, p. 72 (= Æthelweard, p. 51).

was established at Tiddingford (near Leighton Buzzard), with the East Angles and the Northumbrians. There was, of course, a close relationship between diplomacy and military action – a successful campaign would be likely to lead to a 'peace' such as this and, although the *Anglo-Saxon Chronicle* has no information for the immediately preceding years, we can surmise that such a campaign had taken place. The fact that the resulting submission happened deep in West Saxon-held territory suggests that the West Saxon victory had been decisive for the time being, since the Vikings had had the humiliation of making it there rather than in some place on the border of the lands they dominated.[5]

Our information is too sketchy for us to form any real picture of what happened next, but relations with the Vikings in the Viking kingdom of York were evidently difficult. In 909 we find that a combined West Saxon and Mercian army spent five weeks ravaging 'the territory of the northern army'. King Edward was apparently resorting to the kind of punitive military action that William the Conqueror was to use in the 'harrying of the north' of 1069–70. In Edward's case, we cannot be sure whether or not his goal was military conquest, but there is no suggestion that it was – this was probably a raid, designed to intimidate the Vikings and curb their enthusiasm for attacking the lands to the south. As such, it was a failure, for in 910 'the army in Northumbria broke the peace, and scorned every privilege that King Edward and his councillors offered them, and ravaged over Mercia'. Admittedly, the Vikings had underestimated the speed of King Edward's response within Mercia; overtaken at Tettenhall (Staffordshire), they were defeated and routed, with the death of their kings Halfdan, Eowils and Ivar.[6]

Despite such a victory, the West Saxon rulers and their allies seem to have had to limit themselves to making general and probably insubstantial agreements with those who wielded power in the Northumbrian 'successor states'. Under the same year (918) in which the *Annals of Ulster* place the battle which may have been the second battle of Corbridge, the chronicle known as the Mercian Register reports that 'the people of York had promised [Æthelflæd, ruler of Mercia] . . . that they would be under her direction', information which follows on a report that the army which 'belonged' to Leicester had peacefully submitted to her. We can only guess that the military threat posed by her increasingly commanding position had induced her northerly

[5] *ASC*, *s.a.* 897 (*recte* 896), 906; *ASC* D, *s.a.* 901 (*recte* 899). On Æthelwold, see above, pp. 216–17.
[6] *ASC*, *s.a.* 910 (*recte* 909), 911 (*recte* 910); see also, *Sources*, p. 66.

and easterly neighbours to submit to her, although what this meant in practice it is not possible to say. The submission was in any case probably not of lasting significance since Æthelflæd died in the same year. Under the year 924 (probably more correctly 920) we read in the *Anglo-Saxon Chronicle* of a more comprehensive submission to the West Saxon king involving all the power groups in the former kingdom of Northumbria. According to this annal, King Edward went:

> into the Peak district to Bakewell and ordered a *burh* to be built in the neighbourhood and manned. And then the king of the Scots, and Ragnall, and the sons of Eadwulf [earl of Bamburgh] and all who live in Northumbria, both English and Danes, Norsemen and others, and also the king of the Strathclyde Britons and all the Strathclyde Britons, chose him as father and lord.

It is far from clear what degree of submission was really involved in this. First, we should note that Bakewell can only just have been inside the frontier of Mercia. It lay admittedly to the south of Dore where King Egbert had extorted by military intervention the submission of the Northumbrians in 829; but nevertheless it must have represented a sort of frontier location, so that submission there was much less humiliating to the northerners than submission in the heart of West Saxon-held territory as had been the case with the peace established at Tiddingford in 906. Secondly, it is hard to believe that the words 'chose him as father and lord' have any very precise significance in the context of the early tenth century. In a later period, they might well have conveyed the sort of feudal dominance which King Edward I claimed and attempted so forcefully to establish over the kings of Scots, but at this period there was no precedent for any such relationship. It seems likely that something much more in the way of a general but rather meaningless submission was involved, and that the *Anglo-Saxon Chronicle* deliberately inflated its importance to enhance King Edward the Elder's prestige. Thirdly, we should perhaps be most struck by Edward's failure to penetrate farther north than Bakewell.[7]

In 926, we find Edward the Elder's successor Athelstan (924–39) entering into another agreement, this time with the Viking king of York, Sihtric Caoch (920/1–7), to whom Athelstan married his sister at a meeting in Tamworth. The choice of meeting-place may have reflected Athelstan's more powerful position, for it was in the heart of Mercia, an ancient royal palace

[7] *ASC* B C D, *s.a.* 918 (= Mercian Register), *ASC* A, *s.a.* 924 (*recte* 920, Bakewell), and *ASC*, *s.a.* 827 (*recte* 829, Dore – and see above, pp. 26, 27). For a discussion of whether a submission was really made at the Bakewell meeting, see Davidson (2001).

site, far removed from the area of Sihtric's rule. Moreover, Athelstan was able and willing to intervene militarily in the Viking kingdom of York, for in the following year, when Sihtric died, Athelstan expelled the Viking claimant to the kingdom of York, Guthfrith, and razed to the ground a Danish fortification in York, putting himself in a position, to control the city and presumably what had been the Viking kingdom until his death in 939.[8]

How real that control was is hard to assess, for Athelstan's interventions in the Northumbrian 'successor states' were sporadic. In 927, at Eamont Bridge near Penrith (Cumberland), we find him making yet another agreement, this time with Howell, king of the Welsh, Constantine, king of Scotland, Owain, king of Gwent, and Earl Ealdred of Bamburgh, who had presumably retained his independence despite Athelstan's control of York. Like the Bakewell submission, the peace of Eamont Bridge no doubt reflected the growing ambitions of the English kings to rule all Britain – the *Anglo-Saxon Chronicle*, D version, is explicit that Athelstan was bringing under his rule 'all the kings who were in this island'. The claim is impossible to substantiate, and in any case Athelstan may really have been more concerned with the defence of southern and midland England from Viking and Scottish invasion than with extending his power over the Northumbrian 'successor states'. This is hinted at in the *Anglo-Saxon Chronicle*'s statement that the participants in the agreement at Eamont Bridge 'renounced all idolatry'. This may be a reference in coded language to a demand that Constantine, king of Scots, should renounce his alliance with Guthfrith, then Viking king of Dublin, who was at the time the only ruler worthy of the name to be an 'idolator'. King Constantine had married his daughter to Guthfrith's son, Olaf, who was to be king of the Vikings of Dublin from 934 and to seize the Viking kingdom in Northumbria on Athelstan's death in 939. So the danger posed to England by an alliance between the Scots, the Dublin Vikings and potentially the Viking king of York was real, and Athelstan's stance may have been more defensive than aimed at extending real control over the Northumbrian 'successor states'.[9]

Danger to the midlands from the Scots may have prompted Athelstan's expedition into Scotland in 934. This can be seen as a response to Olaf Guthfrithson's accession to power in Dublin, undertaken as a warning to the king of Scots not to collaborate with him against the English. According to the *History of the Kings*, Athelstan 'subdued his enemies, laid waste Scotland

[8] *ASC*, *s.a.* 925 (D, *recte* 926), 926 (E, *recte* 927); *Sources*, pp. 165–6 (V.1.3 = William Malms., *History*, pp. 214–15); see also *Sources*, pp. 67–8.

[9] *ASC*, D, *s.a.* 926 (*recte* 927); for comment, see Smyth (1984), pp. 201–2.

as far as Dunnottar and *Wertemorum* with a land force, and ravaged with a naval force as far as Caithness'. Athelstan was devoting considerable efforts to securing the submission of Scotland, for, although *Wertemorum* is an unidentified place, Caithness is the extreme north-easterly tip of Scotland and Dunnottar is just to the south of Aberdeen. That those efforts were commensurate with the magnitude of the danger of a Viking–Scots alliance is suggested by what is known of the Battle of *Brunanburh* (937). Poetic as is the *Anglo-Saxon Chronicle*'s description, it is clear that Athelstan, in command of Mercian and West Saxon forces, had to face a combined invasion by King Constantine of Scots and Olaf Guthfrithson. Olaf was put to flight leaving 'five young kings' and seven of his earls slain, and Constantine too fled, 'the hoary-haired warrior, . . . shorn of his kinsmen and deprived of his friends, bereaved in the battle – he left his young son on the field of slaughter'. *Brunanburh* was evidently a decisive victory for Athelstan, but assessing its significance for Athelstan's position vis-à-vis Northumbria is made extremely difficult by the fact that its site cannot be located with confidence. The name has not survived in any modern place-name to the extent that a really confident identification can be made, and the issue is further complicated by the fact that in the early twelfth century Symeon of Durham recorded an alternative name for it, *Weondune*. Scholars have located the battlefield at Brinsworth in Yorkshire (near Rotherham), in the Browney Valley to the east of Durham, at Bromborough on the Wirral, and at Bromswold near Huntingdon. If one or other of the first two is correct, this would mean that Athelstan was facing an invasion which had reached to the very heart of the former kingdom of Northumbria. If Bromborough was the place, he was clearly fending off an invasion across the Irish Sea, which is entirely plausible (map 9). It is hard to see what the historical context for a battle near Bromswold might be.[10] At all events, Athelstan's policy appears more defensive than aggressive and, although he was in control of York, he appears to have been unable or unwilling to extend West Saxon power over the Northumbrian 'successor states' as his predecessors had done over the midlands.

The tenuousness of West Saxon control even of York was revealed after Athelstan's death, when Olaf Guthfrithson seized York, and for a time the Five Boroughs also. Only in 942 did Athelstan's successor, Edmund (939–46), regain control of the latter area.[11] The West Saxon response to the reverse

[10] *ASC*, *s.a.* 937; Symeon, *Origins*, bk 2 ch. 18 (see p. 138 n. 105, for a survey of scholarship); see also A. Campbell (1938), pp. 43–80, and specifically for Bromborough, Dodgson (1957).

[11] *Hist. Kings*, *s.a.* 939 (pp. 93–4), and *ASC*, *s.a.* 942; see also Mawer (1923).

was the same mix of diplomacy and military force which we have witnessed previously. After Edmund's reconquest of the Five Boroughs, Olaf Guthfrithson's successor as king of York, Olaf Sihtricson, made peace with him. In the confused political situation of these years, it appears that so too did another Viking claimant to the kingship of York, Ragnall Guthfrithson. Shortly afterwards, however, Edmund expelled both kings and re-established West Saxon control of 'the kingdom of the Northumbrians', by which the Viking kingdom of York was presumably meant. That Edmund's power was also uncertain is suggested by the fact that he could establish control only by means of ravaging to intimidate the inhabitants. Moreover, after his death in 946 West Saxon control of Northumbria lapsed again, for we find his successor Eadred (946–55) having to 'reduce all Northumbria under his rule' – here too it seems likely that the kingdom of York is meant. It is not certain at what point in Eadred's reign this is supposed to have occurred, since in 947 we find him receiving pledges and oaths from Archbishop Wulfstan of York and 'all the councillors of the Northumbrians' at Tanshelf (Yorkshire West Riding) in the Humber area. Not only does this suggest that Wulfstan and the Northumbrians were being treated as having considerable independence, but Tanshelf was surely a frontier location, a suitable place to ratify an agreement between equals rather than the submission of subjects to the ruler of the territory in which they lived.[12] When Wulfstan and the Northumbrians failed to keep their promises and in 948 invited the Viking Eric to become their king, it became clear that Eadred's power over the kingdom of York was no firmer than that of his predecessors. He too ravaged the area, even burning the church of Ripon.[13] After this last event, the army of the kingdom of York intercepted Eadred's at Castleford (Yorkshire West Riding), but the Northumbrians backed down, deserted Eric and compensated Eadred when the latter threatened renewed ravaging.

In the following year (949), the people of the kingdom of York nevertheless accepted as their ruler their former king Olaf Sihtricson, although they expelled him in 952 and once again received Eric. The latter's expulsion and death in 954 marked, as we have seen, the end of the Viking kings, and indeed of independent royal power in Northumbria. But this was not the work of the king of England. Eric was, according to Roger of Wendover's *Flowers of History*, killed 'treacherously' by a certain Earl Maccus on Stainmore, high up on one of the passes across the Pennines from York westwards to the Vale of Eden and the Irish Sea (map 9). Moreover, he was 'betrayed' by

[12] For the texts, see *Sources*, pp. 68–9; for discussion, see Smyth (1975–9), II, pp. 89–126.
[13] *ASC*, D, *s.a.* 948.

Earl Oswulf. The identity of Maccus is not known, although the name is Old Norse; but Earl Oswulf was none other than the then ruling member of the house of Bamburgh.[14] The removal of the last king of York was thus the result as much of political enmities between the kingdom of York and the independent earls of Bamburgh as of the pressure of the West Saxon kings.

Equally tenuous and dependent on recourse to military force was the policy of the West Saxon kings towards Cumbria. According to the annal for 945 in the *Anglo-Saxon Chronicle*: 'King Edmund ravaged all Cumberland (*Cumbra land*) and granted it all to Malcolm, king of Scotland, on condition that he should be his ally both by sea and on land.' Roger of Wendover's *Flowers of History* adds that Edmund blinded the two sons of Dunmail (Donald), king of the Cumbrians, whom he had presumably expelled. The interpretation of the term Cumberland in these annals is a problem. An entry for 946 in the *Welsh Annals* reads: 'Strathclyde (*Strat Clut*) was devastated by the Saxons.' If this refers to the same ravaging as that in the *Anglo-Saxon Chronicle*, admittedly assigning it to a year later, it would seem that 'Cumberland' in the latter source is simply another term for Strathclyde. It can be suggested, however, that 'Cumberland' referred to a much more restricted area than Strathclyde, perhaps limited to the heartland of the kings of Cumbrians in the north of the later county of Cumberland. This does seem more plausible, if only on the grounds of the unlikelihood of Edmund being able to ravage and control Strathclyde, extending as it did northwards to Glasgow and the valley of the Clyde. In either case, it seems likely that Edmund's priority was to prevent Viking incursions into the north-west midlands, even if this was at the cost of the king of Scots enlarging the territory under his control.[15]

After the expulsion of King Eric from York in 954 and the ending of his kingdom, the king of England, as he then was, nominally appointed earls to control the lands of the Northumbrian 'successor states' which were now in his hands. Some of these earls only governed the kingdom of York south of the Tees, while the earls of Bamburgh continued to enjoy effective independence north of it. Some had pretensions to govern both north and south of the River Tees, although the situation was often very confused. To this latter group of earls belonged the first of the earls, Oswulf I (934–66), who had been responsible for Eric's death. He appears earlier as 'high reeve' of Bamburgh, of which house he was a member, and Bamburgh almost certainly continued

[14] *Sources*, p. 69, and Smyth (1975–9), II, pp. 155–90; Wendover, *Flowers*, *s.a.* 950 (*recte* 954).

[15] *ASC*, *s.a.* 945; *Welsh Annals*, *s.a.* 946; Wendover, *Flowers*, *s.a.* 946; see Phythian-Adams (1996), pp. 110–11, and cf. Smyth (1975–9), II, pp. 205–6.

to be his base. In 966, however, control of the former kingdom of York was removed in quite obscure circumstances from the earls of Bamburgh and passed to a certain Oslac (966–75), who was succeeded by Thored (975 × 979–92/3) and Ælfhelm (992/3–1006). The first of these seems to have been ineffective or corrupt, since Archbishop Oswald noted in a memorandum that the lands of York Minster had been despoiled in his time. The earldom of Bamburgh was meanwhile governed by Eadwulf 'Evil-child', and then from 975 by Waltheof, who was certainly a member of the house of Bamburgh. After Ælfhelm, we again find the Northumbrian 'successor states' north and south of the Tees in the hands of the house of Bamburgh, represented now by Uhtred, son of Waltheof. Uhtred sided with Edmund Ironside (d. 1016), son of the king of England, Æthelred the Unready (978–1016), in his struggle against the invasion of the Danish claimant to the throne Cnut the Great (1016–35); but he was compelled to submit to the latter and was murdered by his enemy Thurbrand, probably for political reasons, in Cnut's hall. With a view presumably to tightening his grip over the Northumbrian 'successor states', Cnut then gave the earldom of Northumbria, possibly excluding the earldom of Bamburgh, to one of his generals, Eric of Hlathir (1016–23 × 1033), who was in turn succeeded by another Scandinavian, Siward (1023 × 1033–55). After the latter's death, King Edward the Confessor (1042–66) appointed as earl Tostig (1055–65), brother of Harold Godwinsson, the future King Harold. Once a 'Northumbrian' revolt had expelled him, King Edward appointed the Mercian nobleman Morcar (1065–7), who was later deposed in a 'Northumbrian' revolt against William the Conqueror.[16]

The extent of the control of the kings of England over these earls was often in doubt. Our evidence suggests that the house of Bamburgh held its earldom on an hereditary basis, and there is no evidence that any of its members was appointed as earl of Bamburgh by the king, apart from Oswulf I who may possibly have been at the point when King Eadred assumed control of York after the expulsion and death of King Eric (fig. 18). There is, however, no evidence that Oswulf's predecessors, Eadwulf and his sons, had received any sort of royal commission. Even in the period around 1000, the earl of Bamburgh's relationship to the king was unclear. According to the early twelfth-century text called *The Siege of Durham and the Probity of Earl Uhtred*, at the time of a Scottish siege of Durham, probably in 1006, the then earl of Bamburgh, Waltheof I, 'being too old to lead an army, shut himself up in Bamburgh', and his son Uhtred successfully organized the

[16] *Sources*, pp. 74–6; for Uhtred, see below, n. 17. The memorandum is Robertson (1956), no. 54.

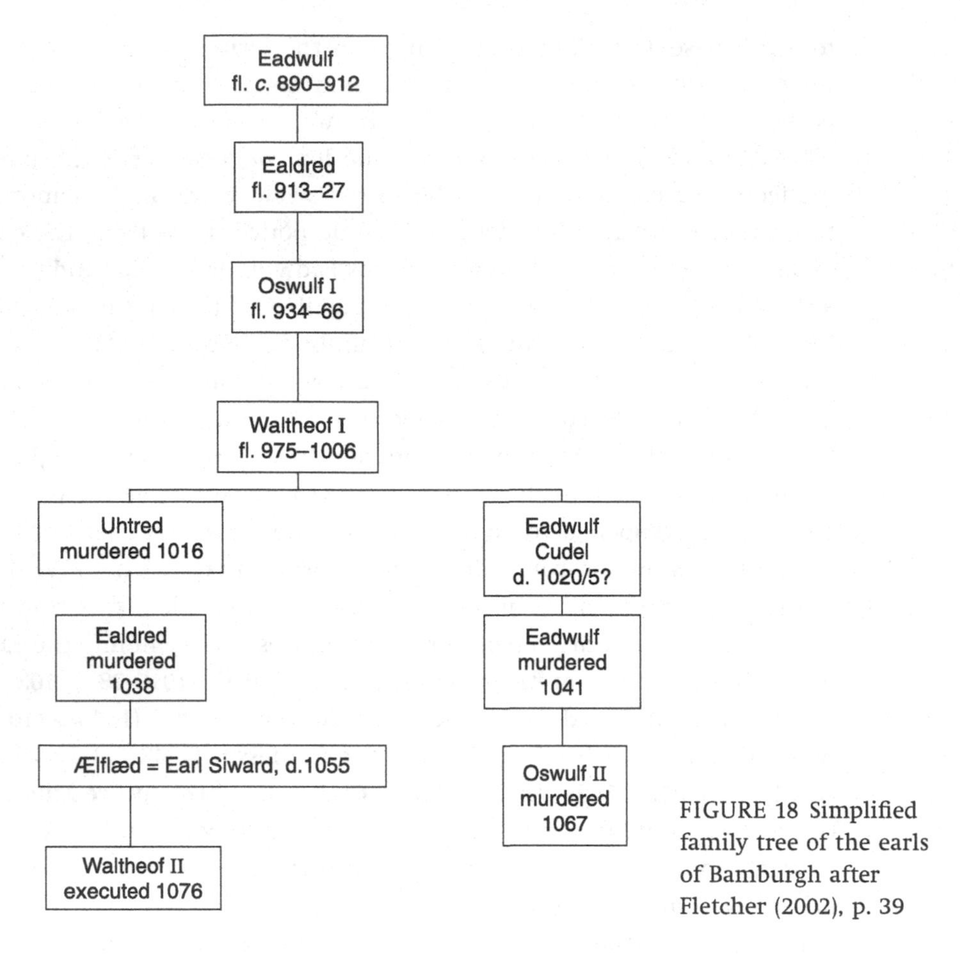

FIGURE 18 Simplified family tree of the earls of Bamburgh after Fletcher (2002), p. 39

defence of Durham, and (the text goes on) 'upon hearing of his victory, King Æthelred rewarded Uhtred's valour by giving him his father's earldom, adding to it the earldom of York'. If this tradition is reliable (and the source is admittedly a late and somewhat romanticized one), this was a royal appointment at least to Northumbria south of the Tees, but it is very unlikely that Uhtred owed his position north of the Tees to royal fiat. In 995, he had taken a prominent part in assisting the Community of St Cuthbert to make their move to Durham, and it is very likely that he developed the site as a fortress in his own right. In particular (according to a tradition recorded by Symeon of Durham which there is no reason to doubt), the peninsula in the River Wear at Durham was cleared of vegetation and made habitable for them by men summoned by Uhtred from the area between the Rivers Coquet and Tees. If this is reliable, it suggests that Uhtred was to all intents and purposes earl north of the Tees, a decade or so before the siege of Durham, assuming 1006 to be the correct date for that event.[17]

Uhtred's independence seems to have been typical. We see the earls' names appearing from time to time in the witness-lists of royal charters, showing that they had been at the court of the king of England wherever it was meeting when the charters in question were drawn up; but we do not see any real indication that they were beholden to the king for their position. No doubt the kings could command, as when King Edgar (957–75) made Earl Oslac responsible for the enforcement in Northumbria of his fourth law-code; but their inability really to control is underlined by the occasions when they resorted to killing the earls or having them killed. Earl Ælfhelm was assassinated and his two sons blinded on the king's orders in 1006, to be replaced by Uhtred, who was himself cut down together with his war-band in King Cnut's hall - and therefore presumably with the king's connivance even if the killing was also part of a personal vendetta as the *Siege of Durham* represents it.[18]

As we have suggested, the dislocations and uncertainties in these relations between the West Saxon kings and the kings of England and the Northumbrian 'successor states' point to their lack of ability or determination really to control the lands of the former kingdom of Northumbria. That they were in a position to have had a consistent impact on the political, ethnic and cultural make-up of those lands seems in view of this unlikely. As regards their political impact, they do appear to have subjected Yorkshire and a small part of Cumbria to geld (taxation), for these areas are included in *Domesday Book*, but they evidently did not achieve this for most of the rest of the former lands of Northumbria, which are omitted from that survey. The system of local administration introduced in the south and the midlands by the West Saxon kings was partially introduced to Yorkshire. At some point the vast shire of Yorkshire had been created, assuming the name of York, but it was the only shire in what had been Northumbria which the Domesday commissioners recognized as such, for the other parts of the area which they surveyed (Craven and Lancashire south of the Sands) they amalgamated with Yorkshire, apparently only for the purposes of the survey. Yorkshire was divided into the three Ridings (that is *trithings* or 'third parts'), and these were in turn subdivided mostly into wapentakes, corresponding to hundreds

[17] *Siege of Durham*; Symeon, *Origins*, bk 3 ch. 2. On the earls of Bamburgh, see above, p. 249 n. 51.
[18] See the tabulated charter witness-lists in Keynes (1998); on Ælfhelm, see *Sources*, p. 75, citing John Worcs., *s.a.* 1006; on Uhtred, see above, n. 17.

in the south (there were in fact some units designated as hundreds in eastern Yorkshire). Like the hundreds, the wapentakes were administrative districts with meeting-places and courts. The terms 'riding' and 'wapentake' are Scandinavian, suggesting that these divisions were made in the period between the arrival of the Vikings and the Domesday survey, but no greater precision is possible. By some point between 1067 and 1069 Yorkshire had a sheriff, Gamel son of Osbert, who received a writ from William I. Moreover, it is just possible that the two reeves of Earl Harold referred to in *Domesday Book* were government officials, although they could equally well have had responsibility for his private estates. In *Domesday Book*, York had four 'lawmen' (*iudices*) who may have constituted a borough court as distinct from a county court, although this is not certain. York certainly had an active mint, the only one producing coins north of the Humber.[19]

Elsewhere in the lands of the Northumbrian 'successor states', there is no indication that the West Saxon kings and the kings of England had an impact on political organization. At the time of *Domesday Book* these lands still had a very long way to go before they would begin to approach the systematic structure of royal adminstration which from the tenth century characterized southern England and involved manageably sized shires divided more or less evenly into hundreds, and centred on the *burhs* which were largely lacking from these lands. Not only did the vast shire of Yorkshire not correspond to this, but there are indications that its organization sat uneasily with what remained of that earlier administrative system of small shires, also known to historians as sokes or multiple estates. Consisting of reasonably consistent numbers of vills focused on a head vill, these shires may have been ancient units of royal administration, although it is true that by the Norman period they may have had little to do with contemporary royal administration.[20]

The kings had only limited land in the Northumbrian 'successor states', to judge at least from the evidence of *Domesday Book*, and they seem to have seen themselves as having little reason to visit that region. At all events, Northumbria was throughout the last century before the Norman Conquest

[19] For subdivisions, see David Hill (1981), nos. 83–98, 174–7, and, for Yorkshire, Palliser (1992), pp. 1–4, Hadley (2000), pp. 104–7; for Gamel son of Osbert, see Bates (1998), no. 32; for the reeves, *Sources*, p. 215 (N.5.2, para. 11); on the lawmen, see *Sources*, pp. 186–7 (N.2.3), and cf. Reynolds (1987), pp. 307–8; for the mint, see *Sources*, pp. 192–3 (N.3.10), on which see also Pirie (1975), pp. xliv–lv; Palliser (1990), p. 14, and Metcalf (1987b), pp. 290–1. For the distribution of mints, see David Hill (1981), map 217.

[20] The most useful treatment is Kapelle (1979), pp. 50–85. See also Geoffrey W. S. Barrow (1973b), pp. 10–11, and, specifically on Yorkshire, K. Mary Hall (1993).

an area which was not visited by the kings of England. King Edgar (957–75) ventured as far north as Chester in 973 where he received the submission of various Scottish and Welsh rulers; Cnut may have been at Wighill in Yorkshire in or around 1016, if that was where the hall in which Uhtred was killed was situated as is possible, and he was certainly at Durham when the gift of Staindrop was made to the church of Durham; but otherwise no king of England is known to have been present in Northumbria in the period between Eadred's ravaging campaign in 948 and Harold's presence in York following his defeat of Harold Hardrada who had invaded Northumbria from Norway in 1066.[21]

There is no indication that the West Saxon kings and kings of England had any impact whatsoever on the ethnic make-up of the Northumbrian 'successor states' even at the level of the political elite. We hear of no settlers, aside from some of the earls of Northumbria themselves. In terms of cultural impact, even their dealings with the church did not involve them in pursuing change. The organization of the church showed little or no sign of the sort of royal influence or intervention which was characteristic of church–state relations further south. From 883 all the land north of the Tees was subject to the bishopric of Chester-le-Street and then, after the move of the Community of St Cuthbert from there to Durham in 995, to that of Durham. We find none of the subdivision of sees into smaller and more manageable units which characterized West Saxon episcopal organization in Wessex (the creation of the sees of Wells and Ramsbury in 909, for example); or indeed any of the movement of sees or the keeping of them vacant which in the east and the midlands may also have been part of royal policy. We find neither any indication of influence from the south of rules for reformed cathedral canons, such as the Rule of Chrodegang; nor much sign of the influence of the tenth-century movement for reforming monasteries according to the Rule of St Benedict, which was so important in royal promotion of the church in the south and the midlands. The only sign of such influence from the south is the presence in the south of the bishop of Chester-le-Street in or around 970, the time when the Council of Winchester instituted the tenth-century monastic 'reformation'. It is just possible that he was involved in some way in that. It appears from inscriptions in the manuscript to have been during that journey that he acquired the Durham Collectar, which contains collects and other texts (notably hymns) representing liturgical practices influenced (as was the south) by the reforming

[21] David Hill (1981), map 179, shows the lands of King Edward in 1066; maps 158–63, 167–9 show royal itineraries.

movements of the continent. But there is no reason to suppose that the king or his circle were involved in this; the acquisition seems to have been entirely fortuitous, and it is not clear that the collectar was actually used in its new home.[22]

Cnut does seem to have made a start at exerting control more directly over the bishopric of Durham through his novel role in the appointment of Edmund, bishop of Durham (1020–42), who had to be made bishop by the king even though he had previously been chosen by the Community of St Cuthbert at Durham. After Bishop Edmund's death, a certain Eadred purchased the office of bishop of Durham from King Harthacnut, although he enjoyed it for only a short time (1042). His successors were Æthelric (1042–56) and Æthelwine (1056–72), monks of the southern abbey of Peterborough, with which Cnut had been particularly closely associated. They had accompanied Edmund on his journey from Cnut's court to take up his see at Durham - another indication, perhaps, of Cnut's determination to control that church. Aside from this, however, there is little sign of royal intervention in the running of the bishopric. The appointment of Æthelric, possibly at King Edward the Confessor's instigation, was unpopular with the Community of St Cuthbert, which expelled him so that he had to be forcibly re-instated by Earl Siward. According to Symeon of Durham, the expulsion was 'because he was an outsider and had been elected against their wishes', and the Community of St Cuthbert accepted him back only because it was 'terrified and overwhelmed by the fearful power of the earl'. The account does not give an impression of either the king or the earl being able to control the church of Durham in any systematic way. They clearly wished to, but were only able to achieve control through the sort of show of force which had reinstated Æthelric.[23]

The most the kings undertook was benefactions to churches which may have been aimed at extending their influence in the region. Athelstan, for example, is supposed to have confirmed to the Community of St Cuthbert the right of sanctuary or asylum for thirty-seven days for those fleeing to the shrine, and granted the estate of South Wearmouth (County Durham), together with books and other treasures; his successor Edmund to have granted it treasures, and King Cnut to have made a barefoot pilgrimage to the church when it was at Durham and (according to *History of St Cuthbert*)

[22] On ecclesiastical reform, see Julia Barrow (1994) and David Hill (1981), map 243; on the bishop of Chester-le-Street, see Bonner (1989), pp. 392–5, and, on the collectar, T. J. Brown (1969) and Correa (1992).

[23] Symeon, *Origins*, bk 3 chs. 6–7, 9; for discussion, see Meehan (1975), and Aird (1998b), pp. 52–3.

granted it the vill of Staindrop (County Durham) with all its appurtenances. It is possible that these kings were deliberately building a relationship with the Community of St Cuthbert to use it as a means of promoting their interests in Northumbria; but the argument is not a wholly convincing one, since is not easy to see what influence they were in fact obtaining. The grant of sanctuary, although not new, could at this time have been seen as a grant of quasi-royal jurisdiction over criminals taking sanctuary in the church which extended the arm of royal justice through this ecclesiastical intermediary; but, if so, no such effect is explicit in our sources. Moreover, there is a long interval between the patronage of Edmund and that of Cnut; and there are good grounds for thinking that the West Saxon kings at least were primarily motivated in their benefactions by piety rather than political calculation – the cult of St Cuthbert was, it seems, well developed in southern England and particularly at the court of King Athelstan, where an office for the saint was being developed, and his *Life* by Bede was being lavishly copied up, as, for example, in the illustrated book which Athelstan is supposed to have given to the saint's shrine (now Cambridge, Corpus Christi College, MS 183), but which may not have been made for that purpose.[24]

York Minster, however, benefited little, as far as we know, from the patronage of the kings of England. Athelstan's gift of the territory known as Amounderness in 934 was of dubious value because the area, in what is now Lancashire on the north-west coast of England, may well have been a buffer area subject to Viking attacks, rather than a worthwhile gift. Late traditions maintain that Athelstan patronized York Minster and Beverley with gifts of sheaves of corn from the harvests in the surrounding shire, established secular canons at Beverley and patronized Beverley and Ripon with grants of sanctuary rights; but the sources are so late that it is hard to believe that they prove more than the fact that Athelstan was regarded as a suitable royal donor by the churchmen who propagated the traditions in their documents. Other recorded grants are those by King Eadwig (955–9) of lands in Nottinghamshire to Archbishop Oscytel of York (950–6), lands which formed the estates of what became Southwell Minster, and King Cnut's grant of Patrington in Holderness to Archbishop Ælfric of York in 1033. In both cases, however, the grants are recorded by charters, the earliest copies of which date from the fourteenth-century register of York Minster, and which might have been fabricated as titles to land by the canons. Otherwise, the only trace of patronage of any Northumbrian church south of the Tees is

[24] *Hist. Cuth.*, paras. 26, 28, 32, on which see Craster (1954). On sanctuary, see David Hall (1989) and, on this aspect of the cult of Cuthbert, Rollason (1989b).

the royal practice (if such it was) of having the archbishop of York be bishop also of a southern see, either Dorchester-on-Thames or Worcester, and thus enjoy the revenues of that see as well as of the archbishopric. This practice may, however, have had more to do with securing the archbishop's loyalty by giving him an interest in the south than with wishing to enrich the church of York through enlarging the resources available to its archbishop.[25]

The West Saxon kings and the kings of England thus had a rather negative impact on the lands of the 'successor states' of Northumbria. They treated them as peripheral to their interests and, although they may have made some progress in the organization of Yorkshire, and they had some role as patrons of certain churches, their impact on the political, ethnic or cultural make-up of the those lands was very slight.

The kings of Scots and the origins of the Scottish Border

The kings of Scots arguably had, in the tenth and eleventh centuries, a greater impact on the Northumbrian 'successor states' than did the West Saxon kings and the kings of England. Already in the early tenth century, probably 914, we read in the *History of St Cuthbert* that Earl Ealdred of Bamburgh, driven out of his earldom by the Viking leader Ragnall, 'went to Scotland, sought the help of King Constantine of Scots and led him into battle against King Ragnall at Corbridge'. The same source describes another battle at the same place a little later, in which Ragnall fought against Eadred, a vassal of the Community of St Cuthbert, whom he killed along with 'a great multitude of English'. After the battle, Ragnall gave Eadred's land to his allies in the battle, Egbrid, son of Eadred, and Count Ælfstan, his brother, who were to judge by their names English. It is possible that the Scots were also involved in this second battle, assuming it to be identical with a battle recorded in some detail in the *Annals of Ulster* under the year 918: 'the men of Scotland moved against them [Ragnall and his allies] and they met on the banks of the Tyne in northern Saxonland'. Setting aside the complication of whether there were two battles or just one which our sources have somehow split up, the important point is the intervention of the king of Scots at such a southerly location as Corbridge, on the

[25] For Athelstan's gifts to churches, see *Hist. Cuth.*, paras. 26–7 (Durham – for discussion; Rollason (1989b), pp. 149–52); *EHD I*, no. 104 (Amounderness = Sawyer (1968), no. 407, and above, p. 230 n. 31); and Hamilton Thompson (1935), p. 97 (Ripon and Beverley). For a summary of evidence relating to Beverley, see Hadley (2000), pp. 239–41, and Cambridge and Morris (1989). On York held in plurality, see Whitelock (1959), pp. 73–6. See also, above, pp. 27, 229 (Southwell), and Sawyer (1968), no. 968 (Patrington).

north bank of the River Tyne. It has even been argued that Earl Ealdred was at this time under some sort of Scottish tutelage; at all events, the *History of St Cuthbert's* account serves to emphasize the remoteness of the West Saxon king who intervened in neither battle (assuming there to have been two).[26]

We have already seen the success of the king of Scots in extending his frontier south to Stainmore in 945, thus taking in most of Cumbria. His success was not an enduring one. In 1092, King William II (Rufus), second of the Norman kings of England, took control of Carlisle, expelled its ruler Dolfin (apparently a scion of the family of the earls of Bamburgh) and constructed a castle there. This moved the Border northwards from Stainmore to the head of the Solway Firth just to the north of Carlisle, where it is today. Nevertheless, Cumbria as it had existed at its fullest extent had been partitioned, with the kings of Scots taking the lands north of the Solway Firth which had pertained to it.

The kings of Scots achieved an equally beneficial partition much earlier on the east side of the country. There the Scottish Border came to be created on the line of the River Tweed, resulting in the area of Lothian (i.e. between the River Tweed and the Firth of Forth) passing under the control of the king of Scots. This Border has endured in broadly the same form throughout the middle ages and beyond (map 10).[27] The process by which it was created is obscure, since our sources give only cursory and inconsistent accounts of it. They suggest three possible scenarios. First, the late eleventh- or early twelfth-century text known as the *First Coming of the Saxons* contains an account of the earls of Northumbria, including the following passage:

> Then under King Edgar [957–75] Oslac ruled as earl of York [966–75] and the places pertaining to it; and Eadwulf called Evil Child ruled over the Northumbrians from the Tees to *Myreford*. These two earls with Ælfsige, who was bishop of St Cuthbert's [?968–?990], led King Kenneth of Scots to King Edgar. When King Kenneth had done homage to him, King Edgar gave him Lothian and sent him back with much honour to his own land.

So, according to this, Lothian passed into the hands of the Scots in the late tenth century, perhaps because Edgar was concerned to consolidate his north-east frontier and to establish in principle his overlordship over Kenneth in return for giving up Lothian to him. Although *Myreford* is

[26] *Hist. Cuth.*, paras. 22, 24; *Annals Ulster*, *s.a.* 918; for comment, Wainwright (1975), pp. 163–79, and Smyth (1984), pp. 196–7, 235.

[27] The best discussion is Geoffrey W. S. Barrow (1973a).

unidentified, it may mean 'the great forth' and so refer to the Firth of Forth, showing that Eadwulf was ruling as far north as Northumbria had extended in Bede's time, and was consequently himself giving up territory to the Scottish king.[28]

The second scenario is presented by the account in the *Siege of Durham* of how, following the killing of Earl Uhtred in the early years of the eleventh century:

> His brother Eadwulf, known as Cudel, a very lazy and cowardly man, succeeded to the earldom. Fearful lest the Scots, whom his brother had slaughtered as aforesaid [at the siege of Durham in 1006] would avenge these deaths upon him, he ceded them by firm treaty (*firmam concordiam*) the whole of Lothian to make amends. In this way Lothian was joined to the kingdom of the Scots.[29]

This passage suggests that the earl of Bamburgh was acting independently of the king of England by disposing of territory at will. A third possible scenario, which may in fact be connected with this second one, is implicit rather than explicit in the sources. According to the *History of the Kings* with reference to the year 1018: 'A great battle between the Scots and the English was fought at Carham between Uhtred, son of Waltheof, earl of Northumbria, and Malcolm, son of Kenneth, king of Scotland.' According to Symeon of Durham's *On the Origins and Progress of this the Church of Durham*, this was an overwhelming Scottish victory and Bishop Ealdhun of Durham died of a broken heart as a result of it. Carham is on the Tweed, and it seems possible that one result of the battle was that Lothian, the lands to the north of that river, passed into Malcolm's control, so that Eadwulf Cudel was only endorsing a *fait accompli*.[30] In fact, all the texts may be over-simplifying as well as confusing the situation, for Lothian may have been in Scottish hands from an earlier date. It is even possible that by the time of the Scottish intervention at the Battle of Corbridge in 914 it was already in the hands of the king of Scotland, which would account for his willingness to assist Ealdred of Bamburgh in that battle as far south as the River Tyne.[31]

[28] For the text of the *First Coming of the Saxons*, Arnold (1882–5), II, p. 382; on *Myreford*, see Geoffrey W. S. Barrow (1973a), p. 153 (identifying it with a ford on the River Tweed), and Duncan (1975), p. 96 (identifying it with the Solway Firth). On William Rufus's seizure of Carlisle, see *ASC* E, *s.a.* 1092; for discussion, see Phythian-Adams (1996), pp. 24–6; on Carlisle, see Summerson (1993).

[29] *Siege of Durham*.

[30] *Hist. Kings*, *s.a.* 1018 (pp. 155–6), and Symeon, *Origins*, bk 3 ch. 5. For discussion, see Meehan (1976), pp. 12–17; cf. Duncan (1976), see also Smyth (1984), pp. 233–7.

[31] Geoffrey W. S. Barrow (1973a), pp. 153–4.

The creation of the Scottish Border had the potential to have had a major impact on the Northumbrian 'successor states', for it was notable for its artificiality, not being a natural frontier or a cultural one or an historic one (map 10). It followed the course of the River Tweed westwards and upstream from Berwick-upon-Tweed; the effect of this was to cut in half the fertile basin of the Tweed. It then left the Tweed downstream from Kelso and took a south-easterly course, turning back on itself to ascend the wild hills of the Cheviots, the ridge of which it then followed south-westwards across the fastness of Kielder Forest to the Kershope Burn. It followed this to Liddesdale and so, via the confluence of the Liddel Water and the River Esk, to the head of Solway Firth, the basin of which it also bisected. (It may be that originally it ran along the River Lyne rather than Liddesdale, but its credentials as a natural frontier would have been in no way improved.)[32]

Moreover, viewed from the perspective of the former kingdom of Northumbria, the Border was not an historic one in the sense that it corresponded to anything within the kingdom of Northumbria or its 'successor states'. The Tweed–Solway line which it followed bore no relationship to any previous frontier known to us. The frontier of the former kingdom of Bernicia was almost certainly on the River Tees; a frontier on the River Tyne does seem to appear in the Viking period with the consolidation of the 'liberty' of the Community of St Cuthbert between Tyne and Tees; nowhere do we see a frontier on the Tweed. The Northumbrian royal heartland in the hinterland of Lindisfarne and reaching north to the Tweed basin and beyond was fractured by that frontier. Nor did the Border respect the ancient estates or ecclesiastical dependencies of the church of Lindisfarne reaching north of the Tweed to Coldingham and Tyninghame as we see them set out in the *History of St Cuthbert*. They too were fractured. The same is true of the west. At no period had the frontier of the kingdom of Northumbria or of its Cumbrian 'successor state' been at the head of the Solway Firth; indeed, when, in the eighth century, King Eadberht conquered the Plain of Kyle in the area of modern Ayrshire the Border presumably lay far to the north, and at an earlier period it embraced the northern coastlands of the Solway Firth westwards to Whithorn. If Carlisle sat at the centre of a north-western heartland, the Scottish Border shattered it.[33] In this sense, the creation of the Scottish Border can be seen as the true end of the kingdom of Northumbria, when its territory was definitively divided, and two of its heartlands were partitioned.

[32] On the Lyne, Phythian-Adams (1996), pp. 133–6. [33] See above, pp. 48–50, 52.

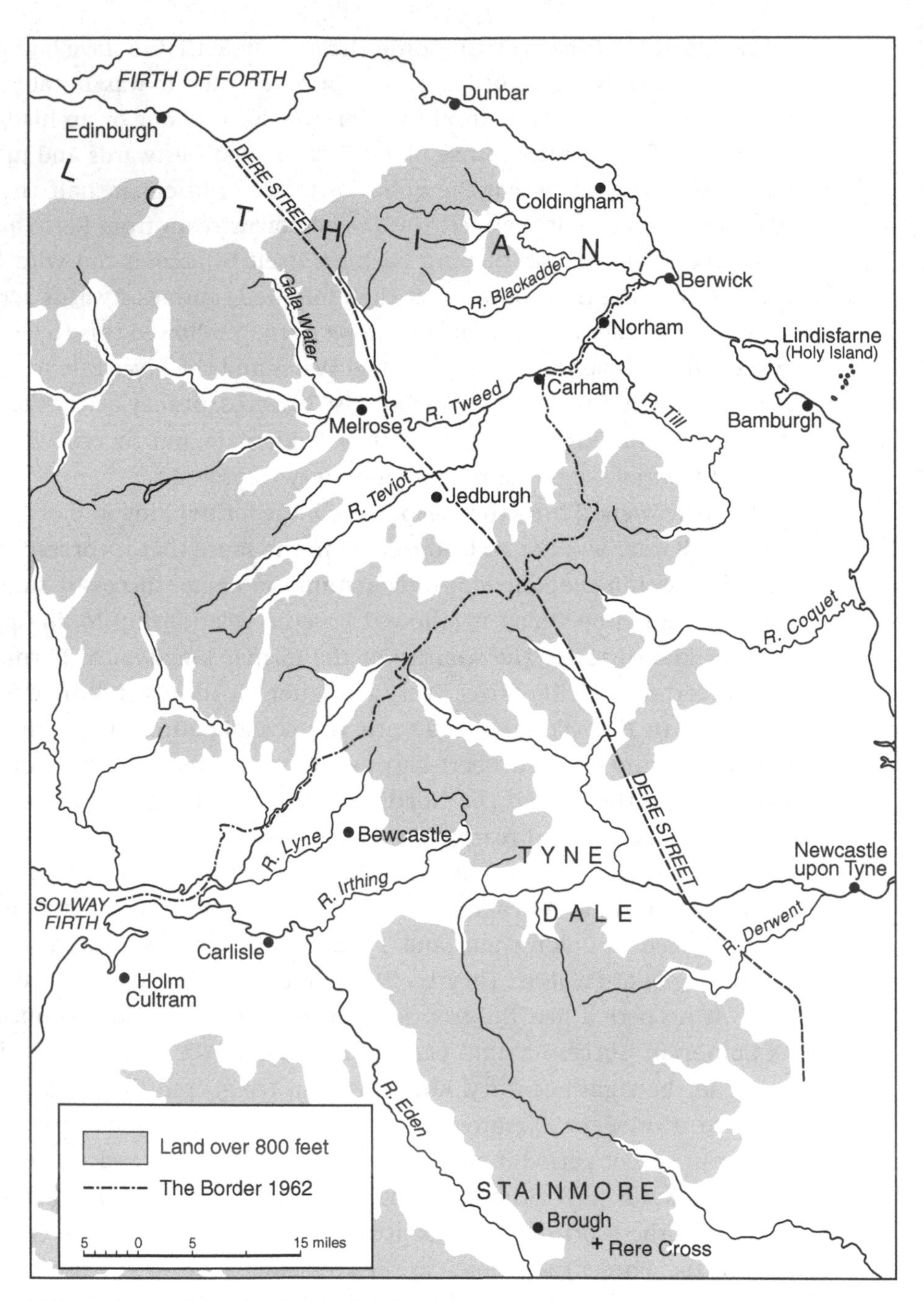

MAP 10 Map of the Anglo-Scottish Border

We should nevertheless not exaggerate the Border's political importance in the tenth, eleventh and even twelfth centuries, for it did little to inhibit the southward ambitions of the Scottish kings, or the anxiety of the kings of England on the contrary to impose at least a general authority over them. Between 1027 and 1031, King Cnut of England led an expedition to Scotland which resulted in King Malcolm II of Scots (1005–34) surrendering to him. In 1035, King Duncan I of Scots (1034–40) mounted an unsuccessful siege of Durham. Such a pattern of relations persisted: in 1072 King William I of England (the Conqueror) led a military expedition to Scotland to force King Malcolm III of Scots (1058–93) to submit to him at Abernethy (Perthshire). In 1093 Malcolm was killed near the River Alne (Northumberland) while leading his fifth military expedition into northern England. Scottish ambitions reached their height during the troubled reign of King Stephen (1135–54) when the bishopric of Durham itself passed for a time into the hands of William Cumin, a nominee of the king of Scots, and when a Scottish army had to be intercepted and defeated by an English force as far south as Lincoln. The ambition of the kings of Scots to rule as far south as the Tees, perhaps even as far south as the Humber, was palpable. Moreover, the Scottish Border had only limited influence over matters such as law and custom and, in the later middle ages at least, distinctive cross-Border traditions and variations developed.[34]

Moreover, the impact of the Border on the ethnic and cultural make-up of the lands of the Northumbrian 'successor states' was arguably not great, and especially in Lothian there are indications that continuity with the kingdom of Northumbria was as great, perhaps greater, north of the Scottish Border, and that the kingdom of Scotland was as much the heir to Northumbria as was northern England. As regards political organization, we see in Lothian the same structure of small shires as in Northumbria; indeed that structure extended farther north into Scotland and may even have been an introduction from England. As regards ethnic make-up, the place-names of Lothian are, as we have seen, predominantly English. The likelihood is that English was the predominant language of that area, as it definitely was (at least in the form of Lowland Scots) in later periods. There are some

[34] In general, see Geoffrey W. S. Barrow (1994a); for the texts, see *ASC* E, *s.a.* 1031 (submission to Cnut), Symeon, *Origins*, bk 3 ch. 9 (1035 siege of Durham), *ASC* E, *s.a.* 1072 (Abernethy), *Hist. Kings*, *s.a.* 1093 (pp. 221–3, incursions of Malcolm III). On Stephen's reign, see Geoffrey W. S. Barrow (1994b); on Cumin, see Symeon, *Origins*, pp. 282–323, and Young (1994); for law, see Neville (1998).

Gaelic place-names in Lothian, presumably indicating the introduction of a Scottish element in the population at whatever time Lothian became part of the kingdom of Scotland. They are not, however, sufficiently numerous to suggest that the area was heavily Gaelicized, reflecting no more than temporary occupation, possibly of a small Gaelic-speaking elite. Amongst the social elite of Lothian in the eleventh and twelfth centuries, our sources suggest the continued importance in Lothian of families of English or Anglo-Scandinavian origin. These included, for example, Liulf, son of Uhtred, who held a large estate around the River Coquet (Northumberland), but whose son, also an Uhtred, was established in southern Scotland, with his daughter (or possibly granddaughter) marrying into the family from whom the Stewart kings of Scots were to be drawn. Moreover, the house of Bamburgh continued to be influential, with the earls of Bamburgh transmuted into earls of Dunbar.[35]

As regards the cultural make-up of Lothian, we see the continuation or revival there of ancient Northumbrian churches – Melrose, Jedburgh, Coldingham in particular – and clear signs of Northumbrian influence on church organization which are less prominent in the essentially reform-dominated church of the post-Conquest period south of the Border. Lothian became part of the bishopric of St Andrews, although there were on-going disputes between that see and Durham, notably over which should be responsible for the district of Teviotdale. It is very striking that St Andrews (Fife), located well to the north of the ancient Northumbrian frontier on the Firth of Forth, retained some much earlier features than did Durham, notably its married clerks, known as 'Culdees', who lived on in St Andrews even when a new reformed cathedral community had been established there. The Culdees' church, the church of St Rule, had a tower notably influenced by Northumbrian architecture, very similar to the tower of Wharram-le-Street in Yorkshire or that of St Mary Bishophill Junior in York, for example (ills. 37–8); and its bishop in the early years of the twelfth century was the Englishman Turgot (1107–15), who had previously been prior of Durham. The orientation of the Scots towards the Northumbrian past was no doubt enhanced by the arrival of English exiles from the Norman Conquest at the court of King Malcolm III, and by the fact that that king had an English

[35] On place-names, see above, p. 61 n. 9, and Nicolaisen (2001), pp. 88–108; and, on families, see Geoffrey W. S. Barrow (1994a), pp. 321–2. On shires, see Geoffrey W. S. Barrow (1973b); and see the very useful survey and discussion of early medieval Scotland in Dodgshon (1981), pp. 26–57.

ILLUSTRATION 37 St Andrews (Fife), St Rule's Church, tower
Note the double openings in the belfry, and the exceptional height of the tower.

bride, Margaret of Scotland, whose *Life* Bishop Turgot was to write; but it was also the result of the heritage of Lothian in particular as an important part of the kingdom of Northumbria.[36] Although in one sense, therefore, the creation of the Scottish Border can be seen as one of the major turning-points in the transformation of the lands of the Northumbrian 'successor states', the real beginning of the process by which they lost their traces of former Northumbrian identity, there are nevertheless good grounds for seeing the continued influence of Northumbrian political organization, ethnic make-up and cultural characteristics in Lothian but also beyond it into more northerly areas of the kingdom of the Scots.

[36] Cowan, Easson and Hadcock (1976), under entries; for the architecture, see Cameron (1994).

ILLUSTRATION 38 Wharram-le-Street (Yorkshire East Riding), St Mary's Church, exterior from the south
Note the belfry openings (compare ill. 37); the pilasters in the masonry on either side of them indicate they have been lowered and would originally have had hood mouldings. See also Taylor and Taylor (1965–78), II, pp. 647–53.

The Norman kings of England

South of the Scottish Border William the Conqueror and his sons were faced with the problem of dominating and controlling what was left to England of the Northumbrian 'successor states'.[37] Throughout much of the eleventh century, the lands in question had appeared to the kings of England as an area which was dangerous to them from both a military and a political point of view. It had been the starting-point, for example, of the campaigns of Edmund Ironside against King Cnut, and the invasion of the Humber by the Norwegian King Harold Hardrada in 1066. Following his installation as king, William came to York in 1068. Edgar the Ætheling, great-nephew of King Edward the Confessor and a claimant to the English throne, had fled to Scotland, the native nobles Edwin and Morcar had left for northern England, and William had good reason to fear revolt. William's tactics elsewhere were to establish castles which were, according to the Norman chronicler Orderic

[37] For what follows, see, for example, the account in Stenton (1971), pp. 581–621.

Vitalis, 'scarcely known in the English provinces' and effective in dampening down English enthusiasm for revolt. In Northumbria, he had constructed a castle in York, probably on the site of what is now Clifford's Tower, and garrisoned it.[38]

William evidently remained anxious about the state of northern England. When Earl Cospatric, scion of the house of Bamburgh, joined the rebel party of Edgar the Ætheling, the king sent north a Norman, Robert Cumin, to be earl of Northumbria. On 28 January 1069, after a bout of ravaging the countryside and killing peasants, he and 700 men entered Durham. They were there savagely killed, all but one, as Symeon of Durham recounted: 'At first light, the Northumbrians who had assembled burst in together through all the gates, and rushed through the whole town killing the earl's companions. So great was the multitude of the slain, that all the streets were full of blood and corpses.'[39] Although the expedition which William dispatched to take vengeance as punishment for this killing turned back (according to Symeon, at any rate) before it reached Durham, there was worse in store. Later in 1069, the Northumbrian aristocrats Merleswein, Cospatric, Archill and 'the four sons of Karl', together with Edgar the Ætheling and 'other powerful rebels', attacked and captured York, and besieged the Norman castle. 'Fealty, oaths and the safety of the hostages were forgotten in their anger at the loss of their patrimonies and the deaths of their kinsmen and fellow countrymen.' The castellan of the castle, William Malet, sent a message to William who made a rapid march to York, and relieved the castle, sparing no man in Orderic's account. His men then plundered the city and 'made St Peter's church a disgrace'.[40]

Later in the year, in September, a Danish fleet anchored in the Humber, and its forces attacked York as part of an army led by the English nobles Edgar the Ætheling, Cospatric and Waltheof. To hinder the attack, the Norman garrison set fire to the city before the army arrived; even so, their efforts at defence were unsuccessful and most of them were killed. The Danish army then withdrew without pressing its advantage. Later in 1069–70, it reentered York as King William was marching north, but he wasted the country to the north and west and the Danish army withdrew again and made peace with the king. According to Orderic Vitalis, it was rumoured that the Scandinavians had intended to keep the midwinter feast in York; but it was in fact King William who kept the Christmas feast there before undertaking

[38] *Sources*, pp. 181–2 (N.1.2 = *Orderic*, bk 4 (II, 216–19)). [39] Symeon, *Origins*, bk 3 ch. 15.
[40] *Sources*, pp. 182–3 (N.1.3 = *Orderic*, bk 4 (II, 222–3)), and *Sources*, p. 188 (N.3.1 = *ASC* D, *s.a.* 1069).

a campaign of ravaging which reached north to the Tees by early 1070 and across the Pennines into Cheshire. This campaign has come to be known as the 'harrying of the north'.[41]

An area north of the River Tees was also ravaged. After the killing of Earl Robert Cumin in 1069, William the Conqueror had sold the earldom of Northumbria to Walcher, a churchman from Lotharingia who had become bishop of Durham in 1071 after the expulsion of the native bishop Æthelwine. In 1080, Walcher was murdered at Gateshead in particularly savage circumstances. His murderer was believed to have been a certain Eadwulf Rus, a scion of the house of Bamburgh. The result of the killing was, according to Symeon of Durham, that: 'Bishop Odo of Bayeux, who was then second only to the king, came to Durham with many of the leading men of the kingdom and a large force of armed men; and in avenging the death of the bishop virtually laid the land waste.'[42]

How far these campaigns of ravaging really resulted in widespread devastation of the land, despite the chroniclers' lurid picture of them, has been questioned. Neither the terminology of *Domesday Book*, nor what is known incidentally of individual manors, really supports that picture. But it seems clear nevertheless that the aristocracy between the Humber and the Scottish Border was to a very large extent killed, exiled or at the least expropriated of its land. South of the Tees, *Domesday Book* seems to show very clearly that new continental land-owners had, to a surprisingly large extent even by the time of the *Domesday* survey in 1086, come to dominate the region, and that the Norman practice of establishing castles with territories to support them was already well advanced as a means of subjugating what had been hostile territory. These were generally intrusive on the pre-existing political structures, in so far as we can perceive them, and resulted in a radically changed political and administrative framework. North of the Tees, there can be little doubt that many of the native Northumbrian aristocracy, including the house of Bamburgh, suffered grave losses to their position. Indeed, we find the family in later years chiefly involved in the affairs of Scotland rather than in those of England, so that continuity with the Northumbrian past in this respect looks greater north of the Tweed than south of it, although

[41] *Sources*, pp. 82, 184–5 (N.1.7 = *Hist. Kings*, *s.a.* 1069 (pp. 187–8)), p. 188 (N.3.1 = *ASC* D, *s.a.* 1069), pp. 193–4 (N.4.1 = Hugh Chanter, pp. 2–3), pp. 188–9 (N.3.2 = *Orderic*, bk 4 (II, pp. 230–3)).

[42] Symeon, *Origins*, bk 3 ch. 24, for this and an account of the murder; but see also p. 213 n. 96 and p. 214 n. 99, citing variant accounts of Walcher's killing in *Hist. Kings*, *s.a.* 1080 (pp. 208–11), and William Malms., *History*, I, pp. 498–501; for Eadwulf Rus, see Arnold (1882–5), II, p. 383.

some families south of the Tweed, such as the Nevilles, survived to prosper in later centuries.[43]

It was, however, in the area of ecclesiastical organization and church reform that the most drastic changes were to come. As we noted earlier, the lands of what had been Northumbria were very little affected by the reformation of monasteries and other churches which so shaped the religious life of southern and midland England. There are some indications that the Community of St Cuthbert at Durham was a little influenced by them. Its sacristan, Elfred, son of Westou, began in the 1030s to assemble a collection of the relics of saints from other churches in the region, an activity strongly reminiscent of that of reformed monasteries in the south. Moreover, as we have seen, Durham had amongst its liturgical books the Durham Collectar imported from the continent via southern England. In general, however, Northumbrian churches seem to have continued, where they did, as communities of secular clerks, certainly married in the case of those of Durham and Hexham and probably in that of other churches also, and therefore no longer in line with the rigorous standards set by the reformers of the late eleventh-century church. To the churchmen installed by the Norman conquerors, this was insupportable. The Community of St Cuthbert at Durham, heir to the ancient Northumbrian church of Lindisfarne, underwent an attempt at reform by the first bishop appointed by William the Conqueror, Walcher, but found the second bishop he appointed, William of St Calais (1080–96), less inclined to take trouble over them. According to Symeon of Durham writing between 1104 and 1107 (and there are no compelling reasons to doubt his account), Bishop William expelled the clerks of the Community of St Cuthbert, and replaced them with monks following the Rule of St Benedict, some at least of whom were from southern England. The same change of personnel (this time for canons following the Rule of St Augustine) occurred at Hexham in the early twelfth century. At York Minster at the end of the eleventh century there was a reform of the chapter, which eventually resulted in it being organized according to the constitution of the Norman cathedral of Bayeux.[44]

[43] On Yorkshire, see Palliser (1992), pp. 34–8, and (1993), and most fully Dalton (1994); on this and north of the Tees, see Kapelle (1979), pp. 120–90; on Cumbria, see Phythian-Adams (1996), pp. 23–43; for the establishment of castelries, see also Pounds (1994), pp. 39–44. For the house of Bamburgh, see above, p. 249 n. 51 and references therein; for the Nevilles, see Offler (1996), no. XIII.

[44] For Durham, see Symeon, *Origins*, bk 4 ch. 3 and nn; on the pre-reform community there, see Rollason (1992). Aird (1998a), pp. 35–40, and (1998b), pp. 126–31, 137–8, disputes Symeon's account, arguing that there was much greater continuity in the personnel of the community. On Hexham, see Rollason (1995b); on York see *Sources*, pp. 195–6 (N.4.2 = Hugh Chanter, pp. 18–21).

Around 1100 and later, the church nevertheless retained some features which derived ultimately from the kingdom of Northumbria. The basic outline of ecclesiastical organization resembled in some respects that of the Northumbrian 'successor states', with York Minster at its head, the 'liberty' of the Community of St Cuthbert dominating the land between the Rivers Tees and Tyne and substantial areas between Tyne and Tweed, and the ancient churches of Ripon and Beverley as subsidiary ecclesiastical centres of the see of York, together with the relatively newly acquired, but nevertheless pre-Conquest, church of Southwell in Nottinghamshire. As the holder of St Cuthbert's liberty, the bishop of Durham had powers which normally belonged to the king, such as holding courts, minting coins and exploiting mines and salvage from wrecked ships. The origin of these powers is obscure and much disputed, but it probably went back to the period of the Northumbrian 'successor states'. An aspect of continuity with the period of the Northumbrian 'successor states' and of the kingdom before them was the rights of sanctuary or asylum which the church of Durham and certain other Northumbrian churches – Ripon, Hexham and Beverley, and also Wetherall near Carlisle – were able to offer to fugitives and other criminals. Although sanctuary rights were general in the early middle ages, these Northumbrian ones were characterized by offering particularly long and wide-ranging protection to those who sought them. In the case of that pertaining to St Cuthbert, the sanctuary afforded by the saint's body is mentioned already in the early eighth century by Bede and, as we have seen, it appears to have been defined and regularized in the late ninth and tenth centuries. There is every reason to suppose that the other sanctuaries also originated before the Norman Conquest, although they need not be as early as St Cuthbert's, and we have noted already the traditional ascription of those of Beverley and Ripon to King Athelstan in the early tenth century. They nevertheless seem to form a strand of continuity with the pre-Norman past.[45] Nevertheless, these elements of continuity were only part of the picture. As we have seen, the churches in question were in fact very different in character by or shortly after 1100; and in any case other churches and other elements of the kingdom of Northumbria had been lost without trace – the position of Hexham as an episcopal see, for example.

[45] For subsidiary ecclesiastical centres, see Nicholl (1964), pp. 123–7; for the liberty, see W. Page (1888) and Scammell (1966); on sanctuary, see David Hall (1989), Cox (1911), Trenholme (1903) and McAleer (2001). On Wetherall, see Phythian-Adams (1996), pp. 118–19.

The shadow of the past

Whatever the reality, however, the church, especially the church of Durham, played an important role in keeping alive the history of, or at any rate traditions about, the Northumbrian past, and in emphasizing or inventing links with it. As we have seen, the church of Durham had been ruthlessly reformed in 1083, but paradoxically the manner of that reformation ensured for two reasons that the former kingdom of Northumbria's history remained a living force in the life of the church of Durham. First, the Benedictine monks with whom Bishop William of St Calais in 1083 replaced the married clerks were drawn from Bede's own churches of Monkwearmouth and Jarrow. These had seemingly been in a ruined and deserted condition in the 1070s when a contingent of monks from the west midlands, led by Aldwine, the future prior of Durham, re-established them as Benedictine monasteries. These monks were, according to Symeon of Durham, inspired by Bede's *Ecclesiastical History* from which they had learned that Northumbria had once been full of monks and monasteries. Although they were no doubt acting in the spirit of the Europe-wide reforming trends of the eleventh century which often emphasized the past as a golden age to be emulated, these monks were clearly specifically influenced by nostalgia for the Northumbrian past, nostalgia which no doubt continued to influence their activities at Durham.[46] Secondly, the position of the monks at Durham depended crucially on claiming that they were the true heirs of the original community of Lindisfarne, which had moved first to Chester-le-Street and then on to Durham, and thus had a right to the status and the possessions (not least in land) of the married clerks who had been expelled. Pursuit of this claim involved historical research and detailed reference to past history. When Bishop William of St Calais undertook to replace the clerks at Durham, he had, Symeon of Durham tells us, consulted both the senior men of his bishopric and Bede's writings, from which he had learned that the church of Lindisfarne had been served by monks. His expulsion of the clerks and their replacement by monks could thus be presented as simply putting the clock back to the golden age of Northumbria. This presentation of the relationship between the past and the present was heightened when in 1104 the new monks of Durham translated the body of St Cuthbert

[46] Symeon, *Origins*, bk 3 chs. 21–2. For discussion, see Knowles (1963), pp. 165–71; on the influence of Bede, see Davis (1989). For a re-appraisal of the background of the reforming monks, suggesting that contemporary reform movements were more important than pre-Conquest English influences, see Dawtry (1982).

into their splendid new cathedral, and discovered (as they thought) that the saint's body remained undecayed just as Bede had described it as being when the saint's grave had first been opened in 698. The body, enshrined in the towering Norman cathedral, was a tangible link, a powerful symbol of continuity with the Northumbrian past. A sense of history was also needed for more mundane matters. The church of Durham's claim to its landed estates was grounded not only in the period of Bede, but also in subsequent periods when the church of Lindisfarne had acquired its wide lands north and south of the River Tyne which the church of Durham still held or claimed.[47]

All this helps to explain why, when between 1104 and 1107 Symeon of Durham came to write the official history of the church, his *On the Origins and Progress of this the Church of Durham*, he began it with the foundation of Lindisfarne in 635, emphasizing in his preface that:

> This venerable church derived its status and its divine religion from the fervent faith in Christ of the former glorious king of the Northumbrians and estimable martyr Oswald. In praise of God and under his perpetual guardianship it preserves those relics of devout veneration, the undecayed body of the most saintly father Cuthbert and the venerable head of that same king and martyr Oswald, both lodged in a single shrine. Although for various reasons this church no longer stands in the place where Oswald founded it, nevertheless by virtue of the constancy of its faith, the dignity and authority of its episcopal throne, and the status of the dwelling-place of monks established there by the king himself and by Bishop Aidan, it is still the very same church founded by God's command.[48]

This practice of referring to the Northumbrian past was not limited to Durham. The Augustinian canons of Hexham translated the remains of the Northumbrian bishops Acca (Bede's own bishop) and Alchmund to shrines of great splendour in their church, and even obtained the services of Ælred, the great abbot of the new Cistercian abbey of Rievaulx in Yorkshire, who was a descendant of one of the pre-reform married clergy of Hexham, to write an account of these saints, apparently as a celebratory sermon.[49] Whitby was re-established by the ex-Norman knight Reinfrid, one of the monks who had refounded Monkwearmouth and Jarrow. Although we know little about the attitude of the new abbey to the Northumbrian past, it is likely that it too shared the same sense of nostalgia and of the importance of history as did Durham. *Domesday Book* records some of Whitby's lands under the name of Abbess Hild of Whitby, who was present at the Synod of Whitby

[47] Symeon, *Origins*, pp. lxxvii–xci. [48] Symeon, *Origins*, bk 1 ch. 1.
[49] Raine (1864–5), I, pp. 173–203.

in 664; and its first abbots chose to refound two ancient Northumbrian monasteries, Hackness and Lastingham, the first of which had been a subsidiary monastery of Whitby in the seventh century, the second a foundation of Bishop Chad's brother Cedd, a pupil of Bishop Aidan of Lindisfarne, at broadly the same period. York was perhaps least affected, although its archbishop Thurstan at the beginning of the twelfth century was directly involved with establishing the house of canons of Nostell, the story of the origin of which focused on veneration of the cult of the Northumbrian saint-king Oswald.[50]

In the hands of Symeon of Durham, however, feeling towards the Northumbrian past took on a more important dimension. He was responsible not only for *On the Origins and Progress* but also for a more general historical compilation, probably incomplete at his death in *c.* 1128, the *History of the Kings*. This too gathered together much information about the early history of Northumbria, and the place of the community of Durham in it; indeed, as we have seen, it preserved our fullest version of the Northern Annals. Another compilation, the *History after the Death of Bede*, was also produced at Durham, possibly as an abbreviation of the *History of the Kings*, but the relationship remains unclear. This considerable achievement in researching and writing history had an immediate influence on history-writing at Hexham, the canons of which church not only produced their own work but also added material to the *History of the Kings* as we now have it. All this historical work on Northumbrian history had a wider influence in England south of the Humber – on John of Worcester, who drew on the *History of the Kings* extensively for his account of northern history, and on Roger of Hoveden, who absorbed into his own *History* the full text of the *History after the Death of Bede*.[51]

This concern with the history of Northumbria had, for us at least, two effects. First, it led to the preservation of materials – Durham not only produced the histories mentioned and collected annalistic compilations, but it also assembled copies of Nennius and Gildas, as well as of Bede.[52] Someone setting out to write this book at Durham in the twelfth century would by

[50] On Whitby, see Burton (1994); on Nostell, see Rollason (1995c), pp. 173–4.

[51] On Durham, see Rollason (1998); on Hexham, see Walterspacher (2002) and Offler (1970); on Roger, see Stubbs (1868–71), I, pp. xxvi–xl. In general, see Gransden (1974).

[52] Durham, Cathedral Library, MS B.II.35 is a copy of Bede's *Ecclesiastical History* given to Durham by Bishop William of St Calais (1080–96). The important historical compilations in Cambridge, Corpus Christi College MSS 66 and 139, and Cambridge, University Library, MS Ff.1.27, which contain *inter alia* Gildas and Nennius, have been convincingly assigned to twelfth-century Durham by Meehan (1994b) and Norton (1998b). On historical interests at Durham Cathedral Priory, see Piper (1998).

no means have been badly placed to do so. Secondly, it led to a particular emphasis being given to the history of Northumbria by medieval historians. Symeon's allegiances and those of contemporary churchmen in northern England were fully engaged with the kingdom of England. These men owed their position to the regime of the Norman kings. They wanted good relations with the kings of Scots, even to the extent in Durham's case of having King Malcolm III present at the laying of the foundations of Durham Cathedral in 1093 and the future King Alexander I present at the translation of St Cuthbert in 1104; but it never seems to have occurred to them to suppose that they were anything but part of the kingdom of England.[53] As we have seen, this was not in fact so self-evident as they seem to have supposed; but the important point for us is that it meant that the post-Conquest histories of Northumbria were focused on its English dimension, viewing its history (as more modern scholars have often done) as one of the preludes to the kingdom of England. It has been the purpose of this chapter to try to demonstrate that the process by which the 'successor states' into which the kingdom of Northumbria had fragmented were absorbed into the kingdoms of England and Scotland was complex and long-drawn-out, that we should not with Symeon and his colleagues give prominence to it as an aspect of English rather than Scottish history and that we should be alert to the continuing influence of what had been the kingdom of Northumbria north and south of the Scottish Border, in the reality of political organization and ethnic and cultural make-up as well as in the mental images and historical researches of the scholar-churchmen of the decades after the Norman Conquest of England.

[53] *Hist. Kings*, *s.a.* 1093 (p. 220), on which see Wall (1994); for the translation, see Battiscombe (1956), p. 105.

References

Abrams, Lesley 2001, 'The conversion of the Danelaw', in James Graham-Campbell, Richard A. Hall, Judith Jesch et al. (eds.), *Vikings and the Danelaw: Select Papers from the Proceedings of the Thirteenth Viking Congress, Nottingham and York, 21–30 August 1997* (Oxford: Oxbow Books), pp. 31–44.

Addyman, Peter 1973–4, 'Saxon Southampton: a town and international port of the eighth to tenth century', in Herbert Jankuhn, Walter Schlesinger and Heiko Steuer (eds.), *Vor- und Frühformen der europäischen Stadt im Mittelalter: Bericht über ein Symposium in Reinhausen bei Göttingen in dem Zeit vom 18.–24. April 1972*, 2 vols. (Göttingen: Vandenhoeck und Ruprecht), I, pp. 218–28.

Addyman, Peter (ed.) 1976–, *The Archaeology of York* (London: Council for British Archaeology).

Aird, W. M. 1998a, 'The political context of the *Libellus de exordio*', in David Rollason (ed.), *Symeon of Durham: Historian of Durham and the North* (Stamford: Paul Watkins), pp. 32–45.

Aird, W. M. 1998b, *St Cuthbert and the Normans: The Church of Durham, 1071–1153* (Studies in the History of Medieval Religion; Woodbridge: Boydell Press).

Alcock, Leslie 1971, *Arthur's Britain: History and Archaeology AD 367–634*, 2nd edn (Harmondsworth: Penguin).

Alcock, Leslie 1975–6, 'A multi-disciplinary chronology for Alt Clut, Castle Rock, Dumbarton', *Proceedings of the Society of Antiquaries of Scotland* 107, 103–13.

Alcock, Leslie 1981, 'Quantity or quality: the Anglian graves of Bernicia', in V. I. Evison (ed.), *Angles, Saxons and Jutes: Essays Presented to J. N. L. Myres* (Oxford: Clarendon Press), pp. 168–86 (repr. in his *Economy, Society and Warfare among the Britons and Saxons* (Cardiff: University of Wales Press, 1987), pp. 255–66).

Alcock, Leslie 1987, *Economy, Society and Warfare among the Britons and Saxons* (Cardiff: University of Wales Press).

Alcock, Leslie 1989, *Bede, Eddius and the Forts of the North Britons* (Jarrow Lecture 1988; Jarrow: St Paul's Church, Jarrow) (repr. in *Bede and his World: The Jarrow Lectures 1958–93*, ed. Michael Lapidge, 2 vols. (Aldershot: Variorum, 1994), II, pp. 773–806).

Alcock, Leslie, Alcock, Elizabeth A. and Foster, Sally M. 1986, 'Reconnaissance excavations on early historic fortifications and other royal sites in Scotland, 1974–84: 1, excavations near St Abb's Head, Berwickshire, 1980', *Proceedings of the Society of Antiquaries of Scotland* 116, 255–79.

Alcock, Leslie, Stevenson, S. J. and Musson, C. R. 1995, *Cadbury Castle, Somerset: The Early Medieval Archaeology* (Cardiff: University of Wales Press).

Anderson, Alan Orr 1922, *Early Sources of Scottish History, AD 500 to 1286*, 2 vols. (Edinburgh: Oliver and Boyd).

Anderson, Alan Orr and Anderson, Marjorie Ogilvie (eds.) 1936, *The Chronicle of Melrose: From the Cottonian Manuscript, Faustina B.IX in the British Museum; A Complete and Full-size Facsimile in Collotype* (London: P. Lund, Humphries).

Anderson, Marjorie Ogilvie 1973, *Kings and Kingship in Early Scotland* (Edinburgh: Scottish Academic Press).

Angenendt, Arnold 1986, 'The conversion of the Anglo-Saxons considered against the background of the early medieval mission', in Anon. (ed.), *Angli e Sassoni al di qua e al di là del Mare* (Settimane di studio sull'alto medioevo 32; Spoleto: Centro Italiano sull'Alto Medioevo), pp. 747–81.

Archibald, Marion M. 1992, 'Dating the Cuerdale hoard: the evidence of the coins', in James Graham-Campbell (ed.), *Viking Treasure from the North West: The Cuerdale Hoard in its Context* (National Museums and Galleries on Merseyside Occasional Papers 5; Liverpool: Liverpool Museum), pp. 21–30.

Armit, Ian 1997, *Celtic Scotland* (London: Historic Scotland/Batsford).

Arnold, Thomas (ed.) 1882–5, *Symeonis monachi Opera omnia*, 2 vols. (Rolls Series; London: Longman).

Aston, T. H. 1983, 'The origins of the manor in England', in T. H. Aston (ed.), *Social Relations and Ideas: Essays in Honour of R. H. Hilton* (Cambridge: Cambridge University Press), pp. 1–43.

Attenborough, F. L. (ed. and trans.) 1922, *The Laws of the Earliest English Kings* (Cambridge: Cambridge University Press).

Backhouse, Janet 1981, *The Lindisfarne Gospels* (Oxford: Phaidon).

Bailey, Richard N. 1980, *Viking Age Sculpture in Northern England* (London: Collins).

Bailey, Richard N. 1985, 'Aspects of Viking-age sculpture in Cumbria', in John R. Baldwin and Ian D. Whyte (eds.), *The Scandinavians in Cumbria* (Edinburgh: Scottish Society for Northern Studies), pp. 53–63.

Bailey, Richard N. 1989, 'St Cuthbert's relics: some neglected evidence', in Gerald Bonner, David Rollason and Clare Stancliffe (eds.), *St Cuthbert, his Cult and his Community to AD 1200* (Woodbridge: Boydell Press), pp. 231–46.

Bailey, Richard N. 1991, 'St Wilfrid, Ripon and Hexham', in Catherine Karkov and Robert Farrell (eds.), *Studies in Insular Art and Archaeology* (American Early Medieval Studies 1; Oxford, Ohio: American Early Medieval Studies), pp. 1–25.

Bailey, Richard N. and Cramp, R. J. 1988, *Corpus of Anglo-Saxon Stone Sculpture*, II: *Cumberland, Westmorland and Lancashire North-of-the-Sands* (Oxford: Oxford University Press).

Bannerman, John 1974, *Studies in the History of Dalriada* (Edinburgh and London: Scottish Academic Press).

Barrow, Geoffrey W. S. 1973a, 'The Anglo-Scottish Border', in Geoffrey W. S. Barrow (ed.), *The Kingdom of the Scots: Government, Church and Society from the Eleventh to the Fourteenth Century* (London: Edward Arnold), pp. 139–61 (repr. from *Northern History* 1 (1966), 21–42).

Barrow, Geoffrey W. S. 1973b, 'Pre-feudal Scotland: shires and thanes', in Geoffrey W. S. Barrow (ed.), *The Kingdom of the Scots: Government, Church and Society from the Eleventh to the Fourteenth Century* (London: Edward Arnold), pp. 7–68.

Barrow, Geoffrey W. S. 1994a, 'The kings of Scotland and Durham', in David Rollason, Margaret Harvey and Michael Prestwich (eds.), *Anglo-Norman Durham 1093–1193* (Woodbridge: Boydell Press), pp. 311–23.

Barrow, Geoffrey W. S. 1994b, 'The Scots and the north of England', in Edmund King (ed.), *The Anarchy of King Stephen's Reign* (Oxford: Clarendon Press), pp. 231–53.

Barrow, Geoffrey W. S. 1998, 'The uses of place-names and Scottish history: pointers and pitfalls', in Simon Taylor (ed.), *The Uses of Place-Names* (Edinburgh: Scottish Cultural Press), pp. 54–74.

Barrow, Julia 1994, 'English cathedral communities and reform in the late tenth and eleventh centuries', in David Rollason, Margaret Harvey and Michael Prestwich (eds.), *Anglo-Norman Durham 1093–1193* (Woodbridge: Boydell Press), pp. 25–39.

Barry, Terry 1995, 'The last frontier: defence and settlement in late medieval Ireland', in T. B. Barry, Robin Frame and Katherine Simms (eds.), *Colony and Frontier in Medieval Ireland: Essays Presented to J. F. Lydon* (London and Rio Grande: Hambledon), pp. 217–28.

Bartlett, Robert 1993, *The Making of Europe: Conquest, Colonization and Cultural Change 950–1350* (London: Allen Lane).

Bately, Janet M., Brown, Michelle P. and Roberts, Jane (eds.) 1993, *A Palaeographer's View: The Selected Writings of Julian Brown* (London: Harvey Miller) (with a preface by Albinia C. de la Mare).

Bates, David (ed.) 1998, *Regesta Regum Anglo-Normannorum: The Acta of William I 1066–1087* (Oxford: Clarendon Press).

Bateson, Mary 1899, 'The origin and early history of double monasteries', *Transactions of the Royal Historical Society*, new series, 13, 137–98.

Battiscombe, C. F. (ed.) 1956, *The Relics of St Cuthbert* (Oxford: Oxford University Press).

Beresford, Maurice W. and Hurst, John 1991, *Wharram Percy: Deserted Medieval Village* (New Haven: Yale University Press).

Biddle, Martin 1975, 'The evolution of towns: planned towns before 1066', in Maurice W. Barley (ed.), *The Plans and Topography of Medieval Towns in England and Wales* (Research Report 14; London: Council for British Archaeology), pp. 19–32.

Binchy, D. A. 1970, *Celtic and Anglo-Saxon Kingship* (Oxford: Clarendon Press).

Binns, J. W., Norton, E. C. and Palliser, D. M. 1990, 'The Latin inscription on the Coppergate Helmet', *Antiquity* 64, 134–9.

Birley, Anthony Richard 1981, *The Fasti of Roman Britain* (Oxford: Clarendon Press).

Birley, Robin 2002, 'Vindolanda', *Current Archaeology* 178, 436–45.

Blackburn, Mark 2001, 'Expansion and control: aspects of Anglo-Scandinavian minting south of the Humber', in James Graham-Campbell, Richard A. Hall, Judith Jesch et al. (eds.), *Vikings and the Danelaw: Select Papers from the Proceedings of the Thirteenth Viking Congress, Nottingham and York, 21–30 August 1997* (Oxford: Oxbow Books), pp. 125–42.

Blackburn, Mark forthcoming, 'The coinage of Scandinavian York', in Richard A. Hall (ed.), *Anglo-Scandinavian York (AD 876–1066)* (Archaeology of York 8; York: York Archaeological Trust).

Blair, John 1988, *Minsters and Parish Churches: The Local Church in Transition 950–1200* (Monograph 17; Oxford: Oxford University Committee for Archaeology).

Blair, John 1992, 'Anglo-Saxon minsters: a topographical review', in John Blair and Richard Sharpe (eds.), *Pastoral Care before the Parish* (Leicester, London and New York: Leicester University Press), pp. 226–66.

Blair, John 1995, 'Anglo-Saxon shrines and their prototypes', *Anglo-Saxon Studies in Archaeology and History* 8, 1–28.

Blair, John 2001, 'Beverley, *Inderauudu* and St John: a neglected reference', *Northern History* 38, 315–16.

Blair, John and Sharpe, Richard (eds.) 1992, *Pastoral Care before the Parish* (Leicester, London and New York: Leicester University Press).

Blair, Peter Hunter 1947, 'The origins of Northumbria', *Archaeologia Aeliana*, 4th series, 25, 1–51 (repr. in his *Anglo-Saxon Northumbria*, ed. Michael Lapidge and Pauline Hunter Blair (London: Variorum, 1984), no. III).

Blair, Peter Hunter 1948, 'The Northumbrians and their southern frontier', *Archaeologia Aeliana*, 4th series, 26, 98–126 (repr. in his *Anglo-Saxon Northumbria*, ed. Michael Lapidge and Pauline Hunter Blair (London: Variorum, 1984), no. IV).

Blair, Peter Hunter 1949, 'The boundary between Bernicia and Deira', *Archaeologia Aeliana*, 4th series, 27, 46–59 (repr. in his *Anglo-Saxon Northumbria*, ed. Michael Lapidge and Pauline Hunter Blair (London: Variorum, 1984), no. V).

Blair, Peter Hunter 1963, 'Some observations on the *Historia Regum* attributed to Symeon of Durham', in N. K. Chadwick (ed.), *Celt and Saxon: Studies in the Early British Border* (Cambridge University Press: Cambridge), pp. 63–118 (repr. in his *Anglo-Saxon Northumbria*, ed. Michael Lapidge and Pauline Hunter Blair (London: Variorum, 1984), no. III).

Blair, Peter Hunter 1970, *The World of Bede* (Cambridge: Cambridge University Press) (repr. with corrections and additional bibliography, Cambridge: Cambridge University Press, 1990).

Bonner, Gerald 1989, 'St Cuthbert at Chester-le-Street', in Gerald Bonner, David Rollason and Clare Stancliffe (eds.), *St Cuthbert, his Cult and his Community to AD 1200* (Woodbridge: Boydell Press), pp. 387–95.

Bonner, Gerald, Rollason, David and Stancliffe, Clare (eds.) 1989, *St Cuthbert, his Cult and his Community to AD 1200* (Woodbridge: Boydell Press).

Breeze, David J. and Dobson, Brian 2000, *Hadrian's Wall*, 4th edn (London: Penguin).

Brentano, R. J. 1953, 'Whithorn and York', *Scottish Historical Review* 32, 144–6.

Brickstock, R. J. 2000, 'Coin supply in the north in the late Roman period', in Tony Wilmott and Pete Wilson (eds.), *The Late Roman Transition in the North: Papers from the Roman Archaeology Conference, Durham 1999* (British Archaeological Reports, British Series 299; Oxford: Archaeopress), pp. 33–7.

Brinklow, D. A. 1984, 'Roman settlement around the legionary fortress at York', in Peter V. Addyman and Valerie E. Black (eds.), *Archaeological Papers from York Presented to M. W. Barley* (York: York Archaeological Trust), pp. 21–7.

Brooke, Christopher N. L. and Keir, Gillian 1975, *London 800–1216: The Shaping of a City* (History of London; London: Secker and Warburg).

Brooke, Daphne 1991, 'The Northumbrian settlements in Galloway and Carrick: an historical assessment', *Proceedings of the Society of Antiquaries of Scotland* 121, 295–327.

Brooks, Nicholas 1971, 'The development of military obligations in eighth- and ninth-century England', in Peter Clemoes and Kathleen Hughes (eds.), *England before the Conquest: Studies in Primary Sources Presented to Dorothy Whitelock* (Cambridge: Cambridge University Press), pp. 69–84.

Brooks, Nicholas 1978, 'Arms, status and warfare in late Saxon England', in David Hill (ed.), *Ethelred the Unready: Papers from the Millenary Conference* (British Archaeological Reports, British Series 59; Oxford: BAR), pp. 81–103.

Brooks, Nicholas 1979, 'England in the ninth century: the crucible of defeat', *Transactions of the Royal Historical Society*, 5th series, 29, 1–20.

Brooks, Nicholas 2000, *Bede and the English* (Jarrow Lecture 1999; Jarrow: St Paul's Church, Jarrow).

Brown, George Hardin 1987, *Bede the Venerable* (Twayne's English Authors Series 443; Boston: Twayne).

Brown, Peter 1997, *The Rise of Western Christendom: Triumph and Diversity, AD 200–1000*, 2nd edn (Oxford: Blackwell).

Brown, T. J. (ed.) 1969, *The Durham Ritual: A Southern English Collectar of the Tenth Century with Northumbrian Additions: Durham Cathedral Library A.IV.19* (Early English Manuscripts in Facsimile 16; Copenhagen: Rosenkilde and Bagger).

Brown, T. J. 1972, 'Northumbria and the Book of Kells', *Anglo-Saxon England* 1, 219–46 (repr. in T. J. Brown, *A Palaeographer's View: The Selected Writings of Julian Brown*, ed. Janet Bately, Michelle P. Brown and Jane Roberts (London: Harvey Miller, 1993), pp. 97–124).

Brown, T. J., Verey, Christopher D. and Coatsworth, Elizabeth (eds.) 1980, *The Durham Gospels together with a Fragment of a Gospel Book in Uncial (Durham Cathedral Library, MS A.II.17)* (Early English Manuscripts in Facsimile 20; Copenhagen: Rosenkilde and Bagger).

Bruce-Mitford, Rupert L. S. 1968, *The Art of the Codex Amiatinus* (Jarrow Lecture 1967; Jarrow: St Paul's Church, Jarrow) (repr. in *Bede and his World: The Jarrow Lectures 1958–1993*, ed. Michael Lapidge, 2 vols. (Aldershot: Variorum, 1994), I, pp. 185–234).

Bruce-Mitford, Rupert L. S. 1974, *Aspects of Anglo-Saxon Archaeology: Sutton Hoo and Other Discoveries* (London: Victor Gollancz).

Bruce-Mitford, Rupert L. S. 1989, 'The Durham-Echternach calligrapher', in Gerald Bonner, David Rollason and Clare Stancliffe (eds.), *St Cuthbert, his Cult and his Community to AD 1200* (Woodbridge: Boydell Press), pp. 175–88.

Brühl, Carl-Richard 1968, *Fodrum, gistum, servitium regis: Studien zu den wirtschaftlichen Grundlagen des Königtums im Frankenreich und in den fränkischen Nachfolgestaaten Deutschland, Frankreich und Italien vom 6. bis zur Mitte des 14. jahrhunderts*, 2 vols. (Kölner historische Abhandlungen 14; Cologne: Böhlau).

Buckland, P. C. 1984, 'The "Anglian Tower" and the use of Jurassic limestone in York', in Peter V. Addyman and Valerie E. Black (eds.), *Archaeological Papers from York Presented to M. W. Barley* (York: York Archaeological Trust), pp. 51–7.

Bullough, Donald A. 1964, 'Columba, Adomnan and the achievement of Iona: part I', *Scottish Historical Review* 43, 111–30.

Bullough, Donald A. 1981, 'Hagiography as patriotism: Alcuin's York poem and the early Northumbrian Vitae Sanctorum', in Anon. (ed.), *Hagiographie, cultures et sociétés: IVe–XIIe siècles: actes du colloque organisé à Nanterre et à Paris, 2–5 mai 1979* (Université de Paris X, Centre de recherche sur l'antiquité tardive et le haut moyen âge; Paris: Etudes augustiniennes), pp. 339–59.

Bullough, Donald A. 1982, 'The missions to the English and the Picts and their heritage', in Heinz Lowe (ed.), *Die Iren und Europa im früheren Mittelalter* (Stuttgart: Klett-Cotta), pp. 80–98.

Bullough, Donald A. 1993, 'What has Ingeld to do with Lindisfarne?', *Anglo-Saxon England* 22, 93–125.

Bullough, Donald A. (2002), *Alcuin: Achievement and Reputation* (Education and Society in the Middle Ages and Renaissance; Leiden: Brill).

Bund, Konrad 1979, *Thronsturz und Herrscherabsetzung im Frühmittelalter* (Bonner Historische Forschung 44; Bonn: Röhrscheid).

Burton, Janet 1994, 'The monastic revival in Yorkshire: Whitby and St Mary's, York', in David Rollason, Margaret Harvey and Michael Prestwich (eds.), *Anglo-Norman Durham 1093–1193* (Woodbridge: Boydell Press), pp. 41–51.

Burton, Janet 1999, *The Monastic Order in Yorkshire, 1069–1215* (Cambridge Studies in Medieval Life and Thought, 4th series, 40; Cambridge: Cambridge University Press).

Byrne, F. J. 1995, *Irish Kingship and High Kingship in the Early Middle Ages* (Woodbridge: Boydell Press).

Cambridge, Eric and Morris, Richard 1989, 'Beverley Minster before the early thirteenth century', in Christopher Wilson (ed.), *Medieval Art and Architecture in the East Riding of Yorkshire* (The British Archaeological Association Conference Transactions for the Year 1983; London: British Archaeological Association), pp. 9–32.

Cameron, Kenneth 1968, 'Eccles in English place-names', in Maurice W. Barley and R. P. C. Hanson (eds.), *Christianity in Britain, 300–700: Papers Presented to the Conference on Christianity in Roman and Sub-Roman Britain held at the University of Nottingham 17–20 April 1967* (Leicester: Leicester University Press), pp. 87–92.

Cameron, Kenneth 1977, *English Place-Names*, 3rd edn (London: Batsford).

Cameron, Neil 1994, 'St Rule's Church, St Andrews, and early stone-built churches in Scotland', *Proceedings of the Society of Antiquaries of Scotland* 124, 367–78.

Campbell, A. (ed.) 1938, *The Battle of Brunanburh* (London: Heinemann).

Campbell, A. (ed.) 1967, *Æthelwulf, De Abbatibus* (Oxford: Clarendon Press).

Campbell, James 1971, 'The first century of English Christianity', *Ampleforth Journal* 76, 12–29 (repr. in James Campbell, *Essays in Anglo-Saxon History* (London and Ronceverte: Hambledon, 1986), pp. 49–67).

Campbell, James 1973, 'Observations on the conversion of England', *Ampleforth Journal* 78, 12–26 (repr. in James Campbell, *Essays in Anglo-Saxon History* (London and Ronceverte: Hambledon, 1986), pp. 69–84).

Campbell, James 1979, 'Bede's words for places', in P. H. Sawyer (ed.), *Names, Words and Graves: Early Medieval Settlement* (Leeds: University of Leeds), pp. 34–54 (repr. in James Campbell, *Essays in Anglo-Saxon History* (London and Ronceverte: Hambledon, 1986), pp. 99–120).

Campbell, James 1980, *Bede's* Reges *and* Principes (Jarrow Lecture 1979; Jarrow: St Paul's Church, Jarrow) (repr. in *Bede and his World: The Jarrow Lectures 1958–93*, ed. Michael Lapidge, 2 vols. (Aldershot: Variorum, 1994), II, pp. 491–505; and James Campbell, *Essays in Anglo-Saxon History* (London and Ronceverte: Hambledon, 1986), pp. 85–98).

Campbell, James 1989, 'Elements in the background to the life of St Cuthbert and his early cult', in Gerald Bonner, David Rollason and Clare Stancliffe (eds.), *St Cuthbert, his Cult and his Community to AD 1200* (Woodbridge: Boydell Press), pp. 3–19.

Carver, M. O. H. 1995, 'Roman to Norman at York Minster', in Derek Phillips, Brenda Heywood and M. O. H. Carver (eds.), *Excavations at York Minster*, I: *From Roman Fortress to Norman Cathedral* (London: HMSO), pp. 177–221.

Carver, M. O. H. 1998, *Sutton Hoo: Burial Ground of Kings?* (London: British Museum Press).

Cassidy, Brendan (ed.) 1992, *The Ruthwell Cross: Papers from the Colloquium Sponsored by the Index of Christian Art, Princeton University, 8 December 1989* (Princeton: Index of Christian Art).

Chadwick, Hector Munro 1905, *Studies on Anglo-Saxon Institutions* (Cambridge: Cambridge University Press) (repr. New York: Russell and Russell, 1963).

Chadwick, Nora Kershaw 1955, *Poetry and Letters in Early Christian Gaul* (London: Bowes and Bowes).

Chadwick, Nora Kershaw 1963a, 'Bede, St Colmán and the Irish abbey of Mayo', in Nora Kershaw Chadwick (ed.), *Celt and Saxon: Studies in the Early British Border* (Cambridge: Cambridge University Press), pp. 186–205.

Chadwick, Nora Kershaw 1963b, 'The conversion of Northumbria: a comparison of sources', in Nora Kershaw Chadwick (ed.), *Celt and Saxon: Studies in the Early British Border* (Cambridge: Cambridge University Press), pp. 138–66.

Chaney, William A. 1970, *The Cult of Kingship in Anglo-Saxon England* (Manchester: Manchester University Press).

Chapman, John 1929, *Saint Benedict and the Sixth Century* (London: Sheed and Ward).

Charles-Edwards, Thomas M. 1983, 'Bede, the Irish and the Britons', *Celtica* 15, 42–52.

Charles-Edwards, Thomas M. 1989, 'Early medieval kingships in the British Isles', in Steven Bassett (ed.), *The Origins of Anglo-Saxon Kingdoms* (Leicester: Leicester University Press), pp. 28–39.

Charles-Edwards, Thomas M. 2000, *Early Christian Ireland* (Cambridge: Cambridge University Press).

Chase, Colin (ed.) 1981, *The Dating of Beowulf* (Toronto: University of Toronto Press).

Chitty, Derwas J. 1966, *The Desert a City: An Introduction to the Study of Egyptian and Palestinian Monasticism under the Christian Empire* (Blackwell: Oxford).

Clanchy, Michael 1979, *From Memory to Written Record: England 1066–1307*, 2nd edn (Oxford: Blackwell).

Coggins, D., Fairless, K. J. and Batey, C. E. 1983, 'Simy Folds: an early medieval settlement in Upper Teesdale, Co. Durham', *Medieval Archaeology* 27, 1–26.

Colgrave, Bertram (ed. and trans.) 1927, *The Life of Bishop Wilfrid by Eddius Stephanus* (Cambridge: Cambridge University Press).

Colgrave, Bertram (ed. and trans.) 1940, *Two Lives of St Cuthbert: A Life by an Anonymous Monk of Lindisfarne and Bede's Prose Life* (Cambridge: Cambridge University Press).

Colgrave, Bertram (ed. and trans.) 1956, *Felix's Life of Saint Guthlac: Introduction, Text, Translation and Notes* (Cambridge: Cambridge University Press).

Colgrave, Bertram and Mynors, R. A. B. (eds. and trans.) 1969, *Bede's Ecclesiastical History of the English People* (Oxford Medieval Texts; Oxford: Clarendon Press).

Correa, Alicia (ed.) 1992, *The Durham Collectar* (Henry Bradshaw Society 107; Woodbridge: Boydell Press).

Cowan, Ian B., Easson, David E. and Hadcock, R. Neville 1976, *Medieval Religious Houses, Scotland: With an Appendix on the Houses in the Isle of Man*, 2nd edn (London: Longman).

Cox, John Charles 1905, *The Royal Forests of Medieval England* (London: Methuen).

Cox, John Charles 1911, *The Sanctuaries and Sanctuary Seekers of Medieval England* (London: George Allen).

Craig, D. J. 1991, 'Pre-Norman sculpture in Galloway: some territorial implications', in R. D. Oram and G. P. Stell (eds.), *Galloway: Land and Lordship* (Edinburgh: Scottish Society for Northern Studies), pp. 45–62.

Cramp, R. J. 1976, 'Monastic sites', in D. M. Wilson (ed.), *The Archaeology of Anglo-Saxon England* (Cambridge: Cambridge University Press), pp. 201–52, 453–62.

Cramp, R. J. 1977, 'Schools of Mercian sculpture', in A. Dornier (ed.), *Mercian Studies* (Leicester: Leicester University Press), pp. 191–233 (repr. in Cramp (1992), pp. 174–216).

Cramp, R. J. 1984, *County Durham and Northumberland* (British Academy Corpus of Anglo-Saxon Stone Sculpture in England 1; Oxford: British Academy).

Cramp, R. J. 1988, 'Northumbria: the archaeological evidence', in Stephen T. Driscoll and Margaret R. Nieke (eds.), *Power and Politics in Early Medieval Britain and Ireland* (Edinburgh: Edinburgh University Press), pp. 69–78.

Cramp, R. J. 1989, 'The artistic influence of Lindisfarne within Northumbria', in Gerald Bonner, David Rollason and Clare Stancliffe (eds.), *St Cuthbert, his Cult and his Community to AD 1200* (Woodbridge: Boydell Press), pp. 213–28 (repr. in Cramp (1992), pp. 329–42).

Cramp, R. J. 1992, *Studies in Anglo-Saxon Sculpture* (London: Pindar Press).

Cramp, R. J. 1993, 'A reconsideration of the monastic site at Whitby', in R. M. Spearman and John Higgitt (eds.), *The Age of Migrating Ideas: Early Medieval Art in Northern Britain and Ireland: Proceedings of the Second International Conference on Insular Art held in the National Museums of Scotland in Edinburgh, 3–6 January 1991* (Edinburgh: National Museums of Scotland), pp. 64–73.

Cramp, R. J. 1994, 'Monkwearmouth and Jarrow in their continental context', in Kenneth Painter (ed.), *'Churches Built in Ancient Times': Recent Studies in Early Christian Archaeology* (Occasional Paper 16; London: Society of Antiquaries), pp. 279–94.

Cramp, R. J. 1999, 'The Northumbrian identity', in Jane Hawkes and Susan Mills (eds.), *Northumbria's Golden Age* (Stroud: Sutton), pp. 1–11.

Craster, E. 1954, 'The patrimony of St Cuthbert', *English Historical Review* 69, 177–99.

Cubitt, Catherine 1995, *Anglo-Saxon Church Councils c. 650–c. 850* (London and New York: Leicester University Press).

Cummins, W. A. 1995, *The Age of the Picts* (Stroud: Sutton).

Cummins, W. A. 1999, *The Picts and their Symbols* (Stroud: Sutton).

Curle, Alexander O. 1914, 'Report on the excavation, in September 1913, of a vitrified fort at Rockcliffe, Dalbeattie, known as the Mote of Mark', *Proceedings of the Society of Antiquaries of Scotland* 48, 125–68.

Curle, Alexander O. 1923, *The Treasure of Traprain: A Scottish Hoard of Roman Silver Plate* (Glasgow: Maclehose, Jackson and Co.).

Dalton, Paul 1994, *Conquest, Anarchy and Lordship: Yorkshire, 1066–1154* (Cambridge: Cambridge University Press).

Daniels, R. 1988, 'The Anglo-Saxon monastery at Church Close, Hartlepool', *Archaeological Journal* 145, 158–210.

Dark, Kenneth R. 1994, *Civitas to Kingdom: British Political Continuity 300–800* (Leicester: Leicester University Press).

Dark, Kenneth R. 2000, *Britain and the End of the Roman Empire* (Stroud: Tempus).

Davidson, Michael R. 2001, 'The (non)submission of the northern kings in 920', in Nicholas J. Higham and David Hill (eds.), *Edward the Elder 899–924* (London and New York: Routledge), pp. 200–11.

Davies, R. R. 1978, *Lordship and Society in the March of Wales, 1282–1400* (Oxford: Clarendon Press).

Davies, R. R. 1987, *Conquest, Coexistence, and Change: Wales, 1063–1415* (The History of Wales 2; Oxford: Clarendon Press).

Davies, R. R. 1994, 'The peoples of Britain and Ireland, 1100–1400: identities', *Transactions of the Royal Historical Society*, 6th series, 4, 1–20.

Davies, R. R. 1995, 'The peoples of Britain and Ireland, 1100–1400: II: names, boundaries and regnal solidarities', *Transactions of the Royal Historical Society*, 6th series, 5, 1–20.

Davies, R. R. 1996, 'The peoples of Britain and Ireland, 1100–1400: III: laws and customs', *Transactions of the Royal Historical Society*, 6th series, 6, 1–23.

Davies, R. R. 1997, 'The peoples of Britain and Ireland, 1100–1400: IV: language and historical mythology', *Transactions of the Royal Historical Society*, 6th series, 7, 1–24.

Davies, Wendy and Vierck, Hayo 1974, 'The contexts of the Tribal Hidage: social aggregates and settlement patterns', *Frühmittelalterliche Studien* 8, 223–93.

Davis, R. H. C. 1989, 'Bede after Bede', in C. Harper-Bill, C. Holdsworth and Janet L. Nelson (eds.), *Studies in Medieval History Presented to R. Allen Brown* (Woodbridge: Boydell Press), pp. 103–16.

Dawtry, A. 1982, 'The Benedictine revival in the north: the last bulwark of Anglo-Saxon monasticism?', *Studies in Church History* 19, 87–98.

Dickinson, Steve 1985, 'Bryant's Gill, Kentmere: another "Viking period" Ribblehead?', in John R. Baldwin and Ian D. Whyte (eds.), *The Scandinavians in Cumbria* (Edinburgh: Scottish Society for Northern Studies), pp. 83–8.

Dodgshon, Robert A. 1981, *Land and Society in Early Scotland* (Oxford: Clarendon Press).

Dodgson, J. McN. 1957, 'The background of Brunanburh', *Saga Book of the Viking Society* 14, 303–16.

Dolley, R. H. M. 1965, *The Viking Coins of the Danelaw and Dublin* (London: British Museum).

Downham, Clare forthcoming, 'Eric Bloodaxe – axed? The mystery of the last Viking-king of York', *Medieval Scandinavia* 14.

Dubois, Jacques 1978, *Les martyrologes du moyen âge latin* (Typologie des sources du moyen âge occidental 26; Turnhout: Brepols).

Dugdale, William 1817–30, *Monasticon Anglicanum: A History of the Abbies and Other Monasteries, Hospitals, Frieries, and Cathedral and Collegiate Churches with their Dependencies in England and Wales*, 8 vols. (London: Longman, Hurst, Rees, Orme and Brown, Lackington, Hughes, Harding, Mavor and Lepard, and Joseph Harding).

Dümmler, Ernst (ed.) 1895, *Epistolae Karolini Aevi, Tomus II* (Monumenta Germaniae Historica, Epistolarum Tomus IV, Karolini Aevi II; Berlin: Weidmann).

Dumville, D. N. 1975–6, 'Nennius and the *Historia Brittonum*', *Studia Celtica* 10–11, 78–95.

Dumville, D. N. 1976, 'The Anglian collection of royal genealogies and regnal lists', *Anglo-Saxon England* 5, 23–50.

Dumville, D. N. 1977, 'Palaeographical considerations in the dating of early Welsh verse', *Bulletin of the Board of Celtic Studies* 27, 246–51.

Dumville, D. N. 1986, 'The historical value of the *Historia Brittonum*', *Arthurian Literature* 6, 1–26 (repr. in his *Histories and Pseudo-Histories of the Insular Middle Ages* (Aldershot: Variorum, 1990), no. VII).

Dumville, D. N. 1987, 'Textual archaeology and Northumbrian history subsequent to Bede', in D. M. Metcalf (ed.), *Coinage in Ninth-Century Northumbria: The Tenth Oxford Symposium on Coinage and Monetary History* (British Archaeological Reports, British Series 180; Oxford: BAR), pp. 43–55.

Dumville, D. N. 1989a, 'The origins of Northumbria: some aspects of the British evidence', in Steven R. Bassett (ed.), *The Origins of Anglo-Saxon Kingdoms* (Studies in the Early History of Britain; London and New York: Leicester University Press), pp. 213–22.

Dumville, D. N. 1989b, 'The Tribal Hidage: an introduction to its texts and their history', in Steven R. Bassett (ed.), *The Origins of Anglo-Saxon Kingdoms* (Studies in the Early History of Britain; London and New York: Leicester University Press), pp. 225–30.

Dumville, D. N. 1997, 'The terminology of over-kingship in early Anglo-Saxon England', in John Hines (ed.), *The Anglo-Saxons from the Migration Period to the Eighth Century: An Ethnographic Perspective* (Woodbridge: Boydell Press), pp. 345–74.

Dumville, D. N. 1999, *A Palaeographer's Review: The Insular System of Scripts in the Early Middle Ages* (Kansai University Institute of Oriental and Occidental Studies Sources and Materials Series 20; Suita: Kansai University Press).

Duncan, Archibald A. M. 1975, *Scotland: The Making of the Kingdom* (Edinburgh: Oliver and Boyd).

Duncan, Archibald A. M. 1976, 'The Battle of Carham, 1018', *Scottish Historical Review* 55, 20–8.

Eagles, B. N. 1979, *The Anglo-Saxon Settlement of Humberside* (British Archaeological Reports, British Series 68; Oxford: BAR).

Eagles, B. N. 1989, 'Lindsey', in Steven R. Bassett (ed.), *The Origins of Anglo-Saxon Kingdoms* (Studies in the Early History of Britain; London and New York: Leicester University Press), pp. 202–12.

Ekwall, Eilert 1928, *English River-Names* (Oxford: Clarendon Press).

Ekwall, Eilert 1960, *The Concise Oxford Dictionary of English Place-Names*, 4th edn (Oxford: Clarendon Press).

Elton, Hugh 1996, *Frontiers of the Roman Empire* (London: Batsford).

Emerton, Ephraim (ed.) 2000, *The Letters of Saint Boniface*, 2nd edn (Records of Western Civilisation; New York: Columbia University Press) with a new introduction and bibliography by Thomas F. X. Noble.

Enright, Michael J. 1985, *Iona, Tara, and Soissons: The Origin of the Royal Anointing Ritual* (Arbeiten zur Frühmittelalterforschung 17; Berlin and New York: De Gruyter).

Etchingham, Colmán 1999, *Church Organisation in Ireland AD 650 to 1000* (Maynooth: Laigin Publications).

Evans, Jeremy 2000, 'The end of Roman pottery in the north', in Tony Wilmott and Pete Wilson (eds.), *The Late Roman Transition in the North: Papers from the Roman Archaeology Conference, Durham 1999* (British Archaeological Reports, British Series 299; Oxford: Archaeopress), pp. 39–46.

Ewig, Eugen 1963, 'Résidence et capitale pendant le haut moyen âge', *Revue historique* 230, 25–72 (repr. in Eugen Ewig, *Spätantikes und fränkisches Gallien: Gesammelte Schriften (1952–1973)*, ed. Hartmut Atsma, 2 vols. (Zurich and Munich: Artemis, 1976–9), I, pp. 362–48).

Ewig, Eugen 1967, 'Descriptio Franciae', in Wolfgang Braunfels (ed.), *Karl der Grosse: Lebenswerk und Nachleben*, I: *Persönlichkeit und Geschichte* (Düsseldorf: Schwann), pp. 143–77 (repr. in Eugen Ewig, *Spätantikes und fränkisches Gallien: Gesammelte Schriften (1952–1973)*, ed. Hartmut Atsma, 2 vols. (Zurich and Munich: Artemis, 1976–9), I, pp. 274–332).

Fairclough, G. 1983, 'Tynemouth Priory and Castle – excavations in the outer court, 1980', *Archaeologia Aeliana*, 5th series, 11, 101–33.

Fanning, S. 1991, 'Bede, *imperium*, and the bretwaldas', *Speculum* 66, 1–26.

Faull, Margaret Lindsay 1977, 'British survival in Anglo-Saxon Northumbria', in Lloyd Laing (ed.), *Studies in Celtic Survival* (British Archaeological Reports, British Series 37; Oxford: BAR), pp. 1–55.

Faull, Margaret Lindsay and Moorhouse, S. A. (eds.) 1981, *West Yorkshire: An Archaeological Survey to AD 1500*, 4 vols. (Wakefield: West Yorkshire Metropolitan County Council).

Feilitzen, Olof von 1937, *The Pre-Conquest Personal Names of Domesday Book* (Acta Universitatis Upsaliensis. Nomina Germanica. Arkiv för germansk namnforskning 3; Uppsala: Almqvist and Wiksell).

Fellows-Jensen, Gillian 1968, *Scandinavian Personal Names in Lincolnshire and Yorkshire* (Copenhagen: I kommission hos Akademisk forlog).

Fellows-Jensen, Gillian 1983, 'Lancashire and Yorkshire names', *Northern History* 19, 231–7.

Fellows-Jensen, Gillian 1985a, 'Scandinavian settlement in Cumbria and Dumfriesshire: the place-name evidence', in John R. Baldwin and Ian D. Whyte (eds.), *The Scandinavians in Cumbria* (Edinburgh: Scottish Society for Northern Studies), pp. 65–82.

Fellows-Jensen, Gillian 1985b, *Scandinavian Settlement Names in the North-West* (Copenhagen: C. A. Reitels).

Fellows-Jensen, Gillian 1992, 'Scandinavian place-names of the Irish Sea province', in James Graham-Campbell (ed.), *Viking Treasure from the North West: The Cuerdale Hoard in its Context* (National Museums and Galleries on Merseyside Occasional Papers 5; Liverpool: Liverpool Museum), pp. 31–42.

Fellows-Jensen, Gillian 1995, 'Scandinavian settlement in Yorkshire: through a rear-view mirror', in Barbara Crawford (ed.), *Scandinavian Settlement in Northern Britain: Thirteen Studies of Place-Names in their Historical Context* (Studies in the Early History of Britain; London and New York: Leicester University Press), pp. 170–86.

Fernie, Eric 1983, *The Architecture of the Anglo-Saxons* (London: Batsford).

Ferris, Iain and Jones, Rick 2000, 'Transforming an elite: reinterpreting late Roman Binchester', in Tony Wilmott and Pete Wilson (eds.), *The Late Roman Transition in the North: Papers from the Roman Archaeology Conference, Durham 1999* (British Archaeological Reports, British Series 299; Oxford: Archaeopress), pp. 1–11.

Fletcher, Richard 1997, *The Conversion of Europe from Paganism to Christianity* (London: HarperCollins).

Fletcher, Richard 2002, *Bloodfeud: Murder and Revenge in Anglo-Saxon England* (London: Penguin).

Foley, W. T. 1992, *Images of Sanctity in Eddius Stephanus' Life of Bishop Wilfrid, an Early English Saint's Life* (Lampeter: Edward Mellen).

Folz, Robert 1984, *Les saints rois du moyen âge en Occident (VIe–XIIIe siècles)* (Subsidia hagiographica 68; Brussels: Société des Bollandistes).

Foot, Sarah 1992, 'Anglo-Saxon minsters: a review of terminology', in John Blair and Richard Sharpe (eds.), *Pastoral Care before the Parish* (Leicester, London and New York: Leicester University Press), pp. 212–25.

Fox, Cyril 1955, *Offa's Dyke: A Field Survey of the Western Frontier-works of Mercia in the Seventh and Eighth Centuries AD* (London: Oxford University Press).

Frazer, James George 1922, *The Golden Bough: A Study in Magic and Religion* (London: Macmillan).

Fry, Timothy (ed.) 1981, *RB 1980: The Rule of St Benedict in Latin and English with Notes* (Collegeville, Minn.: Liturgical Press).

Fryde, E. B., Greenway, D. E., Porter, S. et al. (eds.) 1986, *Handbook of British Chronology*, 3rd edn (Royal Historical Society Guides and Handbooks 2; London: Royal Historical Society).

Gates, T. and O'Brien, C. F. 1988, 'Cropmarks at Milfield and New Bewick and the recognition of grubenhäuser in Northumberland', *Archaeologia Aeliana*, 5th series 16, 1–9.

Gauert, A. 1965, 'Das Itinerar Karls des Großen', in Wolfgang Braunfels (ed.), *Karl der Große: Werk und Wirkung* (Aachen: Schwann), pp. 14–18.

Geake, Helen 1997, *The Use of Grave-Goods in Conversion-Period England, c. 600–c. 850* (British Archaeological Reports, British Series 261; Oxford: BAR).

Geary, Patrick J. 2002, *The Myth of Nations: The Medieval Origins of Europe* (Princeton and Oxford: Princeton University Press).

Gelling, Margaret 1997, *Signposts to the Past: Place-Names and the History of England*, 3rd edn (Chichester: Phillimore).

Gerchow, Jan 1988, *Die Gedenküberlieferung der Angelsachsen mit einem Katalog der Libri Vitae und Necrologien* (Arbeiten zur Frühmittelalterforschung: Schriftenreihe des Instituts für Frühmittelalterforschung der Universität Münster 20; Berlin and New York: Walter de Gruyter).

Gibbs, M. 1973, 'The decrees of Agatho and the Gregorian plan for York', *Speculum* 48, 213–46.

Goffart, Walter 1980, *Barbarians and Romans AD 418–584: The Techniques of Accommodation* (Princeton: Princeton University Press).

Goffart, Walter 1981, 'Rome, Constantinople and the barbarians', *American Historical Review* 86, 275–306 (repr. in Walter Goffart, *Rome's Fall and After* (London and Ronceverte: Hambledon Press, 1989), pp. 1–32).

Graham, A. 1959, 'Guidi', *Antiquity* 33, 63–5.

Graham, Frank 1983, *The Outpost Forts of Hadrian's Wall in the Days of the Romans* (Newcastle-upon-Tyne: Frank Graham).

Gransden, A. 1974, *Historical Writing in England c. 550 to c. 1307* (London: Routledge and Kegan Paul).

Grierson, Philip and Blackburn, Mark 1986, *Medieval European Coinage with a Catalogue of the Coins in the Fitzwilliam Museum, Cambridge*, I: *The Early Middle Ages (5th–10th Centuries)* (Cambridge: Cambridge University Press).

Griffiths, David 2001, 'The north-west frontier', in Nicholas J. Higham and David Hill (eds.), *Edward the Elder 899–924* (London and New York: Routledge), pp. 167–87.

Haddan, Arthur West and Stubbs, William (eds.) 1869–71, *Councils and Ecclesiastical Documents relating to Great Britain and Ireland*, 3 vols. (Oxford: Oxford University Press).

Hadley, Dawn M. 2000, *The Northern Danelaw: Its Social Structure, c. 800–1100* (Studies in the Early History of Britain; London and New York: Leicester University Press).

Hadley, Dawn M. and Richards, Julian D. (eds.) 2000, *Cultures in Contact: Scandinavian Settlement in England in the 9th and 10th Centuries* (Studies in the Early Middle Ages 2; Turnhout: Brepols).

Härke, Heinrich 1990, '"Warrior graves?" The background of the Anglo-Saxon weapon burial rite', *Past and Present* 126, 22–43.

Hall, David 1984, 'The Community of St Cuthbert: its properties and claims from the ninth century to the twelfth', PhD Dissertation, University of Oxford.

Hall, David 1989, 'The sanctuary of St Cuthbert', in Gerald Bonner, David Rollason and Clare Stancliffe (eds.), *St Cuthbert, his Cult and his Community to AD 1200* (Woodbridge: Boydell Press), pp. 425–36.

Hall, K. Mary 1993, 'Pre-Conquest estates in Yorkshire', in H. E. Jean Le Patourel, Moira H. Long, and May F. Pickles (eds.), *Yorkshire Boundaries* (Leeds: Yorkshire Archaeological Society), pp. 25–38.

Hall, Richard A. 1978, 'The topography of Anglo-Scandinavian York', in R. A. Hall (ed.), *Viking Age York and the North* (London: Council for British Archaeology), pp. 31–6.

Hall, Richard A. 1984a, 'A late pre-Conquest building tradition', in Peter V. Addyman and Valerie E. Black (eds.), *Archaeological Papers from York Presented to M. W. Barley* (York: York Archaeological Trust), pp. 71–7.

Hall, Richard A. 1984b, *The Viking Dig: The Excavations at York* (London: Bodley Head).

Hall, Richard A. 1988, 'The making of Domesday York', in D. Hooke (ed.), *Anglo-Saxon Settlements* (Oxford: Blackwell), pp. 233–47.

Hall, Richard A. 1989, 'The Five Boroughs of the Danelaw: a review of present knowledge', *Anglo-Saxon England* 18, 149–206.

Hall, Richard A. 1992, 'Archaeological introduction', in Patrick Ottaway (ed.), *Anglo-Scandinavian Ironwork from 16–22 Coppergate* (Archaeology of York 17.1; York: York Archaeological Trust), pp. 455–63.

Hall, Richard A. 1994, *English Heritage Book of Viking Age York* (London: Batsford/English Heritage).

Hall, Richard A. (ed.) forthcoming, *Anglo-Scandinavian York (AD 876–1066)* (Archaeology of York 8; York: York Archaeological Trust).

Hamilton Thompson, A. (ed.) 1935, *Bede: His Life, Times and Writings: Essays in Commemoration of the Twelfth Centenary of his Death* (Oxford: Oxford University Press).

Harmer, Florence Elizabeth (ed.) 1914, *Select English Historical Documents of the Ninth and Tenth Centuries* (Cambridge: Cambridge University Press).

Harrison, Kenneth 1976, *The Framework of Anglo-Saxon History to AD 900* (Cambridge: Cambridge University Press).

Hart, Cyril R. 1975, *The Early Charters of Northern England and the North Midlands* (Leicester: Leicester University Press).

Hart, Cyril R. 1992, *The Danelaw* (London and Rio Grande: Hambledon Press).

Haughton, Christine and Powlesland, Dominic 1999, *West Heslerton: The Anglian Cemetery*, 2 vols. (Yedingham: The Landscape Research Centre).

Hawkes, S. C. and Wells, C. 1975, 'Crime and punishment in an Anglo-Saxon cemetery?', *Antiquity* 49, 118–22.

Haywood, John 1995, *The Penguin Historical Atlas of the Vikings* (London: Penguin).

Henderson, George 1987, *From Durrow to Kells: The Insular Gospel-Books 650–800* (London: Thames and Hudson).

Henderson, Isabel B. 1967, *The Picts* (Ancient Peoples and Places 54; London: Thames and Hudson).

Herity, Michael 1995, *Studies in the Layout, Buildings and Art in Stone of Early Irish Monasteries* (London: Pindar Press).

Higham, Nicholas J. 1995, *An English Empire: Bede and the Early Anglo-Saxon Kings* (Manchester and New York: Manchester University Press).

Hill, David 1969, 'The Burghal Hidage: the establishment of a text', *Medieval Archaeology* 13, 84–92.

Hill, David 1981, *An Atlas of Anglo-Saxon England* (Oxford: Blackwell).

Hill, David and Rumble, Alex (eds.) 1996, *The Defence of Wessex: The Burghal Hidage and Anglo-Saxon Fortifications* (Manchester: Manchester University Press).

Hill, Peter 1997, *Whithorn and St Ninian: The Excavation of a Monastic Town 1984–91* (Stroud: The Whithorn Trust/Sutton Publishing).

Hill, R. M. T. and Brooke, C. N. L. 1977, 'From 627 until the early thirteenth century', in G. E. Aylmer and Reginald Cant (eds.), *A History of York Minster* (Oxford: Oxford University Press), pp. 1–41.

Hind, J. G. F. 1980, 'Elmet and Deira – forest names in Yorkshire?', *Bulletin of the Board of Celtic Studies* 28, 541–52.

Hines, John 1991, 'Scandinavian English: a creole in context', in P. Sture Ureland and George Broderick (eds.), *Language Contact in the British Isles: Proceedings of the Eighth International Symposium on Language Contact in Europe, Douglas, Isle of Man, 1988* (Linguistische Arbeiten 238; Tübingen: Max Niermeyer), pp. 403–27.

Hines, John (ed.) 1997, *The Anglo-Saxons from the Migration Period to the Eighth Century: An Ethnographic Perspective* (Woodbridge: Boydell Press).

Hirst, Susan M. 1985, *An Anglo-Saxon Cremation Cemetery at Sewerby, East Yorkshire* (York University Archaeological Publications 4; York: University of York).

Hirst, Susan M. 1993, 'Death and the archaeologist', in M. O. H. Carver (ed.), *In Search of Cult: Archaeological Investigations in Honour of Philip Rahtz* (York Archaeological Papers; Woodbridge: Boydell Press), pp. 41–3.

Hlawitschka, Eduard 1967, 'Die Vorfahren Karls des Grossen', in Wolfgang Braunfels (ed.), *Karl der Grosse: Lebenswerk und Nachleben, I: Persönlichkeit und Geschichte* (Düsseldorf: Schwann), pp. 51–82.

Hodges, Richard 1982, *Dark Age Economics: The Origins of Towns and Trade AD 600–1000* (London: Duckworth).

Hodges, Richard and Hobley, B. (eds.) 1988, *The Rebirth of Towns in the West AD 700–1050* (London: Council for British Archaeology).

Hoffmann, Erich 1975, *Die heiligen Könige bei den Angelsachsen und den skandinavischen Völkern: Königsheiliger und Königshaus* (Quellen und Forschungen zur Geschichte Schleswig-Holsteins 69; Neumünster: K. Wachholtz).

Hogg, Richard M. 1992, *The Cambridge History of the English Language, I: The Beginnings to 1066* (Cambridge: Cambridge University Press).

Hooke, Della 1998, *The Landscape of Anglo-Saxon England* (London: Leicester University Press).

Hope-Taylor, Brian 1977, *Yeavering: An Anglo-British Centre of Early Northumbria* (Department of the Environment Archaeological Reports 7; London: HMSO).

Hope-Taylor, Brian 1983, 'Balbridie and Doon Hill', *Current Archaeology* 72, 18–19.

Horn, Walter, White Marshall, Jenny, Rourke, Grellan D. et al. 1990, *The Forgotten Hermitage of Skellig Michael* (California Studies in the History of Art, Discovery Series 2; Berkeley: University of California Press).

Hough, Carole A. 1997, 'The earliest Old English place-names in Scotland', *Nomina* 6, 23–30.

Hough, Carole A. 2001, 'The place-name Penninghame (Wigtownshire)', *Notes and Queries* 48.2, 99–102.

Howe, Nicholas 1989, *Migration and Mythmaking in Anglo-Saxon England* (New Haven and London: Yale University Press).

Howlett, David R. 2000, *Caledonian Craftsmanship: The Scottish Latin Tradition* (Dublin: Four Courts).

Hughes, Kathleen 1966, *The Church in Early Irish Society* (London: Methuen).

Hughes, Kathleen 1971, 'Evidence for contacts between the churches of the Irish and the English', in Peter Clemoes and Kathleen Hughes (eds.), *England before the Conquest: Studies in Primary Sources Presented to Dorothy Whitelock* (Cambridge: Cambridge University Press), pp. 49–67.

Ireland, Colin A. (ed.) 1999, *Old Irish Wisdom Attributed to Aldfrith of Northumbria: An Edition of Bríathra Flainn Fhína maic Ossu* (Tempe: Arizona Center for Medieval and Renaissance Studies).

Isaac, Benjamin 1992, *The Limits of Empire: The Roman Army in the East*, rev. edn (Oxford: Clarendon Press).

Jackson, Kenneth Hurlstone 1953, *Language and History in Early Britain: A Chronological Survey of the Brittonic Languages, First to Twelfth Century AD* (Edinburgh: Edinburgh University Press).

Jackson, Kenneth Hurlstone 1955, 'The Pictish language', in F. T. Wainwright (ed.), *The Problem of the Picts* (Edinburgh and London: Nelson), pp. 129–66.

Jackson, Kenneth Hurlstone 1958, 'The sources of the Life of St Kentigern', in Nora Kershaw Chadwick (ed.), *Studies in the Early British Church* (Cambridge: Cambridge University Press), pp. 273–357.

Jackson, Kenneth Hurlstone 1959, 'Edinburgh and the Anglian occupation of Lothian', in Peter Clemoes (ed.), *The Anglo Saxons: Studies in Some Aspects of their History and Culture, Presented to Bruce Dickins* (London: Bowes and Bowes), pp. 35–42.

Jackson, Kenneth Hurlstone 1963a, 'Angles and Britons in Northumbria and Cumbria', in H. Lewis (ed.), *Angles and Britons* (O'Donnell Lectures; Cardiff: University of Wales Press), pp. 60–84.

Jackson, Kenneth Hurlstone 1963b, 'On the northern British section of Nennius', in Nora Kershaw Chadwick (ed.), *Celt and Saxon: Studies in the Early British Border* (Cambridge: Cambridge University Press), pp. 20–62.

Jackson, Kenneth Hurlstone 1981, 'Varia I: Bede's Urbs Guidi: Stirling or Cramond?', *Cambridge Medieval Celtic Studies* 2, 1–7.

James, Edward 1988, *The Franks* (Oxford: Blackwell).

James, Edward and Heywood, Brenda 1995, 'York in the Roman and early medieval periods: a context for the excavations', in Derek Phillips, Brenda Heywood and M. O. H. Carver (eds.), *Excavations at York Minster* (London: HMSO), I, pp. 1–15.

James, Simon, Marshall, Anne and Millett, Martin 1984, 'An early medieval building tradition', *Archaeological Journal* 141, 182–215.

Jansen, Annemiek 1995, 'The development of the St Oswald legends on the continent', in Clare Stancliffe and Eric Cambridge (eds.), *Oswald: Northumbrian King to European Saint* (Stamford: Paul Watkins), pp. 230–40.

Jarman, A. O. H. (ed.) 1988, *Aneirin: Y Gododdin. Britain's Oldest Heroic Poem* (Welsh Classics 3; Llandysul: Gomer).

Jobey, G. 1965, 'Hill forts and settlements in Northumberland', *Archaeologia Aeliana*, 4th series, 43, 21–64.

John, Eric 1964, *Land Tenure in Early England: A Discussion of Some Problems* (Leicester: Leicester University Press).

John, Eric 1966a, 'English feudalism and the structure of Anglo-Saxon Society', in Eric John, *Orbis Britanniae and Other Studies* (Leicester: Leicester University Press), pp. 128–53.

John, Eric 1966b, 'Folkland reconsidered', in Eric John, *Orbis Britanniae and Other Studies* (Leicester: Leicester University Press), pp. 64–127.

John, Eric 1966c, '"Orbis Britanniae" and the Anglo-Saxon kings', in Eric John, *Orbis Britanniae and Other Studies* (Leicester: Leicester University Press), pp. 1–63.

Johnson South, Ted (ed. and trans.) 2002, Historia de Sancto Cuthberto: *A History of Saint Cuthbert and a Record of his Patrimony* (Anglo-Saxon Texts 3; Cambridge: D. S. Brewer).

Jolliffe, J. E. A. 1926, 'Northumbrian institutions', *English Historical Review* 41, 1–42.

Jones, C. W. (ed.) 1943, *Bedae opera de temporibus* (Cambridge, Mass.: Mediaeval Academy of America).

Jones, C. W. 1994, *Bede, the Schools and the Computus* (Aldershot: Variorum).

Jones, G. J. R. 1975, 'Early territorial organization in Gwynedd and Elmet', *Northern History* 10, 3–27.

Jones, G. J. R. 1979, 'Multiple estates and early settlement', in Peter Hayes Sawyer (ed.), *English Medieval Settlement* (London: Edward Arnold), pp. 9–34.

Jones, G. J. R. 1990, '"Broninis"', *Bulletin of the Board of Celtic Studies* 37, 125–32.

Jones, G. J. R. 1995, 'Some donations to Bishop Wilfrid in northern England', *Northern History* 31, 22–38.

Jones, Putnam Fennell 1929, *A Concordance to the Historia ecclesiastica of Bede* (Cambridge, Mass.: Mediaeval Academy of America).

Kapelle, William E. 1979, *The Norman Conquest of the North: The Region and its Transformation 1000–1135* (London: Croom Helm).

Kemp, R. L. 1999, *Anglian Settlement at 46–54 Fishergate* (Archaeology of York 7/1; London: York Archaeological Trust).

Kendrick, T. D., Brown, T. J., Bruce-Mitford, R. L. S. et al. (eds.) 1956–60, *Evangeliorum Quattuor Codex Lindisfarnensis*, 2 vols. (Olten and Lausanne: Urs Graf Verlag).

Kenyon, Denise 1991, *The Origins of Lancashire* (Origins of the Shire; Manchester: Manchester University Press).

Keynes, Simon 1993, 'The control of Kent in the ninth century', *Early Medieval Europe* 2, 111–31.

Keynes, Simon 1998, *An Atlas of Attestations in Anglo-Saxon Charters, c. 670–1066*, 2nd edn (Cambridge: Department of Anglo-Saxon, Norse and Celtic, University of Cambridge).

Keynes, Simon and Lapidge, Michael (trans.) 1983, *Alfred the Great: Asser's Life of King Alfred and Other Contemporary Sources* (Harmondsworth: Penguin).

Kirby, D. P. 1962, 'Strathclyde and Cumbria: a survey of historical development to 1092', *Transactions of the Cumberland and Westmorland Antiquarian Society*, new series, 62, 77–94.

Kirby, D. P. 1974a, 'The kingdom of Northumbria and the destruction of the Votadini', *Transactions of the East Lothian Ant. and Field Naturalists' Society* 14, 1–13.

Kirby, D. P. 1974b, 'Northumbria in the time of Wilfrid', in D. P. Kirby (ed.), *St Wilfrid at Hexham* (Newcastle upon Tyne: Oriel), pp. 1–34.

Kirby, D. P. 1991, *The Earliest English Kings* (London: Unwin Hyman).

Kitzinger, Ernst 1965, 'The coffin-reliquary', in C. F. Battiscombe (ed.), *The Relics of St Cuthbert* (Oxford: Oxford University Press), pp. 202–304.

Knowles, David 1963, *The Monastic Order in England: A History of its Development from the Times of St Dunstan to the Fourth Lateran Council, 940–1216*, 2nd edn (Cambridge: Cambridge University Press).

Koch, John T. (ed.) 1997, *The Gododdin of Aneirin: Text and Context in Dark-Age North Britain* (Cardiff: University of Wales Press).

Krautheimer, Richard 1980, *Rome: Profile of a City, 312–1303* (Princeton: Princeton University Press).

Krüger, Karl Heinrich 1971, *Königsgrabkirchen der Franken, Angelsachsen und Langobarden zur Mitte des 8. Jahrhunderts: Ein historischer Katalog* (Münstersche Mittelalter-Schriften 4; Munich: Wilhelm Fink).

Kydd, Dafydd 1976, 'Review of *The Anglo-Saxon Cremation Cemetery at Sancton, East Yorkshire* by J. N. L. Myres and W. H. Southern', *Medieval Archaeology* 20, 202–4.

Laing, Lloyd and Longley, David forthcoming, *The Mote of Mark: A Dark Age Hillfort in South-West Scotland* (Edinburgh: Society of Antiquaries of Scotland).

Laistner, M. L. W. 1935, 'The library of the Venerable Bede', in Alexander Hamilton Thompson (ed.), *Bede: His Life, Times and Writings: Essays in Commemoration of the Twelfth Centenary of his Death* (Oxford: Clarendon Press), pp. 237–66.

Lane, Alan and Campbell, Ewan 2000, *Dunadd: An Early Dalriadic Capital* (Oxford: Oxbow).

Lang, James T. 1984, 'The hogback: a Viking colonial monument', *Anglo-Saxon Studies in Archaeology and History* 3, 85–176.

Lang, James T. 1991, *York and Eastern Yorkshire* (British Academy Corpus of Anglo-Saxon Stone Sculpture in England 3; Oxford: Oxford University Press).

Lang, James T. 1999, 'The imagery of the Franks Casket: another approach', in Jane Hawkes and Susan Mills (eds.), *Northumbria's Golden Age* (Stroud: Sutton), pp. 247–55.

Lang, James T. 2001, *Corpus of Anglo-Saxon Stone Sculpture VI: Northern Yorkshire* (Oxford: Oxford University Press).

Lapidge, Michael 1984, 'Gildas's education and the Latin culture of sub-Roman England', in Michael Lapidge and David N. Dumville (eds.), *Gildas: New Approaches* (Woodbridge: Boydell Press), pp. 27–50.

Lapidge, Michael 1986, 'The school of Theodore and Hadrian', *Anglo-Saxon England* 15, 45–72.

Lapidge, Michael 1989, 'Bede's Metrical *Vita sancti Cuthberti*', in Gerald Bonner, David Rollason and Clare Stancliffe (eds.), *St Cuthbert, his Cult and his Community to AD 1200* (Woodbridge: Boydell Press), pp. 77–93.

Lapidge, Michael 1996, *Anglo-Latin Literature 600–899* (London and Rio Grande: Hambledon).

Lapidge, Michael and Bischoff, Bernhard (eds.) 1994, *Biblical Commentaries from the Canterbury School of Theodore and Hadrian* (Cambridge Studies in Anglo-Saxon England 10; Cambridge: Cambridge University Press).

Lapidge, Michael and Herren, Michael (eds. and trans.) 1979, *Aldhelm, The Prose Works* (Cambridge: D. S. Brewer).

Leahy, Kevin 2000, 'Middle Anglo-Saxon metalwork from South Newbald and the "productive site" phenomenon in Yorkshire', in Helen Geake and Jonathan Kenny (eds.), *Early Deira: Archaeological Studies of the East Riding in the Fourth to Ninth Centuries AD* (Oxford and Oakville: Oxbow Books), pp. 51–82.

Lebecq, Stéphane 1983, *Marchands et navigateurs frisons du haut moyen âge*, 2 vols. (Lille: Presses universitaires de Lille).

Leech, R. H. and Newman, R. 1985, 'Excavations at Dacre 1982–4: an interim report', *Transactions of the Cumberland and Westmorland Architectural and Archaeological Society* 85, 87–93.

Lemarignier, Jean-François 1945, *Recherches sur l'hommage en marche et les frontières féodales* (Lille: Bibliothèque universitaire).

Lemarignier, Jean-François 1965, *Le gouvernement royal aux premiers temps capétiens (987–1108)* (Paris: Picard).

Levison, Wilhelm 1940, 'An eighth-century poem on St Ninian', *Antiquity* 14, 280–91.

Levison, Wilhelm 1947, *England and the Continent in the Eighth Century* (Oxford: Clarendon Press).

Liebermann, F. 1903–16, *Die Gesetze der Angelsachsen*, 3 vols. (Halle: Savigny-Stiftung).

Llewellyn, Peter 1971, *Rome in the Dark Ages* (London: Faber and Faber).

Longley, David 2001, 'The Mote of Mark: the archaeological context of the decorated metalwork', in Mark Redknap, Nancy Edwards, Susan Youngs et al. (eds.), *Pattern and Purpose in Insular Art: Proceedings of the Fourth International Conference on Insular Art Held at the National Museum and Gallery Cardiff 3–6 September 1998* (Oxford: Oxbow Books), pp. 75–90.

Loveluck, Christopher 1996, 'The development of the Anglo-Saxon landscape: economy and society "On Driffield", East Yorkshire, 400–750 AD', *Anglo-Saxon Studies in Archaeology and History* 9, 25–48.

Loveluck, Christopher 2002, '"The Romano-British to Anglo-Saxon transition" – social transformations from the late Roman to early medieval period in northern England, AD 400–700', in Catherine Brooks, Robin Daniels and Anthony Harding (eds.), *Past, Present and Future: The Archaeology of Northern England* (Research Report 5; Durham: Architectural and Archaeological Society of Durham and Northumberland), pp. 127–48.

Lowe, C. E. 1991, 'New light on the Anglian "Minster" at Hoddom', *Transactions of the Dumfriesshire and Galloway Natural History and Antiquarian Society*, 3rd series, 66, 11–35.

Lowe, C. E. 1993, 'Hoddom', *Current Archaeology* 135, 88–92.

Loyn, H. R. 1955, 'Gesiths and thegns from the seventh to the tenth century', *English Historical Review* 70, 529–49.

Lucy, Sam 1998, *The Early Anglo-Saxon Cemeteries of East Yorkshire: An Analysis and Reinterpretation* (British Archaeological Reports, British Series 272; Oxford: Archaeopress).

Lucy, Sam 1999, 'Changing burial rites in Northumbria AD 500–750', in Jane Hawkes and Susan Mills (eds.), *Northumbria's Golden Age* (Stroud: Sutton), pp. 12–43.

Lucy, Sam 2000a, *The Anglo-Saxon Way of Death: Burial Rites in Early England* (Stroud: Sutton).

Lucy, Sam 2000b, 'Early medieval burials in east Yorkshire: reconsidering the evidence', in Helen Geake and Jonathan Kenny (eds.), *Early Deira: Archaeological Studies of the East Riding in the Fourth to Ninth Centuries AD* (Oxford and Oakville: Oxbow Books), pp. 11–18.

Lund, Niels, Fell, Christine E., Crumlin-Pedersen, Ole et al. 1984, *Two Voyagers at the Court of King Alfred: The Ventures of Ohthere and Wulfstan, together with the Description of Northern Europe from the Old English Orosius* (York: Sessions).

Lyon, C. S. S. 1987, 'Ninth-century Northumbrian chronology', in D. M. Metcalf (ed.), *Coinage in Ninth-Century Northumbria: The Tenth Oxford Symposium on Coinage and Monetary History* (British Archaeological Reports, British Series 180; Oxford: BAR), pp. 27–38.

McAleer, J. Philip 2001, 'The north portal of Durham Cathedral and the problem of "sanctuary" in medieval Britain', *Antiquaries Journal* 81, 195–258.

Macalister, Robert Alexander Stewart 1945, *Corpus Inscriptionum Insularum Celticarum* (Dublin: Stationery Office; repr. Dublin: Four Courts Press and the Irish Manuscripts Commission, 1996).

McCarthy, Mike 2002, *Roman Carlisle and the Lands of the Solway* (Stroud: Tempus).

McClure, Judith and Collins, Roger (trans.) 1994, *Bede: The Ecclesiastical History of the English People, The Greater Chronicle, Bede's Letter to Egbert* (World's Classics; Oxford: Oxford University Press).

McKitterick, Rosamond (ed.) 1990, *The Uses of Literacy in Early Medieval Europe* (Cambridge: Cambridge University Press) (1990).

McKitterick, Rosamond (ed.) 1995, *The New Cambridge Medieval History, Volume II, c. 700–c. 900* (Cambridge: Cambridge University Press).

McNeill, John T. and Gamer, Helen M. (trans.) 1938, *Medieval Handbooks of Penance: A Translation of the Principal "Libri Poenitentiales" and Selections from Related Documents* (Records of Western Civilisation; New York: Columbia University Press).

McNeill, Peter G. B. and MacQueen, Hector L. (eds.) 1996, *Atlas of Scottish History to 1707* (Edinburgh: Scottish Medievalists and Department of Geography, University of Edinburgh).

MacQueen, John 1990, *St Nynia: With a Translation of the Miracles of Bishop Nynia* (Edinburgh: Edinburgh University Press).

Maitland, Frederick W. 1897, *Domesday Book and Beyond: Three Essays in the Early History of England* (Cambridge: Cambridge University Press).

Malone, K. (ed.) 1977, *Deor*, rev. edn (Exeter: Exeter University Press).

Mann, J. C. 1978, 'Hadrian's Wall: the last phases', in P. J. Casey (ed.), *The End of Roman Britain: Papers Arising from a Conference, Durham, 1978* (British Archaeological Reports, British Series 71; Oxford: BAR), pp. 144–51.

Markus, Robert A. 1976, *Bede and the Tradition of Ecclesiastical Historiography* (Jarrow Lecture 1975; Jarrow: St Paul's Church, Jarrow) (repr. in *Bede and his World: The Jarrow Lectures 1958–1993*, ed. Michael Lapidge, 2 vols. (Aldershot: Variorum, 1994), I, pp. 385–403).

Markus, Robert A. 1997, *Gregory the Great and his World* (Cambridge: Cambridge University Press).

Mattingly, H. (trans.) 1948, *Tacitus on Britain and Germany: A Translation of the 'Agricola' and the 'Germania'* (Harmondsworth: Penguin).

Mawer, A. 1920, *The Place-Names of Northumberland and Durham* (Cambridge: Cambridge University Press).

Mawer, A. 1923, 'The redemption of the Five Boroughs', *English Historical Review* 38, 551–7.

Mayr-Harting, Henry 1991, *The Coming of Christianity to Anglo-Saxon England*, 3rd edn (London: Batsford).

Meehan, Bernard 1975, 'Insiders and property in Durham around 1100', *Studies in Church History* 12, 45–58.

Meehan, Bernard 1976, 'The siege of Durham, the Battle of Carham and the cession of Lothian', *Scottish Historical Review* 55, 1–19.

Meehan, Bernard 1994a, *The Book of Kells: An Illustrated Introduction to the Manuscript in Trinity College Dublin* (London: Thames and Hudson).

Meehan, Bernard 1994b, 'Durham twelfth-century manuscripts in Cistercian houses', in David Rollason, Margaret Harvey and Michael Prestwich (eds.), *Anglo-Norman Durham 1093–1193* (Woodbridge: Boydell Press), pp. 439–49.

Meehan, Denis (ed.) 1958, *Adamnans De locis sanctis* (Scriptores Latini Hiberniae 3; Dublin: Dublin Institute for Advanced Studies).

Metcalf, D. M. (ed.) 1987a, *Coinage in Ninth-Century Northumbria: The Tenth Oxford Symposium on Coinage and Monetary History* (British Archaeological Reports, British Series 180; Oxford: BAR).

Metcalf, D. M. 1987b, 'The taxation of moneyers under Edward the Confessor and in 1086', in J. C. Holt (ed.), *Domesday Studies: Papers Read at the Novocentenary Conference of the Royal Historical Society and the Institute of British Geographers, Winchester, 1986* (Woodbridge: Boydell Press), pp. 279–93.

Miket, R. 1980, 'A re-statement of evidence for Bernician Anglo-Saxon burials', in P. A. Rahtz (ed.), *Anglo-Saxon Cemeteries* (British Archaeological Reports, British Series 82; Oxford: BAR), pp. 289–306.

Miller, Molly 1975, 'Historicity and the pedigrees of the northcountrymen', *Bulletin of the Board of Celtic Studies* 26, 255–80.

Miller, Molly 1976–8, 'The foundation legend of Gwynedd in the Latin texts', *Bulletin of the Board of Celtic Studies* 27, 515–32.

Moisl, Hermann 1983, 'The Bernician royal dynasty and the Irish in the seventh century', *Peritia* 2, 103–26.

Moore, Wilfrid J. 1937, *The Saxon Pilgrims to Rome and the Schola Saxonum* (Fribourg: Society of St Paul).

Moorhead, John 2001, *The Roman Empire Divided, 400–700* (Harlow: Longman).

Morris, Christopher D. 1977, 'Northumbria and the Viking settlement: the evidence for landholding', *Archaeologia Aeliana*, 5th series, 5, 81–104.

Morris, Christopher D. 1981, 'Viking and native in northern England: a case-study', in Hans Bekker-Nielsen, Peter Foote and Olaf Olsen (eds.), *Proceedings of the Eighth Viking Congress, Arhus, 24–31 August 1977* (Mediaeval Scandinavia Supplements 2; Odense: Odense University Press), pp. 223–44.

Morris, John 1974, 'Review of *The Anglo-Saxon Cemeteries of Caistor-by-Norwich and Markshall, Norfolk* by J. N. L. Myres and Barbara Green', *Medieval Archaeology* 18, 225–32.

Morris, Richard 1983, *The Church in British Archaeology* (Research Report 47; London: Council for British Archaeology).

Morris, Richard 1989, *Churches in the Landscape* (London: Dent).

Murison, David 1979, 'The historical background', in A. J. Aitken and Tom McArthur (eds.), *Languages of Scotland* (Edinburgh: Chambers), pp. 2–13.

Mynors, R. A. B., Thomson, R. M. and Winterbottom, Michael (eds. and trans.) 1998, *William of Malmesbury, Gesta Regum Anglorum, The History of the English Kings* (Oxford Medieval Texts; Oxford: Clarendon Press).

Myres, J. N. L. 1969, *Anglo-Saxon Pottery and the Settlement of England* (Oxford: Clarendon Press).

Myres, J. N. L. 1977, *A Corpus of Anglo-Saxon Pottery of the Pagan Period*, 2 vols. (Cambridge: Cambridge University Press).

Myres, J. N. L. and Southern, W. H. 1973, *The Anglo-Saxon Cremation Cemetery at Sancton, East Yorkshire* (Hull Museum Publications 218; Hull: Hull Museums).

Nash-Williams, V. E. 1950, *The Early Christian Monuments of Wales* (Cardiff: University of Wales Press).

Nelson, Janet L. 1973, 'Royal saints and early medieval kingship', *Studies in Church History* 10, 39–44.

Nelson, Janet L. 1987, 'The Lord's anointed and the people's choice: Carolingian royal ritual', in David Cannadine (ed.), *Rituals of Royalty: Power and Ceremonial in Traditional Societies* (Cambridge: Cambridge University Press), pp. 137–80 (repr. in Janet

L. Nelson, *The Frankish World 750–900* (London and Rio Grande: Hambledon), pp. 99–131).

Neuman de Vegvar, Carol L. 1987, *The Northumbrian Renaissance: A Study in the Transmission of Style* (Toronto and London: Associated University Presses).

Neville, Cynthia J. 1998, *Violence, Custom and Law: The Anglo-Scottish Border Lands in the Later Middle Ages* (Edinburgh: Edinburgh University Press).

Nicasie, M. J. 1998, *Twilight of Empire: The Roman Army from the Reign of Diocletian until the Battle of Adrianople* (Amsterdam: J. C. Gieben).

Nicholl, Donald 1964, *Thurstan, Archbishop of York (1114–1140)* (York: Stonegate Press).

Nicolaisen, W. F. H. 2001, *Scottish Place-Names: Their Study and Significance*, rev. edn (Edinburgh: John Donald).

Nordal, S. (ed.) 1933, *Egils Saga Skalla-Grímssonar* (Islenzk Fornrit 2; Reykjavík: Hid Islenzka Fornritafelag).

Nordenfalk, Carl 1977, *Celtic and Anglo-Saxon Painting* (London: Chatto and Windus).

North, J. J. 1994, *English Hammered Coinage, Volume I, Early Anglo-Saxon to Henry III, c. 600–1272*, 3rd edn (London: Spink).

Norton, Christopher 1998a, 'The Anglo-Saxon cathedral at York and the topography of the Anglian city', *Journal of the British Archaeological Association* 151, 1–42.

Norton, Christopher 1998b, 'History, wisdom and illumination', in David Rollason (ed.), *Symeon of Durham: Historian of Durham and the North* (Stamford: Paul Watkins), pp. 61–105.

O'Brien, Colm and Miket, Roger 1991, 'The early medieval settlement of Thirlings, Northumberland', *Durham Archaeological Journal* 7, 57–91.

O Carragáin, Éamonn 1978, 'Liturgical innovations associated with Pope Sergius and the iconography of the Ruthwell and Bewcastle Crosses', in R. T. Farrell (ed.), *Bede and Anglo-Saxon England* (British Archaeological Reports, British Series 46; Oxford: BAR), pp. 131–47.

O Carragáin, Éamonn 1995, *The City of Rome and the World of Bede* (Jarrow Lecture 1994; Jarrow: St Paul's Church, Jarrow).

O Croínín, Dáibhí 1995, *Early Medieval Ireland 400–1200* (London and New York: Longman).

Odegaard, Charles Edwin 1945, *Vassi and Fideles in the Carolingian Empire* (Harvard Historical Monographs 19; Harvard: President and Fellows of Harvard College) (repr. New York: Octagon Books, 1972).

Offler, H. S. 1958, *Medieval Historians of Durham* (Durham: University of Durham) (repr. in H. S. Offler, *North of the Tees: Studies in Medieval British History*, ed. A. J. Piper and A. I. Doyle (Aldershot: Variorum, 1996), no. I).

Offler, H. S. 1970, 'Hexham and the *Historia Regum*', *Trans. Archit. and Archaeol. Soc. of Durham and Northumberland*, new series, 2, 51–62 (repr. in Offler (1996), no. X).

Offler, H. S. 1996, *North of the Tees: Studies in Medieval British History*, ed. A. J. Piper and A. I. Doyle (Collected Studies; Aldershot: Variorum).

Okasha, Elisabeth 1971, *Hand-List of Anglo-Saxon Non-Runic Inscriptions* (Cambridge: Cambridge University Press).

Okasha, Elisabeth 1993, *Corpus of Early Christian Inscribed Stones of South-West Britain* (Studies in the Early History of Britain; London: Leicester University Press).

O'Mahony, Felicity (ed.) 1994, *The Book of Kells: Proceedings of a Conference at Trinity College, Dublin 6–9 September 1992* (Aldershot: Scolar Press).

Oram, Richard D. 1995, 'Scandinavian settlement in south-west Scotland with a special study of Bysbie', in Barbara E. Crawford (ed.), *Scandinavian Settlement in Northern Britain: Thirteen Studies of Place-Names in their Historical Context* (Studies in the Early History of Britain; London and New York: Leicester University Press), pp. 127–40.

O'Sullivan, Deirdre M. 1996, 'A group of pagan Anglian burials from Cumbria?', *Anglo-Saxon Studies in Archaeology and History* 9, 15–23.

Ottaway, Patrick 1984, '*Colonia Eburacensis*: a review of recent work', in Peter V. Addyman and Valerie E. Black (eds.), *Archaeological Papers from York Presented to M. W. Barley* (York: York Archaeological Trust), pp. 28–33.

Ottaway, Patrick 1993, *English Heritage Book of Roman York* (London: Batsford/English Heritage).

Owen, Gale R. 1981, *Rites and Religions of the Anglo-Saxons* (Newton Abbot and London: David and Charles).

Padel, Oliver J. 1998, 'A new study of the Gododdin', *Cambrian Medieval Celtic Studies* 35, 45–55.

Pagan, H. E. 1969, 'Northumbrian numismatic chronology in the ninth century', *British Numismatic Journal* 38, 1–15.

Page, R. I. 1999, *An Introduction to English Runes*, 2nd edn (Woodbridge: Boydell Press).

Page, W. 1888, 'Some remarks on the Northumbrian palatinates and regalities', *Archaeologia* 51, 143–54.

Page, William (ed.) 1910, *The Victoria History of the County of Nottingham*, II (London: Constable; repr. University of London Institute of Historical Research, 1970).

Palliser, D. M. 1978, 'The medieval street-names of York', *York Historian* 2, 2–16.

Palliser, D. M. 1984, 'York's west bank: medieval suburb or urban nucleus?', in Peter V. Addyman and Valerie E. Black (eds.), *Archaeological Papers from York Presented to M. W. Barley* (York: York Archaeological Trust), pp. 101–8.

Palliser, D. M. 1990, *Domesday York* (Borthwick Papers 78; York: University of York).

Palliser, D. M. (ed.) 1992, *Great Domesday Book: County Edition: Yorkshire*, 2 vols. (London: Alecto).

Palliser, D. M. 1993, 'Domesday Book and the "harrying of the north"', *Northern History* 29, 1–23.

Pálsson, Hermann and Edwards, Paul (trans.) 1976, *Egil's Saga* (Penguin Classics; London: Penguin).

Parsons, David (ed.) 1975, *Tenth-Century Studies: Essays in Commemoration of the Millennium of the Council of Winchester and* Regularis Concordia (London and Chichester: Phillimore).

Parsons, David N. 2001, 'How long did the Scandinavian language survive in England? Again', in James Graham-Campbell, Richard A. Hall, Judith Jesch et al. (eds.), *Vikings and the Danelaw: Select Papers from the Proceedings of the Thirteenth Viking Congress, Nottingham and York, 21–30 August 1997* (Oxford: Oxbow Books), pp. 299–312.

Pearce, Susan 1978, *The Kingdom of Dumnonia: Studies in History and Tradition in South-Western Britain AD 350–1150* (Padstow: Lodenek Press).

Peers, Charles and Ralegh Radford, C. A. 1943, 'The Saxon monastery at Whitby', *Archaeologia* 89, 27–88.

Pennar, M. (trans.) 1988, *The Poems of Taliesin* (Lampeter: Llanerch).

Perros, Helen 1995, 'Crossing the Shannon frontier: Connacht and the Anglo-Normans, 1170–1224', in T. B. Barry, Robin Frame and Katherine Simms (eds.), *Colony and Frontier in Medieval Ireland: Essays Presented to J. F. Lydon* (London and Rio Grande: Hambledon), pp. 117–38.

Pertz, Georg Heinrich (ed.) 1829, *Scriptorum Tomus II* (Monumenta Germaniae Historica; Hanover: Hahn).

Phillips, Derek, Heywood, Brenda and Carver, M. O. H. 1995, *Excavations at York Minster*, I: *From Roman Fortress to Norman Cathedral* (London: HMSO).

Phythian-Adams, Charles 1996, *The Land of the Cumbrians: A Study in British Provincial Origins, AD 400–1120* (London: Scolar).

Piper, A. J. 1998, 'The historical interests of the monks of Durham', in David Rollason (ed.), *Symeon of Durham: Historian of Durham and the North* (Stamford: Paul Watkins), pp. 301–32.

Pirie, E. J. E. 1975, *Coins in Yorkshire Collections: The Yorkshire Museum, York, the City Museum, Leeds, the University of Leeds* (Sylloge of Coins of the British Isles 21; London: Oxford University Press/Spink).

Pirie, E. J. E. 1992, 'The seventh-century gold coinage of Northumbria', *Yorkshire Numismatist* 2, 11–15.

Pirie, E. J. E., Archibald, M. M. and Hall, R. A. 1986, *Post-Roman Coins from York Excavations 1971–1981* (Archaeology of York 18/1; London: Council for British Archaeology).

Plummer, Charles (ed.) 1892–9, *Two of the Saxon Chronicles Parallel with Supplementary Extracts from the Others*, 2 vols. (Oxford: Clarendon Press).

Plummer, Charles (ed.) 1896, *Venerabilis Baedae Opera historica*, 2 vols. (Oxford: Clarendon Press).

Pohl, Walter 1997, 'Ethnic names and identities in the British Isles', in John Hines (ed.), *The Anglo-Saxons from the Migration Period to the Eighth Century: An Ethnographic Perspective* (Woodbridge: Boydell Press), pp. 7–40.

Pohl, Walter, Wood, Ian N. and Reimitz, Helmut (eds.) 2001, *The Transformation of Frontiers: From Late Antiquity to the Carolingians* (The Transformation of the Roman World 10; Leiden, Boston and Cologne: Brill).

Poole, Austin Lane 1946, *Obligations of Society in the XII and XIII Centuries: The Ford Lectures Delivered in the University of Oxford in Michaelmas Term 1944* (Oxford: Clarendon Press).

Pounds, Norman John Greville 1994, *The Medieval Castle in England and Wales: A Social and Political History*, 1st paperback edn (Cambridge: Cambridge University Press).

Power, Daniel and Standen, Naomi (eds.) 1999, *Frontiers in Question: Eurasian Borderlands, 700–1700* (London: Macmillan).

Powlesland, Dominic 1997, 'Early Anglo-Saxon settlements, structures, form and layout', in John Hines (ed.), *The Anglo-Saxons from the Migration Period to the Eighth Century: An Ethnographic Perspective* (Woodbridge: Boydell Press), pp. 117–24.

Powlesland, Dominic 1999, 'The Anglo-Saxon settlement at West Heslerton, north Yorkshire', in Jane Hawkes and Susan Mills (eds.), *Northumbria's Golden Age* (Stroud: Sutton), pp. 55–65.

Powlesland, Dominic 2000, 'West Heslerton settlement mobility: a case of static development', in Helen Geake and Jonathan Kenny (eds.), *Early Deira: Archaeological Studies of the East Riding in the Fourth to Ninth Centuries AD* (Oxford and Oakville: Oxbow Books), pp. 19–26.

Prinz, Friedrich 1965, 'Schenkungen und Privilegien Karls des Grossen', in Wolfgang Braunfels (ed.), *Karl der Grosse: Lebenswerk und Nachleben*, I: *Persönlichkeit und Geschichte* (Düsseldorf: Schwann), p. 488 and map.

Prinz, Friedrich 1988, *Frühes Mönchtum im Frankenreich: Kultur und Gesellschaft in Gallien, den Rheinlanden und Bayern am Beispiel der monastischen Entwicklung (4. bis 8. Jahrhundert)*, 2nd edn (Darmstadt: Wissenschaftliche Buchgesellschaft).

Quentin, H. 1908, *Les martyrologes historiques du moyen âge: étude sur la formation du martyrologe romain* (Paris; repr. Aalen: Scientia Verlag, 1969).

Rahtz, Philip 1988, 'From Roman to Saxon at Wharram Percy', in J. Price and P. A. Wilson (eds.), *Recent Research in Roman Yorkshire: Studies in Honour of Mary Kitson Clark (Mrs Derwas Chitty)* (British Archaeological Reports, British Series 193; Oxford: BAR), pp. 123–37.

Rahtz, Philip 1995, 'Anglo-Saxon and later Whitby', in Lawrence R. Hoey (ed.), *Yorkshire Monasticism: Archaeology, Art and Architecture, from the Seventh to the Sixteenth Centuries* (Leeds: British Archaeological Association), pp. 1–11.

Rahtz, Philip, Watts, Lorna and Grenville, Jane 1996–7, *Archaeology at Kirkdale* (Supplement to the Ryedale Historian 18; Helmsley: Helmsley Archaeological Society).

Rahtz, Philip, Watts, Lorna, Okasha, Elisabeth et al. 1997, 'Kirkdale – the inscriptions', *Medieval Archaeology* 41, 51–99.

Raine, James 1852, *The History and Antiquities of North Durham, as Subdivided into the Shires of Norham, Island, and Bedlington* (London: John Bowyer Nicholas and Son).

Raine, James (ed.) 1864–5, *The Priory of Hexham*, 2 vols. (Surtees Society 44, 46; Durham: Andrews and Co.).

Raine, James (ed.) 1879–94, *The Historians of the Church of York and its Archbishops*, 3 vols. (Rolls Series 71; London: Longman).

Ralegh Radford, C. A. 1956, for 1955, 'Two Scottish shrines: Jedburgh and St Andrews', *Archaeological Journal*, 112, 43–60.

Ray, Roger 1987, 'Bede and Cicero', *Anglo-Saxon England* 16, 1–15.

Ray, Roger 2002, *Bede, Rhetoric, and the Creation of Christian Latin Culture* (Jarrow Lecture 1997; Jarrow: St Paul's Church, Jarrow).

Reuter, Timothy 1991, *Germany in the Early Middle Ages c. 800–1056* (Harlow: Longman).

Reynolds, Susan 1987, 'Towns in Domesday Book', in J. C. Holt (ed.), *Domesday Studies: Papers Read at the Novocentenary Conference of the Royal Historical Society and the Institute of British Geographers, Winchester, 1986* (Woodbridge: Boydell Press), pp. 295–309.

Richards, Julian D. 1999, 'Anglo-Saxon settlements of the Golden Age', in Jane Hawkes and Susan Mills (eds.), *Northumbria's Golden Age* (Stroud: Sutton), pp. 44–54.

Richards, Julian D. 2000a, 'Anglo-Saxon settlement and archaeological visibility in the Yorkshire Wolds', in Helen Geake and Jonathan Kenny (eds.), *Early Deira: Archaeological Studies of the East Riding in the Fourth to Ninth Centuries AD* (Oxford and Oakville: Oxbow Books), pp. 27–39.

Richards, Julian D. 2000b, *Viking Age England*, 2nd edn (Stroud: Tempus).

Richardson, W. 1985, 'The Venerable Bede and a lost Saxon monastery in Yorkshire', *Yorkshire Archaeological Journal* 57, 15–22.

Riché, Pierre 1976, *Education and Culture in the Barbarian West: From the Sixth through the Eighth Century* (Columbia: University of South Carolina Press) (trans. from the 3rd French edn by John J. Contreni; with a foreword by Richard E. Sullivan).

Riché, Pierre 1993, *The Carolingians: A Family who Forged Europe*, trans. Michael Idomir Allen (Philadelphia: University of Pennsylvania Press).

Ritchie, Anna 1989, *Picts: An Introduction to the Life of the Picts and the Carved Stones in the Care of Historic Scotland* (Edinburgh: HMSO).

Rivet, A. L. F. and Smith, Colin 1979, *The Place-Names of Roman Britain* (London: Batsford).

Roberts, Brian K. 1977, *The Green Villages of County Durham: A Study in Historical Geography* (Durham County Library Local History Publication 12; Durham: Durham County Council).

Robertson, A. J. (ed. and trans.) 1925, *The Laws of the Kings of England from Edmund to Henry I* (Cambridge: Cambridge University Press).

Robertson, A. J. (ed. and trans.) 1956, *Anglo-Saxon Charters*, 2nd edn (Cambridge: Cambridge University Press).

Robinson, Joseph Armitage 1923, *The Times of Saint Dunstan: The Ford Lectures Delivered in the University of Oxford in the Michaelmas Term, 1922* (Oxford: Clarendon Press).

Rollason, David 1982, *The Mildrith Legend: A Study in Early Medieval Hagiography in England* (Studies in the Early History of Britain; Leicester: Leicester University Press).

Rollason, David 1983, 'The cults of murdered royal saints in Anglo-Saxon England', *Anglo-Saxon England* 11, 1–22.

Rollason, David 1987, 'The wanderings of St Cuthbert', in David Rollason (ed.), *Cuthbert: Saint and Patron* (Durham: Dean and Chapter of Durham), pp. 45–61.

Rollason, David 1989a, *Saints and Relics in Anglo-Saxon England* (Oxford: Blackwell).

Rollason, David 1989b, 'St Cuthbert and Wessex: the evidence of Cambridge, Corpus Christi College MS 183', in Gerald Bonner, David Rollason and Clare Stancliffe (eds.), *St Cuthbert, his Cult and his Community to AD 1200* (Woodbridge: Boydell Press), pp. 413–24.

Rollason, David 1992, 'Symeon of Durham and the community of Durham in the eleventh century', in Carola Hicks (ed.), *England in the Eleventh Century: Proceedings of the 1990 Harlaxton Symposium* (Stamford: Paul Watkins), pp. 183–98.

Rollason, David 1995a, '995: the beginnings of the diocese of Durham', *Friends of Durham Cathedral*, 23–34.

Rollason, David 1995b, 'Hexham after the Vikings', *Hexham Historian* 5, 6–21.

Rollason, David 1995c, 'St Oswald in post-Conquest England', in Clare Stancliffe and Eric Cambridge (eds.), *Oswald: Northumbrian King and European Saint* (Stamford: Paul Watkins), pp. 164–77.

Rollason, David 1996, 'Hagiography and politics in eighth-century Northumbria', in Paul Szarmach (ed.), *Holy Men and Holy Women: Old English Prose Saints' Lives and their Contexts* (SUNY Series in Medieval Studies; New York: State University of New York), pp. 95–114.

Rollason, David 1998, 'Symeon's contribution to historical writing in northern England', in David Rollason (ed.), *Symeon of Durham: Historian of Durham and the North* (Stamford: Paul Watkins), pp. 1–13.

Rollason, David 1999a, 'Historical evidence for Anglian York', in D. Tweddle, Joan Moulden and Elizabeth Logan (eds.), *Anglian York (AD 410–876): A Survey of the Evidence* (Archaeology of York 7/2; York: York Archaeological Trust), pp. 117–40.

Rollason, David 1999b, 'Monasteries and society in early medieval Northumbria', in Benjamin Thompson (ed.), *Monasteries and Society in Medieval Britain* (Harlaxton Medieval Studies 19; Stamford: Paul Watkins), pp. 59–74.

Rollason, David (ed.) 2000, *Symeon of Durham, On the Origins and Progress of this the Church of Durham* (Oxford Medieval Texts; Oxford: Clarendon Press).

Rollason, David forthcoming, 'Anglo-Scandinavian York: the evidence of historical sources', in Richard A. Hall (ed.), *Anglo-Scandinavian York* (York: York Archaeological Trust).

Rollason, David and Cambridge, Eric 1995, 'The pastoral organization of the Anglo-Saxon church: a review of the "Minster Hypothesis"', *Early Medieval Europe* 4, 87–104.

Rollason, Lynda, Armstrong, Nessie, Baty, Vera et al. 1996, *Four Anglian Monuments in Derbyshire: Bakewell, Bradbourne, Eyam and Wirksworth* (Darlington: Workers Educational Association).

Romilly Allen, J. and Anderson, Joseph 1993, *The Early Christian Monuments of Scotland*, 2 vols. (Balgavies: Pinkfoot Press) (with an introduction by Isobel Henderson, repr. from the edition of 1903, Society of Antiquaries of Scotland).

Ross, Anne 1967, *Pagan Celtic Britain: Studies in Iconography and Tradition* (London: Routledge and Kegan Paul).

Routh, A. E. 1937, 'A corpus of pre-Conquest carved stones in Derbyshire', *Journal of the Derbyshire Archaeological Society*, new series, 11, 1–42.

Royal Commission on Historical Monuments (England) 1972, *An Inventory of the Historical Monuments of the City of York, 2: The Defences* (London: HMSO).

Salway, Peter 1982, *Roman Britain*, 2nd edn (Oxford History of England; Oxford: Clarendon Press).

Samuels, M. Benskin, Laing, M. and Williamson, K. 1986, *A Linguistic Atlas of Late Mediaeval English* (Aberdeen: Aberdeen University Press).

Sawyer, Peter Hayes 1968, *Anglo-Saxon Charters: An Annotated List and Bibliography* (Royal Historical Society Guides and Handbooks 8; London: Royal Historical Society).

Sawyer, Peter Hayes 1971, *The Age of the Vikings*, 2nd edn (London: Edward Arnold).

Sawyer, Peter Hayes 1982, *Kings and Vikings: Scandinavia and Europe AD 700–1100* (London and New York: Methuen).

Sawyer, Peter Hayes 1983, 'The royal *tun* in pre-Conquest England', in Patrick Wormald, Donald Bullough and Roger Collins (eds.), *Ideal and Reality in Frankish and Anglo-Saxon Society* (Oxford: Blackwell), pp. 273–99.

Scammell, J. 1966, 'The origin and limitations of the liberty of Durham', *English Historical Review* 81, 449–73.

Scharer, Anton 1982, *Die angelsächsische Königsurkunde im 7. und 8. Jahrhundert* (Vienna, Cologne and Graz: Böhlau).

Scholz, Berhard Walter and Rogers, Barbara (trans.) 1972, *Carolingian Chronicles: Royal Frankish Annals and Nithard's Histories* (Ann Arbor: University of Michigan Press).

Sharpe, Richard 1984, 'Some problems concerning the organization of the church in early medieval Ireland', *Peritia* 3, 230–70.

Sherlock, Stephen J. and Welch, Martin G. 1992, *An Anglo-Saxon Cemetery at Norton, Cleveland* (CBA Research Report 82; London: Council for British Archaeology).

Sims-Williams, Patrick 1980, 'The settlement of England in Bede and the *Chronicle*', *Anglo-Saxon England* 12, 1–41.

Sims-Williams, Patrick 1990, *Religion and Literature in Western England, 600–800* (Cambridge Studies in Anglo-Saxon England 3; Cambridge: Cambridge University Press).

Smith, A. H. 1956, *English Place-Name Elements*, 2 vols. (English Place-Name Society 25–6; Cambridge: Cambridge University Press).

Smith, A. H. (ed.) with a bibliography by Swanton, Michael J. 1978, *Three Northumbrian Poems: Caedmon's Hymn, Bede's Death Song and the Leiden Riddle*, rev. edn (Exeter: University of Exeter).

Smith, Anthony D. 1986, *The Ethnic Origins of Nations* (Oxford: Blackwell).

Smith, Anthony D. 1998, *Nationalism and Modernism: A Critical Survey of Recent Theories of Nations and Nationalism* (London and New York: Routledge).

Smith, Ian Mervin 1991, 'Sprouston, Roxburghshire: an early Anglo-Saxon centre of the eastern Tweed basin', *Proceedings of the Society of Antiquaries of Scotland* 121, 261–94.

Smyth, Alfred P. 1975–9, *Scandinavian York and Dublin: The History and Archaeology of Two Related Viking Kingdoms*, 2 vols. (Dublin: Irish Academic Press).

Smyth, Alfred P. 1977, *Scandinavian Kings in the British Isles 850–880* (Oxford: Oxford University Press).

Smyth, Alfred P. 1984, *Warlords and Holy Men: Scotland AD 80–1000* (London: Edward Arnold).

Snyder, Christopher A. 1996, *Sub-Roman Britain (AD 40–600): A Gazetteer of Sites* (British Archaeological Reports, British Series 247; Oxford: Tempus Reparatum).

Snyder, Christopher A. 1998, *An Age of Tyrants: Britain and the Britons AD 400–600* (Stroud: Sutton).

Southern, Patricia and Dixon, Karen R. 1996, *The Late Roman Army* (London: Batsford).

Stafford, P. 1980, 'The farm of one night and the organization of King Edward's estates in Domesday', *Economic History Review*, 2nd series, 33, 491–502.

Stancliffe, C. 1983, 'Kings who opted out', in Patrick Wormald, Donald Bullough and Roger Collins (eds.), *Ideal and Reality in Frankish and Anglo-Saxon Society: Studies Presented to J. M. Wallace-Hadrill* (Oxford: Blackwell), pp. 154–76.

Stancliffe, Clare 1989, 'Cuthbert and the polarity between pastor and solitary', in Gerald Bonner, David Rollason and Clare Stancliffe (eds.), *St Cuthbert, his Cult and his Community to AD 1200* (Woodbridge: Boydell Press), pp. 21–44.

Stancliffe, Clare 1995a, 'Oswald, "most holy and most victorious king of the Northumbrians"', in Clare Stancliffe and Eric Cambridge (eds.), *Oswald: Northumbrian King to European Saint* (Stamford: Paul Watkins), pp. 33–83.

Stancliffe, Clare 1995b, 'Where was Oswald killed?', in Clare Stancliffe and Eric Cambridge (eds.), *Oswald: Northumbrian King to European Saint* (Stamford: Paul Watkins), pp. 84–96.

Stenton, Frank Merry 1941, 'The historical bearing of place-name studies: Anglo-Saxon heathenism', *Transactions of the Royal Historical Society*, 4th series, 23, 1–13 (repr. in Doris Mary Stenton (ed.), *Preparatory to 'Anglo-Saxon England': Being the Collected Papers of Frank Merry Stenton* (Oxford: Clarendon Press, 1970), pp. 280–97).

Stenton, Frank Merry 1955, *The Latin Charters of the Anglo-Saxon Period* (Oxford: Clarendon Press).

Stenton, Frank Merry 1970a, 'Lindsey and its kings', in Doris Mary Stenton (ed.), *Preparatory to 'Anglo-Saxon England': Being the Collected Papers of Frank Merry Stenton* (Oxford: Clarendon Press), pp. 127–35.

Stenton, Frank Merry 1970b, 'Pre-Conquest Westmorland', in Doris Mary Stenton (ed.), *Preparatory to 'Anglo-Saxon England': Being the Collected Papers of Frank Merry Stenton* (Oxford: Clarendon Press), pp. 214–23.

Stenton, Frank Merry 1970c, 'The supremacy of the Mercian kings', in Doris Mary Stenton (ed.), *Preparatory to 'Anglo-Saxon England': Being the Collected Papers of Frank Merry Stenton* (Oxford: Clarendon Press), pp. 48–66 (repr. from *English Historical Review* 33 (1918), 433–52).

Stenton, Frank Merry 1971, *Anglo-Saxon England*, 3rd edn (Oxford: Clarendon Press).

Stevens, Wesley M. 1986, *Bede's Scientific Achievement* (Jarrow Lecture 1985; Jarrow: St Paul's Church, Jarrow) (repr. in *Bede and his World: The Jarrow Lectures 1958–1993*, ed. Michael Lapidge, 2 vols. (Aldershot: Variorum, 1994), II, pp. 645–88).

Stokes, Whitley (ed.) 1993, *The Annals of Tigernach*, 2 vols. (Felinfach: Llanerch).

Story, Joanna 1998, 'Symeon as annalist', in David Rollason (ed.), *Symeon of Durham: Historian of Durham and the North* (Stamford: Paul Watkins), pp. 202–13.

Story, Joanna 2003, *Carolingian Connections: Anglo-Saxon England and Carolingian Francia c. 750–870* (Studies in Early Medieval Britain; Aldershot: Ashgate).

Stubbs, William (ed.) 1868–71, *Chronica Magistri Rogeri de Houedene*, 4 vols. (Rolls Series 51; London: Longmans, Green, Reader and Dyer).

Summerson, Henry 1993, *Medieval Carlisle: The City and the Borders from the Late Eleventh to the Mid-Sixteenth Century*, 2 vols. (Cumberland and Westmorland Antiquarian and Archaeological Society, extra series 25; Kendal: Cumberland and Westmorland Antiquarian and Archaeological Society).

Swanton, Michael J. (trans.) 1996, *The Anglo-Saxon Chronicle* (London: Dent).

Tait, James 1936, *The Medieval English Borough: Studies on its Origins and Constitutional History* (Manchester: Manchester University Press).

Tangl, Michael (ed.) 1916, *Die Briefe des heiligen Bonifatius und Lullus* (Epistolae Selectae in usum scholarum ex Monumentis Germaniae Historicis separatim editae 1; Berlin: Weidmannsche Buchhandlung).

Taylor, H. M. 1969, 'Corridor crypts on the continent and in England', *North Staffordshire Journal of Field Studies* 9, 17–52.

Taylor, H. M. and Taylor, Joan 1965–78, *Anglo-Saxon Architecture*, 3 vols. (Cambridge: Cambridge University Press).

Tenhaken, Hans P. 1979 (ed.), *Das nordhumbrische Priestergesetz: Ein nachwulfstanisches Pönitential des 11. Jahrhunderts* (Düsseldorf: Stern-Verlag Janssen).

Thacker, Alan T. 1977, 'The social and continental background to early Anglo-Saxon hagiography', PhD Dissertation, University of Oxford.

Thacker, Alan T. 1981, 'Some terms for noblemen in Anglo-Saxon England *c.* 600–900', in D. Brown (ed.), *Anglo-Saxon Studies in Archaeology and History* (British Archaeological Reports, British Series 92; Oxford: BAR), pp. 201–36.

Thacker, Alan T. 1983, 'Bede's ideal of reform', in Patrick Wormald, Donald Bullough and Roger Collins (eds.), *Ideal and Reality in Frankish and Anglo-Saxon Society: Studies Presented to J. M. Wallace-Hadrill* (Oxford: Blackwell), pp. 130–53.

Thacker, Alan T. 1985, 'Kings, saints, and monasteries in pre-Viking Mercia', *Midland History* 10, 1–25.

Thomas, Charles 1971, *The Early Christian Archaeology of North Britain* (Oxford: Oxford University Press).

Thomas, Charles 1981, *Christianity in Roman Britain to AD 500* (London: Batsford).

Thomas, Charles 1984, 'Abercorn and the *Provincia Pictorum*', in Roger Miket and Colin Burgess (eds.), *Between and Beyond the Walls* (Edinburgh: John Donald), pp. 324–37.

Thomas, Charles 1994, *And Shall These Mute Stones Speak? Post-Roman Inscriptions in Western Britain* (Cardiff: University of Wales Press).

Thompson, James Westfall 1935, *The Dissolution of the Carolingian Fisc in the Ninth Century* (University of California Publications in History 23; Berkeley: University of California Press).

Thorpe, Lewis (ed.) 1974, *Gregory of Tours: The History of the Franks* (Harmondsworth: Penguin).

Timby, J. 1993, 'Sancton I Anglo-Saxon cemetery: excavations carried out between 1976 and 1980', *Archaeological Journal* 150, 243–365.

Trenholme, Norman Maclaren 1903, *The Right of Sanctuary in England* (University of Missouri Studies 1.5; Columbia: University of Missouri).

Tugène, Georges 1982, 'L'Histoire "ecclésiastique" du peuple anglais: réflexions sur le particularisme et l'universalisme chez Bède', *Recherches Augustiniennes* 17, 129–72.

Tugène, Georges 2001a, *L'idée de nation chez Bède le Vénérable* (Collections des Etudes Augustiniennes, Série Moyen Age et Temps Modernes 37; Paris: Institut d'Etudes Augustiniennes).

Tugène, Georges 2001b, *L'image de la nation anglaise dans l'Histoire ecclésiastique de Bède le Vénérable* (Strasbourg: Presses Universitaires de Strasbourg).

Turville-Petre, E. O. Gabriel 1964, *Myth and Religion of the North: The Religion of Ancient Scandinavia* (London: Weidenfield and Nicholson).

Tweddle, Dominic 1992, *The Anglian Helmet from 16–22 Coppergate*, 2 vols. (Archaeology of York 17/8; London: York Archaeological Trust).

Tweddle, Dominic and Moulden, Joan 1992, 'The seventh-century gold coins from York', *Yorkshire Numismatist* 2, 17–20.

Tweddle, Dominic, Moulden, Joan and Logan, E. 1999, *Anglian York: A Survey of the Evidence* (Archaeology of York 7/2; York: York Archaeological Trust).

Verey, Christopher D. 1989, 'The gospel texts at Lindisfarne at the time of Cuthbert', in Gerald Bonner, David Rollason and Clare Stancliffe (eds.), *St Cuthbert, his Cult and his Community to AD 1200* (Woodbridge: Boydell Press), pp. 143–50.

Vince, Alan 1990, *Saxon London: An Archaeological Investigation* (London: Seaby).

Vince, Alan (ed.) 1993, *Pre-Viking Lindsey* (Lincoln Archaeological Studies 1; Lincoln: City of Lincoln Archaeology Unit).

Wainwright, F. T. 1948, 'Nechtansmere', *Antiquity* 22, 82–97.

Wainwright, F. T. (ed.) 1955, *The Problem of the Picts* (Edinburgh and London: Nelson).

Wainwright, F. T. 1975, *Scandinavian England: Collected Papers of F. T. Wainwright*, ed. H. P. R. Finberg (Chichester: Phillimore).

Wakelin, Martyn 1988, *The Archaeology of English* (London: Batsford).

Wall, J. 1965, 'Christian evidences in the Roman period: the northern counties: Part I', *Archaeologia Aeliana*, 4th series, 43, 201–25.

Wall, J. 1966, 'Christian evidences in the Roman period: the northern counties: Part II', *Archaeologia Aeliana*, 4th series, 44, 147–64.

Wall, Valerie 1994, 'Malcolm III and the foundation of Durham Cathedral', in David Rollason, Margaret Harvey and Michael Prestwich (eds.), *Anglo-Norman Durham 1093–1193* (Woodbridge: Boydell Press), pp. 325–37.

Wallace-Hadrill, John Michael 1960, 'Rome and the early English church: some questions of transmission', in Anon. (ed.), *Settimane di studio sull'alto medioevo* (Spoleto: Presso la Sede del Centro), vol. VII (2), pp. 519–48 (repr. in John Michael Wallace Hadrill, *Early Medieval History* (Oxford: Blackwell, 1975), pp. 115–37).

Wallace-Hadrill, John Michael 1962, 'The blood-feud of the Franks', in J. M. Wallace-Hadrill (ed.), *The Long-Haired Kings and Other Studies in Frankish History* (London: Methuen), pp. 121–47.

Wallace-Hadrill, John Michael 1971, *Early Germanic Kingship in England and on the Continent* (Oxford: Clarendon Press).

Wallace-Hadrill, John Michael 1988, *Bede's Ecclesiastical History of the English People: A Historical Commentary* (Oxford Medieval Texts; Oxford: Clarendon Press).

Walterspacher, Ralph 2002, *The Foundation of Hexham Priory, 1070–1170* (Papers in North-East History 11; Middlesbrough: North-East England History Institute).

Ward, Benedicta 1990, *The Venerable Bede* (Harrisburg: Morehouse).

Waterman, D. M. 1959, 'Late Saxon, Viking and early medieval finds from York', *Archaeologia* 97, 59–105.

Watts, Victor E. 1995, 'Northumberland and Durham: the place-name evidence', in Barbara Crawford (ed.), *Scandinavian Settlement in Northern Britain: Thirteen Studies of Place-Names in their Historical Context* (Studies in the Early History of Britain; London and New York: Leicester University Press), pp. 206–13.

Watts, Victor E. and Insley, John 2002, *Dictionary of County Durham Place-Names* (English Place-Name Society Popular Series 3; Nottingham: English Place-Name Society).

Webster, Leslie 1999, 'The iconographic progamme of the Franks Casket', in Jane Hawkes and Susan Mills (eds.), *Northumbria's Golden Age* (Stroud: Sutton), pp. 227–46.

Webster, Leslie and Backhouse, Janet (eds.) 1991, *The Making of England: Anglo-Saxon Art and Culture AD 600–900* (London: British Museum).

Welsby, Derek A. 1982, *The Roman Military Defence of the British Provinces in its Later Phases* (British Archaeological Reports, British Series 101; Oxford: Archaeopress).

Wenham, L. P., Hall, Richard A., Briden, C. M. et al. 1987, *St Mary Bishophill Junior and St Mary Castlegate* (The Archaeology of York 8/2; London: Council for British Archaeology).

Werner, Karl Ferdinand 1979, 'Important noble families in the kingdom of Charlemagne', in Timothy Reuter (ed.), *The Medieval Nobility: Studies on the Ruling Classes of France and Germany from the Sixth to the Twelfth Century* (Europe in the Middle Ages: Selected Studies 14; Amsterdam, New York and Oxford: North Holland), pp. 137–202 (trans. from 'Bedeutende Adelsfamilien im Reich Karls des Grossen: Ein personengeschichtlicher Beitrag zum Verhältnis von Königtum und Adel im frühen Mittelalter', in *Karl der Grosse: Lebenswerk and Nachleben*, I: *Persönlichkeit und Geschichte*, ed. Wolfgang Braunfels (Düsseldorf: Schwann, 1965), 83–142; repr. in Karl Ferdinand Werner, *Vom Frankenreich zur Entfaltung Deutschlands und Frankreichs:*

Ursprünge, Strukturen, Beziehungen: ausgewählte Beiträge: Festgabe zu seinem sechzigsten Geburtstag (Sigmaringen: J. Thorbecke, 1984), pp. 108–56).

Werner, Karl Ferdinand 1980, 'Missus-marchio-comes: entre l'administration centrale et l'administration locale de l'empire carolingien', in Werner Paravicini and Karl Ferdinand Werner (eds.), *Histoire comparée de l'administration (IVe–XVIIIe siècle)* (Beihefte der Francia 9; Munich: Artemis), pp. 191–239 (repr. in Karl Ferdinand Werner, *Vom Frankenreich zur Entfaltung Deutschlands und Frankreichs: Ursprünge, Strukturen, Beziehungen: ausgewählte Beiträge: Festgabe zu seinem sechzigsten Geburtstag* (Sigmaringen: J. Thorbecke, 1984), pp. 22–81).

Whitelock, Dorothy 1959, 'The dealings of the kings of England with Northumbria in the tenth and eleventh centuries', in Peter Clemoes (ed.), *The Anglo-Saxons: Some Aspects of their History and Culture: Studies Presented to Bruce Dickins* (London: Bowes and Bowes), pp. 70–88.

Whittaker, Charles R. 1994, *Frontiers of the Roman Empire: A Social and Economic Study* (Baltimore and London: Johns Hopkins University Press).

Williams, Ann, Smyth, Alfred P. and Kirby, D. P. 1991, *A Biographical Dictionary of Dark Age Britain: England, Scotland and Wales c. 500–c. 1050* (London: Seaby).

Wilmott, Tony 2001, *Birdoswald Roman Fort: 1800 Years on Hadrian's Wall* (Stroud: Tempus).

Wilson, David 1992, *Anglo-Saxon Paganism* (London and New York: Routledge).

Wilson, David M. and Hurst, D. Gillian 1966, 'Medieval Britain in 1965', *Medieval Archaeology* 10, 175–6.

Wilson, P. A. 1966, 'On the use of the terms "Strathclyde" and "Cumbria"', *Transactions of the Cumberland and Westmorland Antiquarian and Archaeological Society*, new series, 66, 57–92.

Wilson, P. R., Cardwell, P., Cramp, R. J. et al. 1996, 'Early Anglian Catterick and Catraeth', *Medieval Archaeology* 40, 1–61.

Wolfram, Herwig 1997, *The Roman Empire and its Germanic Peoples* (Berkeley: University of California Press).

Wood, I. N. 1990, 'Ripon, Francia and the Franks Casket', *Northern History* 26, 1–19.

Wood, I. N. 1994, *The Merovingian Kingdoms 450–751* (London: Longman).

Wood, I. N. 1995, 'Northumbrians and Franks in the age of Wilfrid', *Northern History* 31, 10–21.

Wood, P. N. 1996, 'On the little British kingdom of Craven', *Northern History* 32, 1–20.

Wormald, Patrick 1976, 'Bede and Benedict Biscop', in Gerald Bonner (ed.), *Famulus Christi: Essays in Commemoration of the Thirteenth Centenary of the Birth of the Venerable Bede* (London: SPCK), pp. 141–69.

Wormald, Patrick 1978, 'Bede, Beowulf and the conversion of the Anglo-Saxon aristocracy', in R. T. Farrell (ed.), *Bede and Anglo-Saxon England: Papers in Honour of the*

1300th Anniversary of the Birth of Bede, Given at Cornell University in 1973 and 1974 (British Archaeological Reports, British Series 46; Oxford: BAR), pp. 32–95.

Wormald, Patrick 1983, 'Bede, the bretwaldas and the origins of the *gens anglorum*', in Patrick Wormald, Donald Bullough and Roger Collins (eds.), *Ideal and Reality in Frankish and Anglo-Saxon Society: Studies Presented to J. M. Wallace-Hadrill* (Oxford: Blackwell), pp. 99–129.

Wormald, Patrick 1985, *Bede and the Conversion of England: The Charter Evidence* (Jarrow Lecture 1984; Jarrow: St Paul's Church, Jarrow) (repr. in *Bede and his World: The Jarrow Lectures 1979–1993*, ed. Michael Lapidge, 2 vols. (Aldershot: Variorum, 1994), pp. 611–44).

Wormald, Patrick 1991, 'In search of King Offa's "law-code"', in Ian N. Wood and Niels Lund (eds.), *People and Places in Northern Europe 500–1600: Essays in Honour of Peter Hayes Sawyer* (Woodbridge: Boydell Press), pp. 25–45.

Wormald, Patrick 1998, *The Making of English Law: King Alfred to the Twelfth Century*, I: *Legislation and its Limits* (Oxford: Blackwell).

Yorke, Barbara 1981, 'The vocabulary of Anglo-Saxon overlordship', in David Brown, James Campbell and Sonia Chadwick Hawkes (eds.), *Anglo-Saxon Studies in Archaeology and History*, II (British Archaeological Reports, British Series 92; Oxford: BAR), pp. 171–200.

Yorke, Barbara 1990, *Kings and Kingdoms of Early Anglo-Saxon England* (London: Seaby).

Young, Alan 1994, 'The bishopric of Durham in Stephen's reign', in David Rollason, Margaret Harvey and Michael Prestwich (eds.), *Anglo-Norman Durham 1093–1193* (Woodbridge: Boydell Press), pp. 353–68.

Index